WATER SUPPLY, WASTE WATER TREATMENT AND SEWAGE DISPOSAL

WATER SUPPLY, WASTE WATER TREATMENT AND SEWAGE DISPOSAL

(in SI Units)

[*A Text book for Degree, Diploma and Industrial Engg. Students*]

Dr. M.N. Maulik

B.Sc. (Cal), B.Sc. Engineering (Civil) (London)
Ph.D (Ind.)
Assistant Professor
Civil Engineering Department
Jalpaiguri Govt. Engineering College
Jalpaiguri, West Bengal

STANDARD BOOK HOUSE

unit of : **RAJSONS PUBLICATIONS PVT. LTD.**

1705-A, Nai Sarak, PB.No. 1074, Delhi-110006 Ph.: +91-(011)-23265506
Show Room: 4262/3, First Lane, G-Floor, Gali Punjabian, Ansari Road, Darya Ganj,
New Delhi-110002 Ph.: +91-(011)43551085 Tel Fax : +91-(011)43551185,
Fax: +91-(011)-23250212
E-mail: sbh10@hotmail.com www.standardbookhouse.com

Published by:
RAJINDER KUMAR JAIN
Standard Book House
Unit of: Rajsons Publications Pvt. Ltd.
1705-A, Nai Sarak, Delhi - 110006
Post Box: 1074
Ph.: +91-(011)-23265506 Fax: +91-(011)-23250212

Showroom:
4262/3. First Lane, G-Floor, Gali Punjabian
Ansari Road, Darya Ganj
New Delhi-110002
Ph.: +91-(011)-43551085, +91-(011)-43551185
E-mail: sbhl0@ hotmail.com
Web: www.standardbookhouse.com

Twelfth Edition : 2018
Thirteenth Edition *(Reprint)* : 2024

Price : **₹ 340.00**

ISBN: 978-81-89401-38-2

Typeset by:
C.S.M.S. Computers, Delhi.

Printed by:
R.K. Print Media Company, New Delhi

Preface

An attempt has been made in this book to explain the fundamentals of Sanitary Engineering, Sewage, Lab. Testing Treatment and Disposal of Industrial Waste Water. The subject as a whole is a complicated one. But it is believed that the basic ideas are exposed in this book, the reader will be able to have a clear idea of the subject. A new topic on Water Supply has been introduced in this edition according to the need of the subject matter.

This book is written in SI units, although the metric units also given for the comprehension of the students. The subject-matter explained in simple, easy and comprehensive language assisted by self-explanatory and neatly drawn sketches where necessary.

This book covers the syllabus prescribed by the various university of India— B.E. College Shibpur, Jadavpur Universities of India, Burdman University, North Bengal University, Bombay University etc. This book will therefore be useful to students preparing for Degree, Diploma and Industrial Engineering examination or for examinations governed by the various professional bodies.

Suggestions to improve the utility of the book will be gratefully acknowledged by the Publishers in the forthcoming editions.

Dr. M.N. Maulik

2010

Contents

Conversion of MKS Units to S.I. Units

WHAT IS S.I. UNITS ?

This is known as System International Units to avoid the consequent confusion in International trade and to introduce uniformity over the entire world, it was decided during the 4th General Conference on Weights and Measures held in Paris in the year 1960 that a new system called S.I. units be adapted all over the globe. But, however, our country still largely follow the M.K.S. system.

It is fact that the S.I. unit system is somewhat similar to M.K.S system because of the fact the basic units of length, mass and time and the same in both the systems, i.e., the length is measured by metre (m), the mass is measured by kilogram (kg), and the time by seconds (s). This system is rather a rationalisation of M.K.S. system, as it removes any confusion between kg and kg wt (also usually expressed as kg), as adapted in M.K.S system.

S.I. UNIT OF FORCE

The only basic difference between M.K.S. units and S.I. units is in the unit of force.

In M.K.S. system–the unit of force is kg wt or kgf (kg force) usually written as kg.

But in S.I. unit, the unit force is Newton (N).

One Newton (N) : The force required to move a body of mass equal to 1 kg, as to generate in it an acceleration equal to 1 m/sec^2.

Since, Force = mass × acceleration, by definition of Newton

$$1 \text{ Newton} = 1 \text{ kg} \times 1 \text{ m/sec}^2$$

or $$\mathbf{N = 1 \text{ kg m/sec}^2}$$

Hence, Newton (N) will have the dimensions of kg m/sec^2.

Relation between two force units kgf and N

$$1 \text{ kgf} = 1 \text{ kg wt.}$$
$$= \text{kg.g}$$

where g is acceleration due to gravity = 9.807 m/sec^2

$$= 1 \text{ kg} \times 9.807 \text{ m/sec}^2$$
$$= 9.807 \text{ kg m/sec}^2 = 9.807 \text{ N}$$

or $$\mathbf{1 \text{ kgf} = 9.807 \text{ N}}$$

As Newton is seems to be very small unit of force (approximately) 1/10th of kg wt. So, it is customary to use a bigger unit. Such as kilo-Newton (kN) or mega Newton (MN) for expressing force. Kilo Newton = 10^3 N and mega Newton = 10^6 N as usual.

S.I. UNITS OF STRESS OR PRESSURE

As the force per unit area is stress or pressure, we would evidently have S.I. unit of pressure as Newton/m^2 (N/m^2), and this is known as Pascal (Pa).

But the M.K.S. unit of pressure being kgf/m^2 usually written as kg/m^2 or its bigger units of kg/cm^2. The relation between these two units can be derived as follows :

$$\text{kg. f/m}^2 = 9.807 \text{ N/m}^2$$

or $$9.807 \text{ N/m}^2 = 1 \text{ kg/m}^2$$

(kgf/m^2 being usually expressed as **kg/m^2**)

or $$9.807 \text{ N/m}^2 = \frac{1}{10^4} \text{ kg/cm}^2$$

or $$1 \text{ kg/cm}^2 = 9.807 \times 10^4 \text{ N/m}^2$$

or $$\mathbf{1 \text{ kg/m}^2 = 98.07 \text{ kN/m}^2}$$

Kilo Pascal (kPa) : The bigger S.I. units of pressure will be

$$\text{KPa} = \text{kN/m}^2 = \frac{100}{98.07} \text{ kg/cm}^2$$

or $$100 \text{ KPa} = \frac{1}{98.07} \text{ kg/cm}^2$$

$$= 1.02 \text{ kg/cm}^2$$

$$= 1.02 \times 10 \text{ m head of water}$$

$$= 10.2 \text{ m of water head.}$$

$$\therefore \quad \mathbf{100 \text{ KPa} = 1.02 \text{ kg/m}^2 = 10.2 \text{ m of water head}}$$

S.I. UNIT OF TEMPERATURE

In S.I. unit—the unit temperature is Kelvia (K) and is equal to the temperature interval which is equal to temperature interval corresponding to 1°C. The relation between K and °C is given by

$$\mathbf{°C + 273° = K}$$

CONVERSION TABLE OF SI UNITS

Basic SI Units

Length	Meter	m
Mass	Kilogram	kg
Time	Seconds	s

Current	Ampere	A
Temperature	Kelvin	K
Derived Units		
Force	Newton	N
Energy	Joule	J
Power	Watt	W
Frequency	Hertz	Hz
Pressure stress	Pascal	Pa

Conversion

1kg/m^3 = 1 × 10^{-3} Mg/m^3

1gm/c^3 = 1 g/ml

1t/m^3 = 9.8 kN/m^3

1g/cm^3 = 9.8 kN/m^3

1tf = 1kgf = 9.8 N

1kg/cm^2 = 98 kN/m^2

1t/m^2 = 9.8 kN/m^2

1m^2/kg = 102.041 m^2/kN

Conversion Table for c.g.s., and S.I. Units to M.K.S Units

Conversion of c.g.s units into M.K.S. units	*Conversion of S.I. units into M.K.S. units*
For length	
1 cm = 0.01 m 1 micron (μ) = 10^{-8} m	S.I. unit is same as in M.K.S. i.e., m
For Area	
1 cm^2 = 10^{-4} m^2	S.I. unit is same as in M.K.S. i.e. = m^2
For Volume	
1 cm^3 = 10^{-4} m^3	S.I. unit is same as in M.K.S. i.e. m^3
For Discharge	
1 cm^3/sec = 10^{-6} m^3/sec	S.I. unit is same as in M.K.S. i.e. m^3/sec (cumecs)

For Force

$1\ gm^\circ = 10^{-3}\ kg^\circ$

$1\ dyne = \frac{1\ gm^\circ}{g} = \frac{1}{98.07} gm^\circ$

$= 0.102 \times 10^{-5}\ kg^\circ$

S.I. unit is Newton, where

$1\ Newton\ (N) = \frac{1}{9.81} kg^\circ$

$= 0.102\ kg$

$= 10^5\ dynes$

For Pressure

$1\ gm^\circ/cm^2 = 10^4\ kg^\circ/m^2$

$= 10$ m of water head

S.I. units is Pascal,

where 1 Pascal (Pa)

$= 1\ N/m^2$

$= 0.102\ kg^\circ/m^2$

1 KPa = 1000 Pa

$= 102\ kg^\circ/m^2$

= 0.102 m of water head

100 KPa = 10.2 m of water head

For Density (ρ)

$1\ gm/cm^2 = 1000\ kg/m^3$

S.I. unit is same as in M.K.S. i.e. kg/m^3

For unit wt. (r or w = ρg)

$1\ gm^\circ/cm^3 = 1000\ kg^\circ \times m^3$

S.I. unit is N/m^3 where

$1\ N/m^3 = 0.102\ kg^\circ/m^2$ and

$1\ kN/m^3 = 102\ kg^\circ/m^3$

For Unit wt (γ or ω = ρ.g)

$1\ gm^\circ/cm^2 = 1000\ kg^*/m^2$

S.I. unit is N/m^2, where

$1\ N/m^2 = 0.102\ kg^*/m^2$ and

$1\ kN/m^2 = 102\ kg^*/m^2$

For Energy

$1\ gm^*\ cm = 10^{-6}\ kg^*\ m$

S.I. unit is Joule, where

1 Joule = 1 N.m

$= 0.102\ kg^*\ m$

For Power

1 gm* cm/sec = 10^{-5} kg* m/sec

S.I. unit is watt, where
1 watt (W) = 1 J/s
= 1 N m/s
= 0.102 kg* m/sec
=

For Absolute Viscosity (μ)

c.g.s units gm*′ cm/sec = 1 Poise
= kg*/m sec

S.I. unit is N/m sec
1 N/m sec = 0.102 kg*/m sec

1 centi Poise = 10^{-2} Poise
= 10^{-3} kg* m/sec

For Dynamic Viscosity

(c.g.s unit is gm sec/m^2)
1 centi Poise (cgs) = 1.02 × 10^{-4} kg sec/m^2 (M.K.S.)

S.I. unit is N.s/m^2
1 N/m^2 = 0.102 kg* sec/m^2

For Kinematic Viscosity

1 cm^2/sec = 10^{-4} m^2/sec

S.I. unit is same as in M.K.S. i.e., m^2/sec

*gm is here gm *w*, and kg here is kg wt.
For water = 100 kg/m^3 = 98 kN/m^2

Chapter 1

Water Demand

CHAPTER OBJECTIVE

In this chapter, the demand of water has been delt with in a very simple way, to make it understandable to any one. As much as possible the informations are given. It is expected that will be good enough to have a clear conception about the subject method of this chapter. To make the things clear, same stages through the Solved Examples, the things will be cristal clear.

1.1 DEFINITION

No life in this World can survive without water, as water is as essential for life as air is. Actually, two-third of human body is constituted of water. Water is formed by a chemical reaction of hydrogen and oxygen. (H_2O). It can be formed in liquid forms, in the form of gas as well as in solid form. Water is not only essential for human to survive but also for animals, plants and other living beings.

But this water need should be good and should not have any impurities which is not wanted for the human body. So to get the sufficient quantities of good quality of water, water supply system has been developed. With the help of modern knowledge and technique the water supply system has been developed to give the society a healthy water which is quite harmless.

Water is used for drinking, cooking, bathing, washing, etc., and it also use to help the farming and it is used huge in amounts in the industry, and also for sanitation as well. It also used for beautification of surrounding, that will reduce the environmental pollution.

It is quite clear that from morning to the end of night man using water. Whenever you need water, you open the tap to get it, this water is quite safe for use. This is the achievement of the modern water supply system.

1.2 WATER DEMAND

It is very essential to know the particulars about the scheme of water supply. It is most important to know the community to serve, whether it is a city or a small town. Because the water demand varies according to the needs.

The needs are such as follows :

1. Domestic need of water
2. Need for industries
3. Water demand for institutions
4. Water demand for fire
5. Unaccounted water

To get an accurate assessment all the above items should be counted to get a true requirements.

1.3 DOMESTIC NEED OF WATER

This is purely for a private use in a building for drinking, cooking, bathing, lawn sprinkling, gardening, sanitary purposes, etc. This also depends on the living style of the consumers. As per IS : 1172–1993. The minimum consumption for a town or a city. The amount of water need per capita per day for domestic used is 200 l/h/d.

Use	*Consumption in litres per head per day (l/h/d)*
1. Drinking	1.5
2. Cooking	2.5
3. Bathing	3.75
4. Washing of clothes	4.25
5. Washing of utensils	5.15
6. Washing and cleaning of houses and residences	6.15
7. Lawn watering and gardening	7.45
8. Flashing of water closets etc.	8.45
Total	200

Water consumption for colonies in small Indian towns and cities (for weaver section)

Use	Consumption in $l/h/d$
1. Drinking	1.5
2. Cooking	2.5
3. Washing of clothes	3.55
4. Washing of utensils	4.20
5. Washing and cleaning of houses and residence	5.10
6. Flushing of water closets etc.	6.10
Total	135

According to I.S. Code a limit is pointed out on domestic water consumption is between 135–225 l/h/d (with 200 l/h/d being minimum under ordinary circumstances with flushing system).

1.2 INDUSTRIAL WATER DEMAND

For ordinary industries per capita consumption on account of industrial needs of a city is generally taken as 55 liters/person/daily; this is only to meet the demand of small settered industries in the town or city. Separate provision will be made for a large scale industries.

S. No.	Nature of industries	Units of production	Water requirement litres
1.	Automobiles	Vehicle	40,000
2.	Distillary	Litre of Protochol	122–170
3.	Fertilizer	Tonne	90,000–2,50,000
4.	Leather	kg (tanned)	45
5.	Paper	kg	200–500
6.	Special variety paper	kg	500–1000
7.	Straw Board	kg	80–100
8.	Petroleum refinery	Tonne (crude)	1500–2000
9.	Steel	Tonne	2,00,000–3,00,000
10.	Sugar	Tonne	1000–2500
11.	Textile	kg	90–150
12.	Wood pulp	Tonne	2,40,000
13.	Coke	Tonne	14,000
14.	Ice	Tonne	2,000
15.	Woo scouring	kg 1	5–20
16.	Thermal powder	Mega Watt	1,50,000–1,60,000

Big industries will have their own water supply system. Almost all the industries are nowaday a little away from town or cities.

1.3 WATER DEMAND FOR INSTITUTIONS

This institutions demand will also cover the commercial establishment. Generally the water demand for institutions will cover railway stations, hospitals, hostels, offices, factories etc. This type of water demand should be treated separate from domestic and industrial demand. This is quite true that the quantities will vary with the nature of the city, type of commercial establishment, type of hospitals, etc. The Table 1.1 shows the requirement for individual establishment.

Table 1.1 *Institutional demand of water*

S. No.	*Types of Institution*	*Recomended rate LPCD*
1.	Hospital and laundry	350
	(a) Not more than 100 beds	450
	(b) More than 100 beds	185 (per bed)
2.	Hotels	140
3.	Hostels	70 (per cent)
4.	Restaurant	75
5.	Air ports and sea ports	75
6.	Junction Stations	75
7.	Terminal and intermediate station	50
8.	Day school and colleges	50
9.	Offices	50
10.	Factories	50
11.	Cinemas/theatres	20
12.	Medical quarters and nurses homes	160

1.4 DEMAND FOR PUBLIC PLACES

This is not so essentials to public demand. But still in some cities or towns they have provision for watering public parks, gardening, washing and sprinkling on roads, fountain etc. But for this purpose a nominal of 50% of the total consumption will be added. This is on an arbitrary basis only.

Fire outbreak in a big city or town, say more than 50,000 of population is a big problem. Say in a big cities fire outbreak at four or five places at a same time in that case need of water should be essential at a high rate. Provision should be made at the time of designing the whole system. The high rate of water consumption during a fire effects highly to the design of distribution system. Sometimes the rate of fire demand treated as a function of population. A formula to calculate the amount is given below.

According to National Board of under writers formation, as follows :

(a) **For a central congested high valued city–**

(i) Population less than or equal to 200,000

$$Q = 4637\sqrt{P}[1-0.01\sqrt{P}]$$

where Q = discharge

P = Population in thousands.

(ii) Population is more than 2 lakhs, provision for 54,600 litres/minute may be make with an extra additional provision for 9,100 to 36,400 litres/minute for second fire.

(b) **For residential city–**

(i) Small or low buildings = 2,200 litres/minute

(ii) Large and high buildings = 4,500 litres/minute

(iii) High value residences apartments = 4,500 litres/minute

(iv) Three storeyed buildings in densely built up sections = 27,000 litres/minute

1.5 CALCULATION FOR PER CAPITA DEMAND OF WATER

This is an approximate estimation for calculate the requirement of water demand of a town. For different heads the need of water as shown before is made a total of that and then that will be divided by the total population of that area or town, that will give an equivalent water demand per capita. As domestic and non domestic, in both. The cases population is involved. To get a guidance of that a break up is shown as follows.

1.	Fire demand	–	10 lpcd
2.	Institutions/public requirement	–	35 lpcd
3.	Industrial demand	–	35 lpcd
4.	Domestic demand	–	100 lpcd
5.	Unaccounted water	–	40 lpcd
	Total		**250 lpcd**

That will show the total water consumption per capita per day.

1.6 AFFECTING FACTORS OF WATER CONSUMPTION

These are depending on the following factors such as :

1. Standard of Living. For high standard of living people the consumption of water will be high, but for poor people, the consumption will be low. Naturally, in cities the consumption will be higher than the villagers.

2. Climatic Condition. Living in hot climate, people need more water, comparing to the people living cold climate.

3. Quality of Water. Good quality of water will be consumed more by people than that of the bad quality of water. So the consumption of water will be higher.

4. Industries. Water consumption in the industrial area, depends on the size and number of industries. Naturally, increase of industry will increase the water demand.

5. Water Pressure in the Distribution System. Due to high pressure in the distribution system the rate of flow will be high at the same time the leakages also will be high, that means that the demand of water per capita will be increases.

1.6.1 System of Sanitation

Specially in town the sanitation system is generally water carriage system is provided, due to that the water demand per capita is also increase.

The above mentioned factor's will effect the water consumption.

1.6.2 Forecasting of Population

Population forecasting is a very important factor for design purposes. Generally the design for the project is done on the basis of present population. But in the near future this design may not adequate for increasing population. There are lots of methods for population forecasting are accepted in practice. Some of them are most common and useful methods are described here.

1. Geometrical increase method.
2. Rate of growth decrease method.
3. Arithmetical/increase method.
4. Incremental increase method.
5. Graphical method.

1. Geometrical Increase Method. On the basis of growth of the town the percentage population increased is assumed. For further increase of population an average increase of population is used. But this method will give the higher values and quite good for applicable to newly develop industrial town for two or three decades. After that it is adviceable not to continue this. As for example, this method of geometric increase may not working at present. As the geometric method have existed in Durgapur during 1956–1976, in Ranchi 1960–1980 and steel/city Bokaro it was expected since 1970–1990, after that.

2. Rate of Growth Decrease Method. This method may called the modified method of geometric increase method. In this method, after the geometric increase method failed after two or three decades, then this method is applied to average size cities growing population under normal condition. In decrease rate of growth method the percentage fall increase is considered and after that the future population is calculated. Actually after the geometric increase method this will well covered the water demand.

3. Arithmetical Increase Method. For old and large city, the population is calculated on the basis of census report for past decade, then the increase is added to the present population to get the population for the next decade. Though this may not be so accurate, but it will be very close to get the increase population.

4. Incremental Increase Method. Practically this is the modified method of the arithmetical increase method. In this method the calculations are made as such.

The growth rate of population is found to be increased. It may found that an incremental increase in each decade. The incremental increase is taken to calculate for future population. Truely said that this may improve the arithmetical increase method, which may given the low result.

5. Graphical Method. Population Curve (Population vs. Year) in drawn and a smooth extension will be able to get the future population. The population care is drawn from the result of past census result. Comparing the other cities or towns curve with the some type of population and growth. This curve can be extended to calculate the future population. This method may be called as uniform increase method.

An idea and shape of the curve is given in Fig. 1.1.

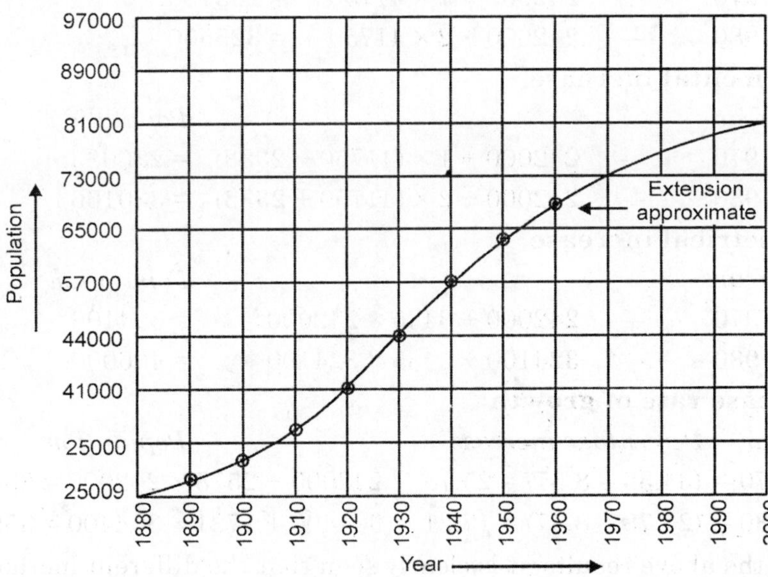

Fig. 1.1

Example 1.1 *The population data for a certain town is given blow. Find out the population in the year 1970 and 1980 by arithmetical, geometrical, incremental increase and decrease rate of growth methods respectively.*

Year :	*1920*	*1930*	*1940*	*1950*	*1960*
Population :	*75,000*	*1,10,000*	*1,50,000*	*2,00,000*	*2,42,000*

(CU)

Solution

Year	*Population*	*Increase*	*Invermental increase*	*Percentage increase*	*Decrease in percentage increase*
1920	75000	—	—	—	—
1930	110000	35000	–	46.70	10.30
1940	150000	40000	+ 5000	36.40	3.10
1950	200000	50000	+ 5000	33.30	12.30
1960	242000	42000	– 8000	21.00	–
Total		167000	7000	137.40	25.70
Average		41750	2333	34.35	8.57

Incremental increase

Year			*Population*
1970	—	242000 + 1 × 41750	= 283750
1980	—	242000 + 2 × 41750	= 325500

Incremental increase

Year			*Population*
1970	—	242000 + 1 × (41750 + 2333)	= 286083
1980	—	242000 + 2 × (41750 + 2333)	= 330166

Geometrical increase

Year			*Population*
1970	—	242000 + 3435 × 242000	= 324100
1980	—	324100 + 3435 × 324100	= 436000

Decrease rate of growth

Year	*Percentage increase*	*Population*
1970	—(34.35 – 8.57) = 25.78	242000 + 2578 × 242000 = 304400
1980	—(25.78 – 8.57) = 17.21	304400 + 1721 × 304400 = 356800

From the above results, it is clearly seen that the different methods gives the different values. So it will be better to select the proper method by considering into the nature and growth of the city or town.

Example 1.2 *The population of a locality as obtained from a census report is given below :*

Census year	*Population*
1880	*8,000*
1890	*12,000*
1900	*17,000*
1910	*22,500*
1920	*29,000*
1930	*37,500*
1940	*47,000*
1950	*15,000*
1960	*66,500*

Estimate the population of the locality in 1980 by the incremental method.

(U.P.S.C. Engg. Services)

Solution

Census year (*1*)	*Population* (*2*)	*Increase* (*3*)	*Incremental* (*4*)
1880	8,000		
1890	12,000	4,000	1,000
1900	17,000	5,000	
1910	22,500	5,500	(+) 500
1920	29,000	6,500	(+) 1,000
1930	37,500	8,500	(+) 2,000
1940	47,000	9,500	(+) 1,00
1950	57,000	10,000	(+) 500
1960	66,500	58,500	(–) 500
Total		58,500	(+) 5,500

$$\text{Average} = x = \frac{58,500}{8} \qquad y = \frac{5,500}{7}$$

$$= 7312.5 \qquad = 785.714$$

$$\Omega\ 7313 \qquad \Omega\ 786.0$$

∴ The population in the year 1980 will be

$$P = P_1 + 2x + \frac{2(2+1)}{2} \times y$$

$= 66{,}500 + 2\ (7313) + 3\ (7861)$

$= \mathbf{83.489}$

Example 1.3 *With the help of the following data, estimate by incremental increase method, the population of a city for the year 2000 AD.*

Year	*Population*
1880	*25,000*
1890	*27,000*
1900	*34,100*
1910	*41,500*
1920	*47,500*
1930	*54,050*
1940	*61,000*

(U.P.S.C. Engg. Services)

Solution

In the tabular form of the average of the net incremental increase is shown.

Year (1)	*Population* (2)	*Population increase* (3)	*Incremental increase* (4)
1880	25,000	2500	+ 4,100
1890	27,500	6000	+ 800
1900	39,100	74000	– 1,850
1910	41,500	55000	+ 1,900
1920	47,050	7450	– 950
1930	54,500	6500	
1940	61,000		
Total		36,000	6,800 – 2800 = 4000

Average decade $x = \dfrac{36000}{6}$ $\qquad y = \dfrac{4000}{5} = 800$

At the end of 2000 the expected population

$$P = P_1 + 6x + \frac{6 \times 7}{2} \times y$$

$$= 61{,}000 + 6 \times 600 + 21 \times 800$$

$$= \mathbf{1{,}13{,}800}$$

N.B.

where

P_1 = Population in the year 1940

Increase population = 6 (times)

Seven decade × increase population in between the decreases (i.e. 6 times)

$$= \frac{7\times6}{2}=\frac{42}{2}=21$$

6. Logistic Curve Method. Actually at normal situation the population growth i.e., like by birth, death, and migration, that does not make a big changes. But considering the situation like, coming of refugees, epidemic and war in the nearest country to us takes place and also an earthquake takes place in other provinces in that case the growth of population is going to be changed and that will effect the demand. In that case within the limited economic opportunity the growth curve will be followed to face the problem.

This growth curve is almost shaped like 'S' as shown in Fig. 1.2.

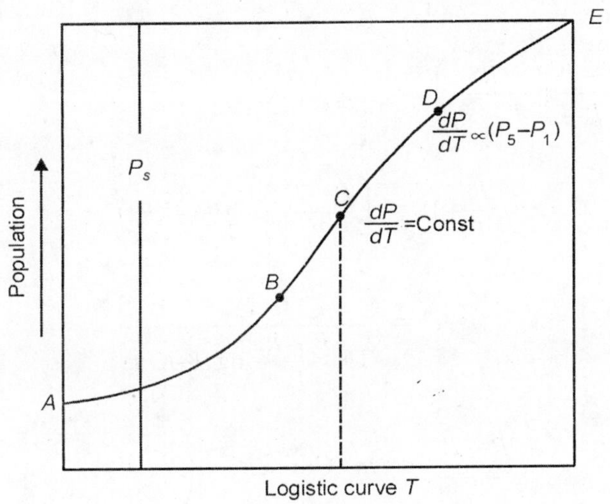

Fig. 1.2

At an early growth AB = increase rate i.e., Log growth $\frac{dP}{dt}\propto P$

At increasing rate DE = Late growth i.e., Ist order curve $\frac{DP}{dt}\propto(P_5-P)$

P_s is approached the situation population. BD = arithmetic increase i.e., $\frac{dP}{dt}$ = constant.

Considering the above equation, forecasting is done. According to Verhaulst a methemetical solution is done. The total curve AE = auto catalytic first order

$$\log_e\left[\frac{P_s - P}{P}\right] - \log_e\left[\frac{P_s - P_0}{P_0}\right] = KP_s t \qquad ...(1)$$

As P_0 = Point A population at start
P_s = Population saturation
P = Population at any time.
K = Const.

Then, from Eq. (1) we get

$$\log_e\left[\left(\frac{P_s - P}{P}\right)\left(\frac{P_0}{P_s - P_0}\right)\right] = -KP_s t$$

or
$$\left(\frac{P_s - P}{P}\right)\left(\frac{P_0}{P_s - P_0}\right) = \log_e^{-1}(-KP_s t)$$

or
$$\frac{P_s}{P} - 1 = \left(\frac{P_s - P_0}{P_0}\right)\log_e^{-1}(-KP_s t)$$

or
$$P = \frac{P_s}{1 + \frac{P_s - P_0}{P_0}\log_e(-KPt)}$$

If
$$\frac{P_s - P_0}{P} = m \text{ (Const.)}$$

and
$$KP_s = n \text{ (Const.)}$$

Then we will get
$$P = \frac{P_s}{1 + m\log_e^{-1}(nt)} \qquad ...(2)$$

Equation (2) will be the logistic curve equation. According to Melean's suggestion that if P_0, P_1, P_2 at time $t = 0$, t_1 and $t_2 = 2t$, if this is extended over a certain period, then the saturation value P_s, and the constants m and n can be calculated from the following equation :

$$P_s = \frac{2P_0 - P_1 P_2 - P^2 (P_0 + P_2)}{P_0 P_2 + P_1^2}$$

where $m = \frac{P_s - P_0}{P_0}$

and $$n = \frac{1}{t} \log_e \left[\frac{P_0 (P_s - P_1)}{P_1 (P_s - P_0)} \right]$$

$$= \frac{2.303}{t} \log_{10} \left[\frac{P_0 (P_s - P_1)}{P_1 (P_s - P_0)} \right]$$

Example 1.4 *The population of a town is, 40,000, 10,000 and 1,30,000 at three decades. Find out the following :*

(*1*) *Population at saturation*

(*2*) *Logistic curve equation*

(*3*) *Population expected in the next decade*

Solution

We get from the following equation

$$P_s = 2P_0 P_1 P_2 - \frac{P_1^2 (P_0 + P_2)}{P_0 P_2 - P_1^2}$$

Now,

$$P_0 = 40{,}000 \; t = 0$$
$$P_1 = 1{,}00{,}000 \; t = 10 \text{ years}$$
$$P_2 = 1{,}30{,}000 \; t = 20 \text{ years}$$

Then $$P_s = \frac{2 \times 40000 \times 100000 \times 130000 - (100000)^2 (40000 + 130000)}{40000 \times 130000 - 100000 \times 10000}$$

$= 1{,}37{,}500$

Then the saturation population = **1,37,500**

Again $m = \frac{P_s - P_0}{P_0} = \frac{137500 - 40{,}000}{40{,}000} = 2{,}4375$

$$n = \frac{2.3}{t} \log_{10} \left[\frac{P_0 (P_s - P_1)}{P_1 (P_s - P_0)} \right]$$

$$= \frac{2.3}{t}\log_{10}\left[\frac{40000(137500-100000)}{100000(137500-40,000)}\right]$$

$$= -0.18697$$

In the next decade (i.e., 20 years after) will be

$$P = \frac{137500}{1+24375\log_{10}^{-1}(-0.18697\times t)}$$

$$P = \frac{137500}{1+24375\log_{e}^{-1}(-0.18697\times 30)}$$

$$= \frac{137500}{1+24375\log_{e}^{-1}}(-5.6091)$$

Substituting $\log_e^{-1}$ (–5.6091) = x

or $\log_e x = -5.6091$

or $2.3\log_{10} x = 5.6091$

or $\log_{10} x = \frac{-5.6091}{2.3} = -2.43874$

or $x = 0.00364$

Therefore $P = \frac{137500}{1+24375\times 0.00364}$

$= \mathbf{126291}$

1.7 PLANNING AND DESIGN PERIOD

A city or a town may be divided into zones. Some part of the city is thickly populated and some part is highly populated; and other part may be thinly populated. This is an important factor of planning.

On the other hand the other point is to Waters Supply Planning and the Design Period. The design is to be done in such a manner that it should have the capacity to cope up the increase of population. But from the economic point of view this period cannot be too long or too short. If it is too long, the present capacity will remain idle and that will be uneconomical, i.e., the investment at present may not be useful till the long time to come. On the other hand a water supply system cannot be for a short period, that will trouble within a short period create which is not desirable. So the design period should be such that it will work satisfactorily within a reasonable period.

For example, some normal design period of various water supply system normally accepted and used in the planning.

Name of units	*Design period*
1. Pumping station	10 – 15 years
2. Pipe lines	20 – 25 years
3. Water treatment plant	10 – 15 years
4. Impounding reservoir	30 – 35 years
5. Distribution line for maximum hourly demand	5 – 10 years
6. Wells for underground source	5 – 10 years

1.8 VARIATION IN DEMAND PER CAPITA

This demand of water is not a constant factor. It will vary according to season i.e., seasonal variation, hourly variation and also diurnal variation. It is not possible to get an annual average daily consumption. During certain months of the year water demand is found to be high. Similarly, certain period or the day the water demand is maximum. Generally, it is found that, during the day say from 7 A.M. to 12 Noon in a day demand of water found maximum. Rates of hourly demand should be considered in design of distribution system. On the other hand the daily variation is found to be useful for pumping station and treatment plant. An hourly demand curve is shown in Fig. 1.5.

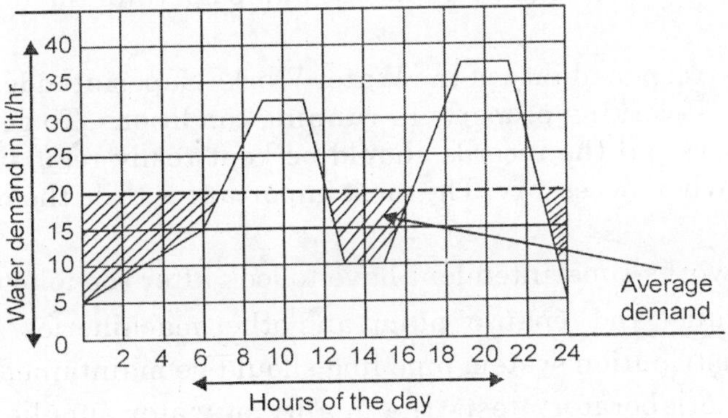

Fig. 1.5 *Hourly rate of demand*

From the curve it is clearly seen that in the early hours of the morning say 0–6 hours is generally small, then increase sharply as the day advances. Ultimately it will reach the peak value at about 8 to 11 AM. After that it will decreases sharply at about 10 PM it will remain constant to 4 PM. After that will increases sharply in the evening and reaches peak between 7–9 PM. Finally falling to a low value in the late hours in night.

1.9 NORMAL VARIATIONS IN DEMAND OF WATER

1. Generally the maximum daily consumptions take 180 percent of the average.

Maximum daily demand

$$= 1.8 \text{ (i.e. 80\%)} \times \text{Average daily demand}$$
$$= 1.80\%$$

2. Generally the maximum hourly demand is taken 150% of its average.

∴ Max. hr. demand i.e. peak demand

$$= 1.5 \text{ (i.e. 150\%)} \times \text{Av. hr. demand of the max. days}$$

$$= 1.5 \times \left(\frac{\text{Max. daily demand}}{24}\right)$$

$$= 1.5 \times \left(\frac{1.8 \times \text{av.}}{24}\right) = 2.7\left(\frac{\text{av.}}{24}\right)$$

$$= 2.7 \text{ (Annual average hourly demand)}$$

1.10 MANAGEMENT OF WATER WORKS

Generally in a water works a Superintendent or a Manager, whatever it may be will be the full incharge of the water works. It will be his duty to look after whole water works should be properly working. In this connection he will have other personals to help him. Such as pump operator, meter reader and helpers etc.

It will be the responsibility of the Water Works Superintendent to see the supply project is working properly in running conditions. To get the proper efficiency of work, all the records should be kept ready regularly and take prompt action when necessary. The most important is to do the maintenance work regularly.

Mainly the works superintendent have to look after the following :

(i) Maintenance and repair of pump and other machineries.

(ii) Water distribution system pipe line should be maintained properly.

(iii) Using the laboratory tests the quality of water supplied should be regularly checked.

(iv) Assesment detection and prevention of water losses.

Wastage and losses of water sometimes makes a great trouble and this is due to lack of preventive maintenance; attention should be paid to this carefully. The leakages mainly from, reservoir losses, from treatment plant, water mains are corroded, joints are faulty, and also from consumers end.

To reduce the water wastage, a systematic waste and leakage survey should be done and action should be taken promptly to reduce the wastage.

To find out the waste in water mains, survey work is done in zone wise. From the ratio of the lost amount to the average daily demand of flow, the percentage loss can be found out. On the other hand the losses at the consumers end can be find out by local enquiries.

To find out the leakage waste if necessary, the following methods should be taken :

(i) By visual examination the leakage in the pipe may be find out.

(ii) By sounding rod, or electronic leak locator for pinpoiting of leak of pipes.

The sounding rod is about 1.2 m long and 12 mm hollow mild steel rod, flat on pointed in one end the other end and is fixed with a cup shape grass cap of 50 mm diameter. The rod is carried over the surface alone the centre line of pipe noise due to water leaking is picked up. The ear thus locating the possible leaks.

On the other hand the electronic leak detector is a very sophisticated instrument consists of a pick up, amplifier and headphone. The vibrations of round created by water leakage are selected and magnified by the magnetic pick up and converted into electrical impulse. These are very sensitive and pin point the position of leaks.

Losses due to leakage in valves, joints and pipes should be reported promptly and take prompt action to stop the leakage. The percentage of water loss due to the leakage should not be more than 20 percent.

PROBLEMS

1.1. Enumerate the methods you can adopted for determining the total water requirement of a community. **(Ran, U.P.S.C. Poona)**

1.2. Explain how per capita rate is arrived in a water supply scheme of an industrial town. **(AMIE)**

1.3. Suggest method to minimise the wasteage of water in your halls of residence. **(LIT)**

1.4. Explain three different methods of forecasting the future population of a town. Comment on their relative merits. **(U.P.S.C.)**

1.5. Discuss how the future population of a town is estimated ? **(Ran)**

1.6. What are the assumptions made for forecasting the population of a city ? **(I. I.T., P.U.)**

1.7. Estimate the future population of a town in 2001. Given

Year	1911	1921	1931	1941	1951	1961
Population	350000	466000	994000	1560000	1623000	1839000

Justify the method you have used. **(I.I.T.)**

1.8. Why is population forecast necessary in the design of public water supply scheme ? **(Pat)**

1.9. What rate of supply would your recommended for a town with a population of 50,000 ? **(Pat, A.M.I.E.)**

1.10. Write short notes on the wastage of water in public water supplies. State various methods of detection and prevention of wastage. Enumerate the cause of such wastage. **(U.P.S.C.)**

1.11. A town water supply is given at 1351 (30 gallons) per head per day. Suggest a possible break up of the consumption on different heads.

What allowance should be made in India for loss and wastage in water supply ? Give reasons. **(U.P.S.C.)**

1.12. What are the average consumption of water per capita per day for most of towns in Bihar ? Explain daily and hourly variation in consumption of water and state how the consumption quantity for town is assessed. **(Bha.)**

1.13. Mention and describe the purposes for which water is supplied to a modern industrial town. What do you consider by per capita day water supply ? What factors determine the rate per capita water supply to a city? What are the common values of rate per capita/day of water supply in India with varying local conditions? **(U.P.S.C.)**

1.14. What is meant by the term "Per Capita demand" ? How is it estimated ? What values of per capita would you recomended for :

(i) a small town with a population at 50,000.

(ii) a small city with a population of 1–2 lakhs.

(iii) a large city with a population of 5 lakhs.

State the reasons in support of your answer.

1.15. Describe the incremental method by which the population of a locality on a future date can be estimated. Also state its advantages over arithmetical and geometrical methods. **(U.P.S.C.)**

Chapter **2**

Sources of Water

CHAPTER OBJECTIVE

In this chapter, the important examples are useful from practical point of view has been discussed in a simple way. This is vast portion to be explained. This is not useful at this stage to explain. The only portion which has been discussed, which will be helpful to anyone to understand the real problem.

2.1 QUALITY OF WATER

Water quality greatly depends upon the climate conditions, geographic and geological conditions. In natural water many factors are playing a part to vary their quality. As in natural condition, some water having a suspended particles and some of them are very clear. Some of the natural water having soluble minerals in it which is good for health and some of them are not so good for health, because they are having different types of virus which is creating trouble to the human health.

2.1.1 Hydrological Cycle

From the hydrological cycle the natural water found as a continuing cycle. As water evaporates from oceans, river, and ponds, and then it is precipitated in the form of rain, snow and hail. Then some part of the precipitation goes into the water surfaces and the other parts fall upon the land. From the land some parts will be used by the vegetation and the other parts will run off through the streams and lakes, and for the river which will lead it to the oceans, and some portions will sinks into the ground. The main idea of water supply is to develop by the interception of surface run off or by the collection of percolated water from the ground.

From the water supply point of view it will be a great importance in the development and purification of water for supply. This will not only effect the quantity of water but also effect the quality also. Considering hydrological cycle available sources are classified as :

1. Rain and snow
2. Surface water
   (a) Streams
   (b) Natural ponds and lake
   (c) Impounding reservoirs
3. Ground Water
   (a) Springs
   (b) Shallow wells
   (c) Deep wells
   (d) Artesian wells
   (e) Infiltration galleries and wells.

1. Rain and Snow

Water vapour condensed in to form cloud and then falls into rain and in high altitude where the temperature is very low, it forms snow. When the rain and snow are high above the earth they are practically pure in nature. As they falls down, they are absorbing oxygen, carbon dioxide and many other gases from the air, and also absorb dust, smoke and fumes. At the time of falling rain and snow, they will carry bacteria and spores of minute plants.

Rain and snow falls in the country side is much more clear than that of a cities. But in the drought season and in a drought area, rain contain the more dust than falling in the rainy season.

Before using the rain water for supply system, they should be stored in a reservoir and should be properly treated to see that it should quite safe for consumption.

2. Surface Water

At the time of rain some parts of water run off to the stream, ponds, river and ocean. But at the high altitude snow evaporates and in winter season snow melts less and supply water to the stream in small amount, but as the temperature rises. Snow melt more and more. Thus the stream gets flooded. This will give the more quantity of water but the quality may not be so good.

So the quality of water from surface is depending upon the character of the area where it is available. It depends upon the geology and topography of the area. It has been found that the water from stream is more variable and less satisfactory than ponds and lakes. Water passing from lime stone is hard and corrosive than water passing from granite rocks. Sometimes the surface water may be heavily polluted by the effect of sewage and industrial wastes.

(a) *Streams.* Due to the direct rainfall water flows in the stream over the surface of the ground at that time it overflows from lakes, swamps and also sweeping the ground from the high land and valleys. This proportion of flow will vary from season and will effect geology and drainge system of the area.

During the time of heavy melting of snow and flood the flow will be muddy and will contain large amount of bacteria.

Stream water is generally polluted by man of their activities because highly objectionable because of over loading with putrescible organic matter.

Soluble minerals are present in the stream water which is coming the soil. The type of minerals are vary with the geological formation of the earth.

But ultimately the man made, and natural pollution will change the colour and turbidity, tests, and odours, and so it effect the hardness bacteria and other micro-organism in the supplied water.

(b) *Natural Ponds and Lakes.* Water collecting in ponds and lakes are from streams. As the water in ponds and lakes are quite in nature comparing to the stream, a change of quality is observed and they are self purified. This purification also depend on the shape and size of the ponds and lakes. Due to the long storage, the suspended materials are settled at the bottom, colour will be bleached; also the bacteria will be removed. Usually the stored water will have uniform quality than water taken directly from stream. Sometimes the micro-organisms may grow heavily in such water. Also flooded water coming into it may cause pollution, also pollution may occur by the surrounding population. In a large lake due to dillution and self purification make sure to get good quality water. Comparing to the small lake, purification is less. In a small pond algae growths are likely to occur. Many times both ponds and small lakes suffer from spring and fall "overturns", which will stirring-up of bottom sediment temporarily.

2.2 IMPOUNDING RESERVOIR

Impounding reservoir is more or less like a natural lakes; and pounds. It is formed by a dam across the valleys cut by a stream. At the time of building a reservoir the surface soil was stripped of all the vegitation and the soil to avoid the effect of decomposing organic matter. Modern practice to omits soil stripping and relie on the selection of point of intake to secure satificatory quality of water.

At the mid-depth the quality of water will be best. At the top depth there will have some algae. And at the bottom depth, there will have carbon-dioxide, iron, manganese, etc. It is found that in deep of lakes and reservoir, the water at the bottom remain cold throughout the year; because a permanent zone of relative stagnation occurring at depth below six metres.

2.3 GROUND WATER

At the time of rain fall some portion of water flows away and some portion will percolates into the ground and becomes ground water. As it passes through the ground will have to meet many sustances both organic and inorganic. Some of them are dissolved in water such as carbon dioxide, and other organic matter decomposed in the soil. Oxygen will be removed from water by the decomposed organic matter in the soil. This water free from oxygen and rich in carbon-dioxide, will dissolve the iron and manganese from the soil. Water having dissolved iron and manganese is favourable for the development of crenothrix similary organisms in collected ground water supplies.

At the time of rainfall some bacteria and living organisms on the ground surface may be picked up by water will be removed when the water percolated through the soil, due to filtration. Only exception is found when fissured rock such as lime stone is found near the surface.

The sanitary condition around the ground water, should be avoided, because there will be a great chance to have the sea-soil water polluted. To avoid any pollution due to the sanitary system, it is better to avoid that spot, if it is not possible, in that case precaution should be taken to guard the pollution.

Generally the ground water are clear, cold, and colourless than the surface water, but will be harder than the surface water. From the point of view of bacteria, the ground water is much safe than the surface water; where the surface pollution may be present.

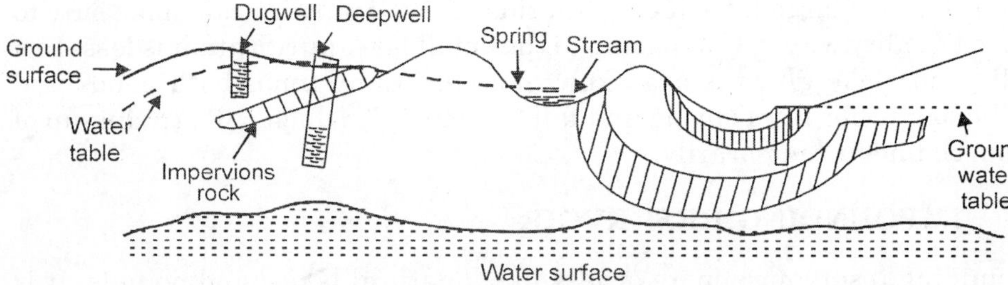

Fig. 2.1

(a) *Springs.* Water flowing over an impervions strata comes out into the surface is called spring. This will happen when the impervious strata out crops below an ulpand area of pervious strata. Spring occur from a hollow strata is more likely to be polluted than the spring occurs from a deep strata. Generally, their quality largily depends on the geologic formation as well as their locality. Water supply from spring is only worked for small communities.

Hot springs are found in India, this is due to the presence of sulphur in them. As springs in Sohana in Haryana, and at the banks of Satluj river at TattaPani near Simla, at Manikaran near Manali on Parvati river in H.P., hot spring in Rajgirh in Bhiar is well known. Also the hot spring at Bakreswar in West Bengal may have helium. These chemicals like sulphur etc. mixed with water to produce hot water at the top.

(b) *Shallow Wells and Infiltration Galleries.* Generally the shallow wells are developed in the surface deposite of pervious deposite laying in top of a impervious deposite. Generally a well deeper than 20 metre is called a deep well. The suction lift used to obtain water from them. The nearby catchment area will influence the quantity of water of a shallow well. Properly protected well may have the good quality of water.

The shallow water bearing strata near to the springs and ponds the infiltration galleries will form. Generally the water in the galleries will have the same quality of water as the shallow well in that area.

Infitration wells are more or less to that of infiltration galleries. Generally they are underground structure. But it is not necessary to construct them with the form of a trench, the water can be collected from wells.

(c) *Deep Wells.* Depending upon the nature of the underground strata, the deep well may be driven or drilled. Quite often they reach the water bearing strata before the desired depth. In the deep well also the nature and quality of water also depend on the nearest catchment area; tributary and also the geological formation through which the water have passed.

Actually the too deep well is expensive. In this case to get into water long way to go under the ground and may have to overcome some rock layer. But the deep well water are generally to be more highly mineralised than the shallow well water. Again the deep well waters are clean, colourless and may have iron and manganese. Deep well waters are usually good from bacterial point of view. Due to the driving or drilling operation may be contaminated temporary at the early stage. But after sometime the good quality of water will be available from it.

(d) *Artesian Wells.* Due to the geological formtion an aquifer may be such that it will behave like a pressure tube. Some parts of the aquifer will have below the hydraulic line. When the hydraulic pressure is produced in some region the pressure goes very high, that may even cause the water to rise up and may be even above the ground, this condition is called the artesian condition. So, if a well is dug in this condition to get the water supply purpose that will be known as artesian well.

2.4 SOURCE SELECTION

For the construction of a water supply scheme, the following points should be considered :

1. Quality of water of the source
2. Quantity
3. Source location

2.4.1 Quality of Water

Water quality can be determined by physical, chemical and bacteriological test. Depending on the basis of these results, the source of water can be classified into different groups.

Group 1

Best available water found in nature no need to have any other treatment except disinfectation. This is called the highly satisfactory source of water and many deep wells may be formed.

Group 2

Water need only the treatment of filtration and disinfectation is known as standard, and in this category, water from shallow wells, springs, and lakes are covered.

Group 3

Water requires the treatment coagulation, filtration, and some time aeration, is classified as specious water.

Group 4

Water containing toxic substances, highly polluted by sewage and industrial wastes. Practically impossible to purify by the normal treatment. So this type of water is not satisfactory at all.

2.4.2 Quantity

Yield in the same thing as the quantity of water available from the source. Depending upon the type of source, there are number of methods are applied to measure the quantity of water. The units of measurement are hectare-water per day or m^3 per second etc.

This can be determined by the following methods :

1. In practice
   (a) Pumping directly
   (b) Recuperating test
2. Theoretically
3. Combination of both.

The discussion of all these will be seen under the heading of "Ground Water".

To get the surface source quantity the following methods are generally adopted.

1. Stream gauging
2. Method of cross-section velocity
3. Chemical method
4. Radiological method.

Discussion of all these methods are found under the heading of "Rain fall run off".

2.5 LOCATION OF SOURCES

The source should be choosen in such a way that it should be as close as possible to the town. But care should be taken to see that the source should be as far as possible from point of waste disposal. On the other hand in case of a river the source should be well off in the upstream since.

2.6 RAIN FALL RUN OFF

The rainfall is composed of showers and steady rainfall. The winds carrying good amount moisture from Bay of Bengal and strikes the south slope of Himalayas causes intense rain. Mainly rain causes by a wind carrying heavy amount of moisture, when obstructed by mountain, hills, etc. causes rain.

2.7 DISTRIBUTION OF RAINFALL

At the beginning of rain fall, it will be obstructed on intercepted by big buildings, bushes, trees etc. This way some portion of rain is prevented to reach in the ground. If the total rain is denoted by R the intercepted rain will be denoted as R_i. But on the other hand the difference between the total rain and the intercepted rain is called the ground rainfall; and is denoted by R_g i.e. $R_g = R - R_i$. But at the time when rate of rainfall (r) exceeds the interception rate (r_i) at that time water will start to reach the ground, infiltration or percolation starts, this is denoted by P. The soil at a given condition can absorb the maximum amount of water is known as infiltration capacity (f) or percolation capacity. When the infiltration capacity is reached after that, the water starts to overflow on top of the ground. This quantity is called the storage depression S_d. But all this depressed water will either evaporate or used by the vagitation or percolated into the soil after some time. None of this appears as surface run-off.

In the soil, when the storage depression is completed, and the intensity of rain exceeds the infitration capacity of the soil, the differences is called the excess rainfall (R_e) and the rate of excess rainfall is called (r_e).

Then $$R_e = r - f$$

Accumulation of water on the ground surface due to the excess water is called surface detention (*D*) and flows over the slopy land towards the channel is called surface runoff or direct runoff (D.R.). But the channel storage (S_c) is a permanent stream channel which will have entire water in it.

Initial rain is the rain which falls at the beginning of a storm at this stage the storage depression is just completed.

Residual rain is called to this rain which falls near the end of storm at a rate less than the infiltration capacity. But both of these two types of rainfall does not have any contribution to the surface run-off.

The intervening period of these two is called the net supply interval.

The residual infiltration is known as the infiltration occurs after the supply interval.

2.8 RAINFALL MEASUREMENT

It is necessary to estimate precipitaion and also its effect on the ground surface. Generally the precipitations are measured as the vertical depth of water will collect on a level surface. For a given period, the total amount of precipitation will express the depth to which it collect on the horizontal surface of the earth to get accurate mearsurement, this should be in mind that there should not be any losses like evaporation or runoff. Two types of precipitations are measured, rain and snow. They have the different type of measuring devices. To get the over all preciptation measurement, the measuring devices should be placed at different parts of the country. The name of the precipitation measurement instrument is Rain Gauge and for snow, is snow gauge.

2.9 RAIN GAUGES

An open face with vertical sides to collect the drop of rain fall can be used as a gauge to measure the rain fall. Two types of rain gauges are used to measure the rain fall. One is non-recording type and the other is recording type.

2.9.1 Non-recording Type Rain Gauge

This type of gauge is most commonly used in India. In this one no record is kept of rainfall. In this one the rain falls are collected, and then measured in a graduated measuring cylinder. So as to directly measure the rainfall in cm of water depth.

$$\text{i.e.,}\quad \frac{\text{Vol. of water collected in cm}^3}{\text{Area of the gauge in cm}^2} = \text{cm depth of rail fall}$$

Lots of different methods are designed in this gauge but one of the most common one is Symon's type.

The polythene bottle and funnel are used in this case. The bottle is as a collector collect rains. The capacity of the collector may have 100, 200, 400 and 1000 mm of rainfall is generally used. When the rain falls it is collected in the bottle. One can come daily at 8.0 A.M. to collect the rain water from

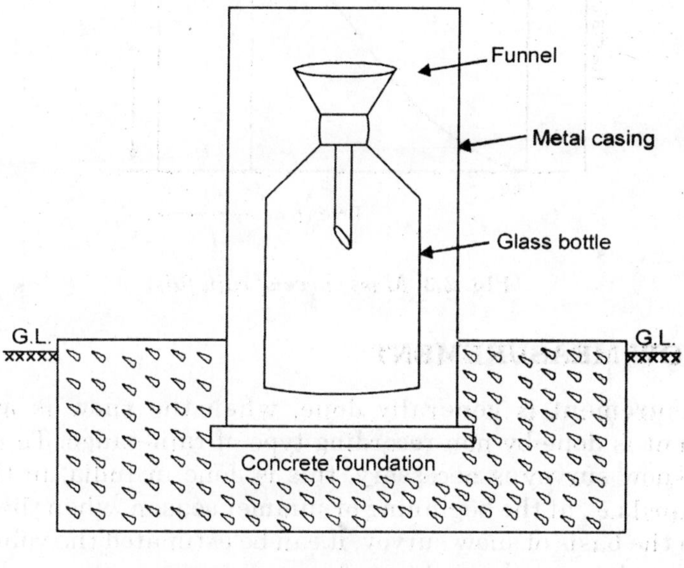

Fig. 2.2 *Simon's rain gauge*

the bottle, this may be recorded. This is the common process in the country. The recorded rain is thus said to be at 8.0 A.M. If the rain is heavy, that may exceed the capacity of the bottle, in that case, the record can be taken two or three times, in a day and the sum of those will said as rainfall in 24 hours.

2.9.2 Recording Type Rain Gauge

This is a type of rain gauge, which records the rainfall automatically by mechanical means without the help of any person the rainfall record will be recovered in a graph paper. The gauge thus produces a record of commulative rain against time, which is called the mass curve of rainfall.

This curve will indicate the time of on set and cessation of rain. So this will give the duration period of rain. The slope of the curve will give the intensity of rain at that period.

This is also called integrating rain gauges. This gauge can give the record of rainfall continuously for number of days. This is very helpful in a hilly and far off areas. Number of different types of model found in the market.

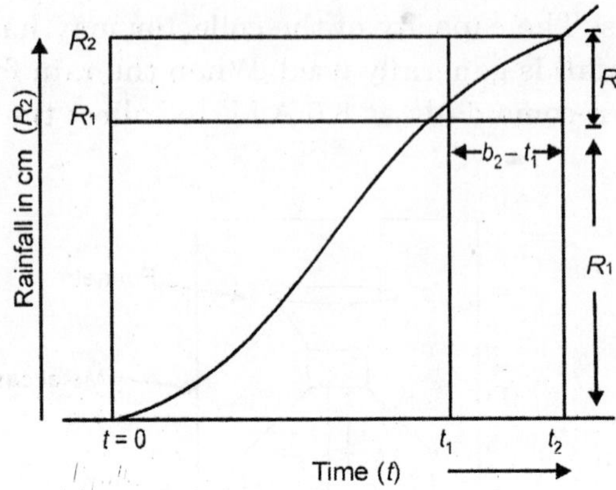

Fig. 2.3 *Mass curve of rain fall*

2.10 SNOW MEASUREMENT

Snow measurement is generally done, when the snow is melted. This measurement is done by non-recording type of rain-gauge. To measure the snow, the snow survey is necessary, this is done in India, in the month of March or April i.e., at the beginning of summer season; when the snow starts to melt. On the basis of snow survey, it can be estimated the volume of water can be expected due to the melting of snow.

2.11 RUN OFF

The rain fall in a catchment area, some portion is immediately appears to flood flow, some of them will be evaporated. Some of them will be intercepted by the vegitation, and some of them will be percolated into the ground. Then the total flow on the ground surface is the stream flow and the flow of underground water is called the baseflow.

2.11.1 Run off Measurement

Most of the times it is necessary to measure the run off, this can be done directly by gauging. By gauging the mean velocity is measured by using current meter or any other method to calculate the discharge. At any stages of water say at high, low and intermidiate stages, in that case the elevation of water level can be measured by absolutely fined staff on the river bed this is also known as a gauge.

To measure the low flow of a stream a masonry wier is constructed across the stream. That may be rectangular, or a triangular notch (i.e., a Vee notch) is fitted there. Then applying the standard formula for them the discharge can be calculated.

(a) Formula for rectangular notch :

$$Q = 1.84 \text{ LH}^{3/4} \text{ m}^3/\text{sec}$$

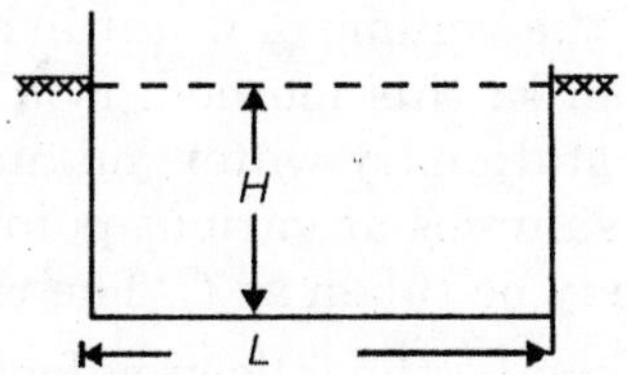

Fig. 2.4 *Rectangular notch*

where Q = discharge in m^3/sec

L = the width of the wier in metre

H = head over crest in metre

(b) Formula for traingular notch :

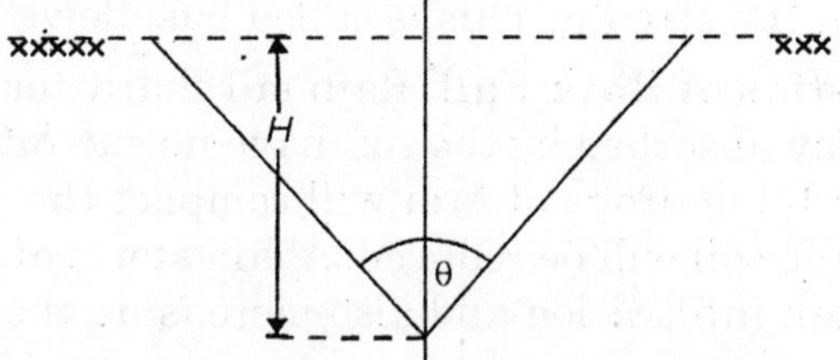

Fig. 2.5 *Triangular notch*

$$Q = 1.4 \tan \frac{\theta}{2} \text{ H}^{5/2} \text{ m}^3/\text{sec.}$$

θ = angle of the notch in degrees

H = head over crest in metres

(c) Chemical and method of radio isotope

This method is very simple and very complicated to work in practice. A chemical salt with known concentration is mixed at a point in the stream at a constant rate, then knowing the concentration of the salt, the water is tested at a point in the downstream.

where q = the rate of flow chemical at cut-stream side

c = concentration of solution

Q = water flow in stream

C = concentration of solution in downstream

This discharged $$Q = \frac{Cq}{C}$$

It is much better to get the volume of water in a lake, the method of radio isotope is the suitable one. In this method, isotope dispersion rate is very high, even in the case of stationary water, mixing is not a problem in this case. Then collecting the samples at various points and positions. Then the average of these values may be taken as C. The rest will be as above.

The runoff will be affected by the following factors :

1. Infiltration
2. Characteristics of rain fall
3. Characteristic of soil
4. Evaporation

1. Infiltration. This is nothing but the percolation of rain water into the ground and forming ground water. This is the same as infiltration. This ground water flow may be exposed in the different forms. This may sometimes comes out and join the stream. This is called baseflow.

2. Characteristics of Rain Fall. Rain fall character will control the run off. Slit rainfall may absorbed by the soil have no run off. On the other hand heavy rain fall with big drops of rain will compact the soil surface and the entry of rain into the soil will be reduced. Again storm of low intensity rate of rain may cause high infiltration and also increasing the soil moisture.

3. Characteristics of Soil. In the surface of the soil if the pore size between the soil grain is low, the infiltration will be low. But the percentage porosity may not be low. The small grain soil size will have the low permeability. So,

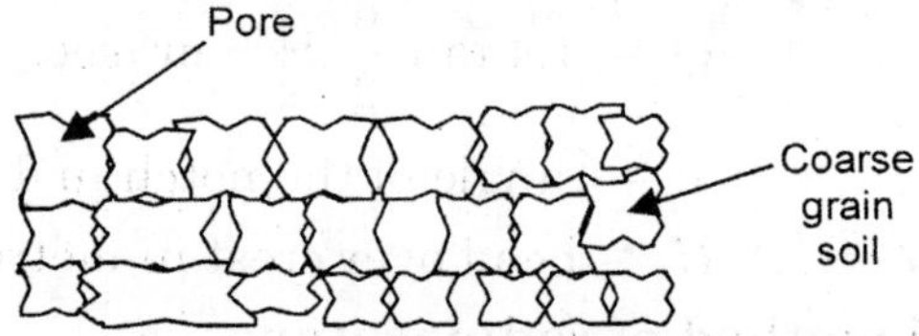

(a) Sandy soil

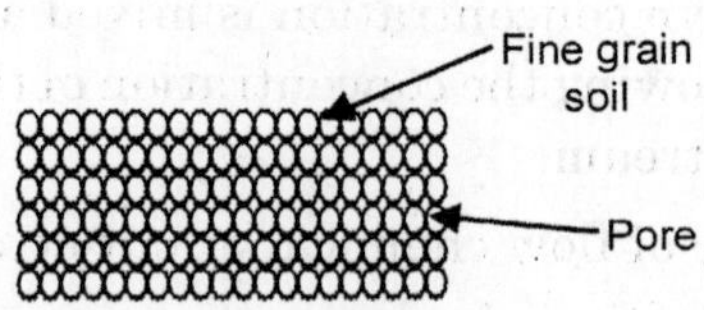

(b) Silty or clay soil

Fig. 2.6

sand bed or sand will have high permeability i.e., high filtration, on the other hand silt and clay will have low permeability i.e., low infiltration. As 0.2 mm or larger size of sand will increase the infiltration on the other hand soil particals having 10^{-6} mm size will have low infiltration.

4. Evaporation. Evaporation of surface water from lakes and reservoirs will effect the calculation, because this evaporation is really be a considerable amount. So the rate of evaporation should be known to get the losses of the stored water. Practically, the losses are due to the temperature, climatic condition and humidity and wind effect cannot be neglected.

An approximate method to find out the rate of evaporation is done in the field. This is done by a pan 1.2 m dia. 25 cm deep. This is set on a 15 cm wood grill in such a way that the water surface is just a bit above the 30 cm above the ground surface. The water is placed 5 cm from the rim of the pan. The result found is multimplied by factor 0.7 to get the equivalent evaporation of reservoir.

2.11.2 Storm Run Off

There are three ways of determining the storm run off.

1. The rational method
2. Empirical formulas
3. Mac Math formula

1. The Rational Method. In this method, all the factors are giving maximum run off to be considered, and then combining them to get the amount expected. The intensity of rainfall can be obtained for that by the duration equal to the time of concentration.

In this case the drainage area should be considered, because it will have direct effect to the run off. The quantity of storm flow can be found by the formula,

$$Q = \frac{1}{36} ACR$$

where Q = storm flow in m^3/sec

A = drainage area in hectares

C = co-efficient of imperviousness of the area

R = rainfall intensity cm/hour

2. Empirical Formulas. The most useful formula Barkli – Ziegler is used in this case.

The formula is

$$Q = \frac{1}{36} (ARC) (S/A)^{1/4}$$

where S = average slope of ground in m/1000 m other terms are same in the rational method.

3. The Mac Math Formula, is stated as

$$Q = \frac{1}{36} (ARC) (S/A)^{1/5}$$

Q will be same as in empirical formuals.

2.11.3 Catchment Yield

Considering the big catchment area which added good amount of water to the stream, other methods are adopted to find out the flood flow. The formulas are as follows :

(1) Ryves Formula

$$Q = CM^{2/3}$$

where Q = discharge in cm/sec.

C = constant ranging from 6.5 to 10

M = catchment area in sq. km

(2) Dicknes Formula

$$Q = CM^{3/4}$$

where Q = discharge in cuseas

C = constant varies from 800 to 1800

M = catchment area in sq. km

(3) Myers Formula

$$Q = 176 \times p \times \sqrt{M}$$

where

Q = flood peak in cum. sec.

p = constant varying from 20 to 150 or so.

M = catchment area in sq. km

(4) Burgers Formula

$$Q = Cal/l^{2/3}$$

where

Q = discharge in m³/sec

l = length of area in km

a = the area in sq. km or in hectares

C = constant 18.4 for a in sq. km

c = constant 0.184 for a in hectares

The following formulas are used to find out the runoff on the annual basis.

1. Inglis Formula

There two separate formula is used, one for hilly areas and the other is used for plane area. This formula is used in South India.

(a) $$F = 0.85\,R - 304.8 \text{ for hilly area}$$

(b) $$F = \frac{R(R-177.8)}{2540} \text{ for plane area}$$

where

F = annual runoff in mm in the catchment

R = annual rainfall in mm in the catchment.

2. Khosla's Formula

From a research paper published in Central Board of Irrigation, this formula based on the mean temperature, which is also depends on the losses by evaporation, transpiration, sunshine and the velocity of wind.

$$F = R - 4.813\,T$$

where

F = Run off in mm

R = Rainfall in mm

T = Temperature in degree centimeter $\because C + 273 = K$

This formula is quite useful to use when the rainfall is uniformly distributed throughout the year. But if the rainfall is not so well distributed throughout the year, on the other hand the rainfall concentrated within a short part of the year, in that case the runoff can be calculated on the monthly basis.

To get the correct results, the formula described above is not much accurate because they are useful for a particular region. In general the application of the formulas are not advisable. To get the more accurate result, the hydrograph method may be applied in general to get more accurate result (Fig. 2.6).

This is a graph, ploted the flow in cubic meter as y-axis and against time in x-axis. Actually it happens, immediately after rainfall the flow starts increasing and the peak value is reached. After that it start decreasing slowly and at the end it comes to its original flow, which called the base flow.

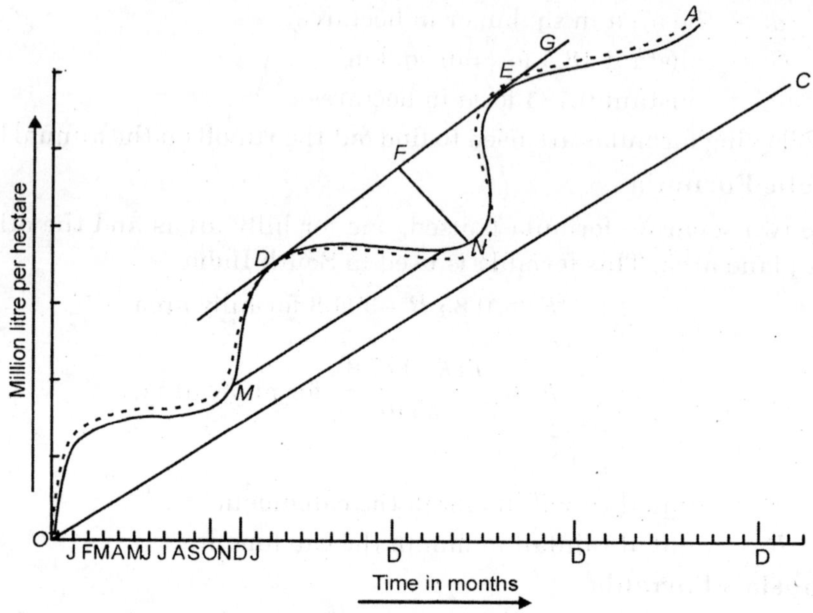

Fig. 2.6 *Integrated hydrograph*

Repeating in the case of another rainfall. Similar curve may be drawn. So for each rainfall, the hydrographs are drawn separately. In the case of high frequency storm, due to subsequent discharge, the curve obtained will overlap to another different curve which will indicate effective sotrm flow. But it should be remember that the hydrograph at a particular point of flow is drawn.

In a catchment when there is no rainfall the stream produced a flow is the indication of a base flow. Generally this is constant. But the fluction obserbed during the rainfall with the rising limb and the falling limb. Considering for a long period a statistical report may help to get a predicting report of a river flow discharge and quite useful for water supply.

It is found sometime necessary to draw a cumulative flow of stream. This is called the integated hydrograph. It is found quite helful to determine capacity of a reservoir. This interated Hydrograph is also called the mass curve or mass diagram.

During the month the accumulated flow is represented by the curve *OA*. The draft is equal to *OC*. A tangent parallel to *OC* are drawn at low points and summits in *OA*. So the line *DE* and *MN* is drawn. So the vertical distance *FN* is the required storage to maintain this particular draft. But at point *D* reservior will be full, so the run off will be less than the draft and the depletion of reservoir starts, and will be maximum at N in October. During the other parts of the year the flow will exceeds the draft, and the reservoir refills again and will be full at E. But the excess water will be wasted by the spillway unless the draft again exceeds the runoff, which will occur at point G. But if the tangent DE exceeds upwards and will not inersect the curve any where, this will have the meaning that the runoff will be the insufficient to supply the expected draft.

For a fixed size of storage reservoir it will be necessary to find out the maximum draft that can be maintained. This can be done by drawing parallel lines at top and low points of OA which will be the ordinates between them will be the available storage.

2.12 METHOD OF UNIT HYDROGRAPH

By using this method, the flood flow can be predicted, this is very useful method. Schnmon has first suggested this and defined as :

The unit hydrograph is the hydrograph of surface runoff excluding ground water runoff on a given basin; due to an effective rain falling for a unit time. Effective rain means, rain produce surface runoff. The unit time means may be done day or a fraction of a day.

Rainfall excess is the same as effective rain, which may be available for direct runoff or the total rainfall minus infiltration and detention losses.

Direct runoff hydrograph plotted for various storms of line duration will have approximately the same duration, but the ordinates will depend upon the total volume of runoff. If the ordinates are divided by the total volume of runoff in cm. There will have a number of graphs, representing 1 cm runoff unit graph.

PROBLEMS

2.1. What are the common sources of water for a water supply scheme ? State the factors that govern the final selection. Explain how you would determine the yield from any three of such sources. **(U.P.S.C.)**

2.2. Give a list of different sources of water, which can be considered for large supplies, bringing out the quality that can be expected in each case, and the treatment that may have to be undertaken in each case. **(U.P.S.C.)**

2.3. (a) Discuss the various methods to control the evaporation losses of a reservoir.

(b) Explain with the help of a sketch the utility of a mass curve.

(A.M.I.E.)

2.4. Write brief notes on any four :

(a) Rainfall and its distribution.

(b) Rainfall measurement, rain gauge.

(c) Average annual rainfall.

(d) Determination of yearly river discharge.

(e) Using empirical formula, determine the years run off from a catchment.

(f) Flow mass curve and their use in determining storage capacity of a reservoir.

2.5. How precipitation is measured at a particular observation ?

2.6. Explain rain intensity, frequency and duration relationship.

2.7. How formulations have been done for measuring the storm runoff ?

2.8. Write short notes on :

(a) Unit Hydrograph method.

(b) Yeild of a catchment.

(c) Infiltration.

(d) Run off

(e) Measurement of precipitation.

2.8. Write a detail notes on the sources from which water is derived for public supply. Give the tabulated statement of the purity and palatability of water derived from various sources. **(U.P.S.C.)**

2.9. Discuss the advantages and disadvantages of underground and surface sources of public water supply. **(A.M.I.E.)**

2.10. What factors are considered in choosing between ground water and river water as a source of water supply to a town ? **(Ran.)**

2.11. Write notes on :

(a) Infiltration and galleries **(A.M.I.E.)**

(b) Artesian Well **(A.M.I.E.)**

(c) Spring as source **(Roorkee)**

2.12. Explain clearly the condition under which ground water is available in nature. What are the different types of well adopted for collection of ground water. Why ground water is preferred to surface water for domestic water supply ? Why ground water is not suitable for some specific purposes ? **(I.I.T.)**

Chapter 3

Ground Water Sources

CHAPTER OBJECTIVE

In this chapter, the most important portions are dealt with from the students point of view as well as from the practical application also. Solved Examples are given to make the theory comprehensive from the practical point of application. In the questions, the problems are given for the practice. This chapter is dealt with carefully as far as possible.

3.1 PRESENCE OF GROUND WATER

Ground water is generally formed by the rainfall that percolates through the ground and comes into contact with the water table. The percolation through the soil will vary with the type of soil present on the ground surface. The permissible capacity of a different type of soil will vary. Some of them will permit the rain water quite quickly and some of them will restrict to the flow. This depends on the structure of the soil in the ground. Underneath the soil some rock will prevent water to flow and some on them may allow them to flow. So the geology of the area will have a great part in the ground water flow. Actually all the materials have variable porosity near the upper portion of earth's crust can be considered as the place of potential storage for ground water ; and may be called as ground water reservoir.

3.1.1 Presence of Ground Water on the Basis of Geological Factors

The geological factors are as—

(1) Porosity of soil.

(2) Permeability of soil.

Porosity. It is mainly the voids present in the soil structure. Due to the presence of voids in the soil, water will be occupied the voids. That is why the occurrence of water is observed. In gerenal the void is defined as the percentage of the voids present in a given volume of soil aggregate. Sometimes it is expressed as :

$$\text{Porosity} = \frac{\text{Vol. water required to saturated the dry sample } v_u}{\text{Total vol. of soil aggregate } V}$$

$$\therefore \qquad \text{Porosity} = \eta = \frac{v_v}{V} \times 100\%$$

It depends upon the shape, packing and degree of sorting of component grains in a given materials.

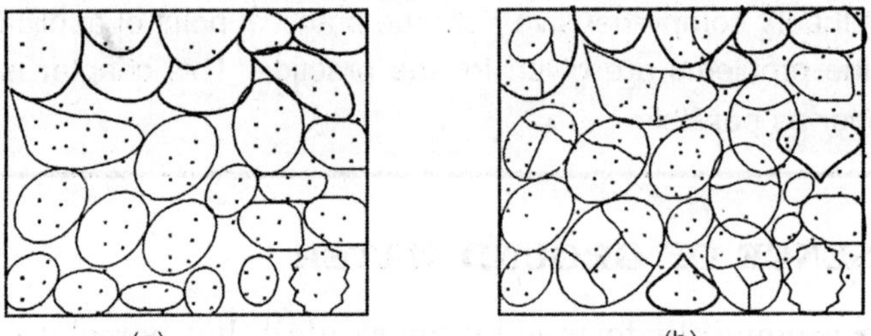

(a) Well sorted (b) Poorly sorted

Fig. 3.1

Porosity of (a) is higher than (b) (Refer Fig. 3.1) because the grain in (a) is more or less is of the same size. But in (b) the smaller grains fills the voids between the larger grains. So the void space is reduced.

In case of unconsolidated materials may vary considerably. It may be less or more than 50%. But generally, it will not exceed 40% except in very poorly compacted materials.

Table 3.1 shows values of porosity of few rock formation.

Table 3.1 *Values of porosity of few rock formation*

S.No.	*Types of rock formation*	*Porosity*
1.	Granite, quartzite	1.5%
2.	Slate, shale	4%
3.	Limestone	5 to 12%
4.	Sanastone	10 to 18%
5.	Sand and gravel	20 to 35%
6.	Only gravel	28%
7.	Only sand	About 40%
8.	Clay and soil	47%

3.2 PERMEABILITY

Water can enter into the rock only if the rocks permit them to flow through it. It depends whether the rock is permeable or not. It is clear that the rock may be porous, but may or mạy not be permeable. So the permeability may be defined as the ability of a rock or unconsolidated sediment to pass water through itself. The permeability measured in terms of coefficient of permeability.

3.3 TRANSMISSIBILITY

This is the more or less the same physical meaning but only the mathematical difference is there. The explanation is as, the total soil has the capability of full width and depth is represented by permeability. But a soil having unit width and full depth i.e, $b = 1$, $d = d$ so $A = d$ is known as transmissibility.

3.4 UNDERGROUND WATER ZONES

Going down below the earth's surface towards the centre of the earth, it will be observed that water will appear at different layers in different forms. On that basis earth's curst can be divided into two zones :

(i) Rock fracture zone.

(ii) Rock flowage zone.

1. Rock Fracture Zone. This zone lies above the rock flowage zone and have the stresses within the elastic limit and also interstices are also present. Depending on the porosity of the rock water stored in the voids. The depth of the zone generally varies from 100 m to 1000 m or more. This zone may be divided into two zones. One that is below the water table is known as zone of saturation, and the other that lies above the water table is called the zone of aeration.

2. Rock Flowage Zone. In this zone rock permit the determination, actually the depth is not known accurately. Interstices are most probably absent in this zone, reason for that the stresses are beyond the elastic limit and the rock remain in the plastic flow. The water found in this zone is known as internal water.

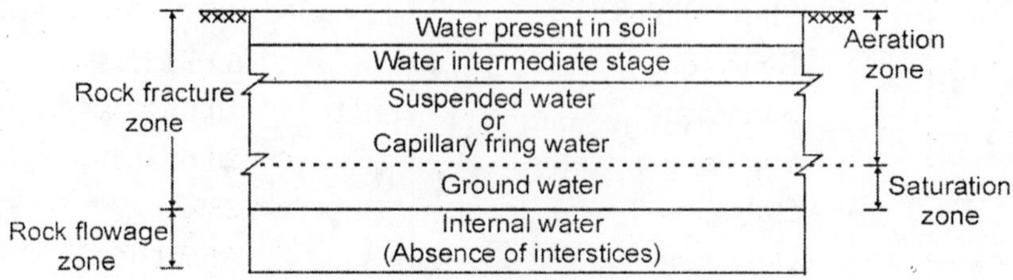

Fig. 3.2

3.5 VELOCITY OF GROUND WATER

In the underground the water table is not in equilibrium state. It will flow from high point to low points as the water table is not maintain a straight line. But this is also depend upon the permeability of soil as well as the hydraulic gradient of the surface. The difference between the two points is expressed as a ratio of the elevation (H) and the length (L) which is the distance between the two points. In this case Darcy's Law is applied to get the Velocity of Ground Water.

3.6 DARCY'S LAW FOR VELOCITY OF GROUND WATER

According to Darcy's Law, the discharge is proportional to the loss of head i.e., (ΔH), and the cross sectional area (A) of the soil and also inversely proportional to the length (L) of the soil sample.

$$Q \propto \frac{\Delta H}{L} \times A$$

But the rate of heat loss $\frac{\Delta H}{L}$, this is the hydraulic gradient (i)

$\therefore \quad Q \propto iA$

or $$Q = KiA \qquad ...(1)$$

where K = Const. but as it changes with the type of soil, hence this called the co-efficient of permeability. Here K will have the units of $\frac{L}{T}$ i.e., cm/sec., unit of velocity.

Now dividing the side of Eq. (1) by A,

$$\frac{Q}{A} = Ki$$

where A = area of soil through which the water flow

or $$v = Ki$$

where v = discharge velocity but the actual velocity v_a occurs through. The voids of the cross-sectional area i.e., A_v and through A. The permeability of the soil can be found as a supercritical flow velocity.

Under a unit hydraulic gradient

So $$A_v \times v_a = A.v$$

or $$v = \frac{A_v V_v}{A} \quad ...(1)$$

Assuming that void area A_v can be taken as A, because the ratio of area of voids to the total area A is very small, Similarly the volume of voids. V_v to be taken a time total vol. V.

$$\therefore \quad \frac{A_v}{A} = \frac{V_u}{V} = \text{porosity } n$$

Substituting this into each in Eq. (1) above

We will get $v = (\eta . v_a)$

Then knowing the value of v from equation $v = Ki$ and divided this by η (porositys) acutal flow velosity v_a through the soil can be found.

3.7 GROUND WATER VELOCITY ESTIMATED BY USING EMPIRICAL FORMULA

The empirical formulas were used before Darcy's formula came into this field. The common formulas are as follows.

1. Slichter's formula

$$v_a = K'i\frac{D_{10}^2}{\mu}$$

where v_a = ground water flow velocity in m/day

K'' = constant having a value about 400

D_{10} = effective particle size in aquifer in mm

μ = viscosity of water depending on temperature

2. Hazen's formula

$$v_a = \frac{K'.i.D_{10}^2}{60} \times (1.8T + 42)$$

where V = Velocity in m/day

T = Temperature in °C or (°K) $\because K = C + 273$

K' = Constant whose value about 1000

Example 3.1 *The data is given below :*

Viscosity coefficient of water of ground water temperature 10°C = 1

Aquifier's particles effective size = 0.1mm

Hydraulic gradient = 1 in 70

Slichter constants = 400

Hazen's constants = 800

Find the velocity of the ground water flow, by both the methods i.e., Slilchter and Hazen's.

Solution

By Slichter's method ; we have

$$v_a = \frac{K'i\ D_{10}^2}{\mu} = \frac{400\times\frac{1}{70}\times(0.1)^2}{1} = \frac{400\times0.01}{70}$$

$$= \mathbf{0.056\ m/day}$$

By using Hazen's method, we have

$$v_a = \frac{K'.i.D_{10}^2}{60}(1.8\times10+42)$$

$$= \frac{800\frac{1}{70}\times(0.1)^2}{60}(1.8\times10+42)$$

$\therefore\ T = 10°C$ or (283°K)

$$= \frac{800\times\frac{1}{70}\times\frac{1}{100}\times60}{60} = \mathbf{0.11 m/day}$$

3.8 GROUND WATER DRAINAGE

Water drainage is not a natural phenomenon except some springs and natural reservoir. Actually this is happening when by artificially water taken out from the ground below the water table by means of a well or pumps. This drained water may be used for municipal needs.

3.9 YIELD OF GROUND WATER

In a particular formation in the ground get filled up by water replenishment. When all the voids are completely fill up by water then the formation will be saturated. When a well is active, due to the gravitational action water comes up; it is found that the water is so drained is less than the space of the volume of voids space which is found by its porosity. This is due to the fact that whole water cannot comes out, but some of them will be retained by these interestices due to their molecular attraction.

3.10 SPECIFIC YIELD

By the gravitational force the water extracted from a saturated water bearing material is known as the yield of water. This is expressed by a ratio of volume of the total material drained is known as specific yield.

$$\therefore \text{ Specific yield} = \frac{\text{Vol. of water drained by gravitational force}}{\text{Total vol. of the dewatered}} \times 100$$

3.11 TYPES OF AQUIFERS

Due to the geological formation in the under ground a preamble strata capable of yielding an appreciable quantities of water by gravitational force is known as aquifers. But in the under ground where the ground water is found with a great difficulty though the very fine grained materials containing less quantities of ground water is called the principal aquifers.

But when a confined bed of impervious materials are on top of an aquifer then a confined bed of over burden is called aquiclude. When a well excavated through the aquifer, then the water yield will driven upon many factors as diameter of the wells, the thickness of the aquifer and also the permeability.

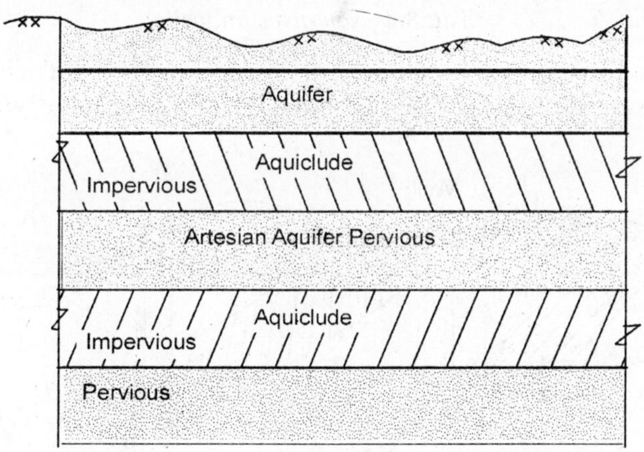

Fig. 3.3

But equifers are in lies any one of the two categories.

1. Non artesian aquifers
2. Artesian aquifers

1. Non Artesian Aquifers. A non artesian aquifer is such that the top most water bearing strata having no confined impermeable over burden on top of it.

Generally the ordinary gravity well say 2 to 4 m dia. are constructed to tap water from the top most water bearing strata, are called the non artisian wells. This type of wells are called gravity well or may be water-table wells.

2. Artesian aquifers are formed, when both the top and bottom surface of an aquifer is contained by impervious rock formations and also have a good inclination so as to expose the aquifer any where to the catchment at a high level to create a sufficient hydraulic head, then this will become an artesian aquifer.

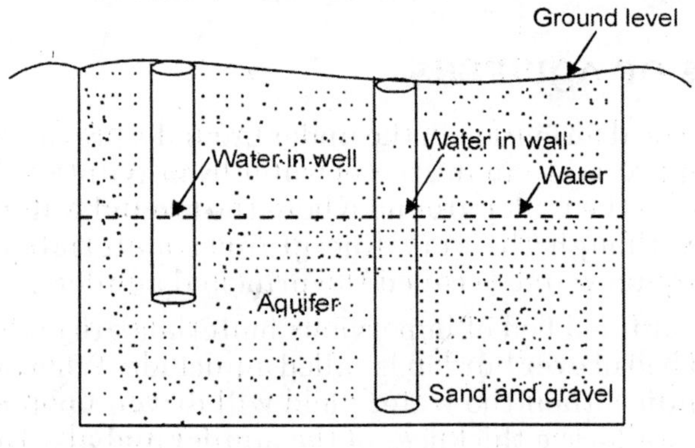

Fig. 3.4 *Non artesian wells*

Any well exacavated in this zone water will yield and flow starts automatically, under hydraulic pressure. Even the rise of water may rush

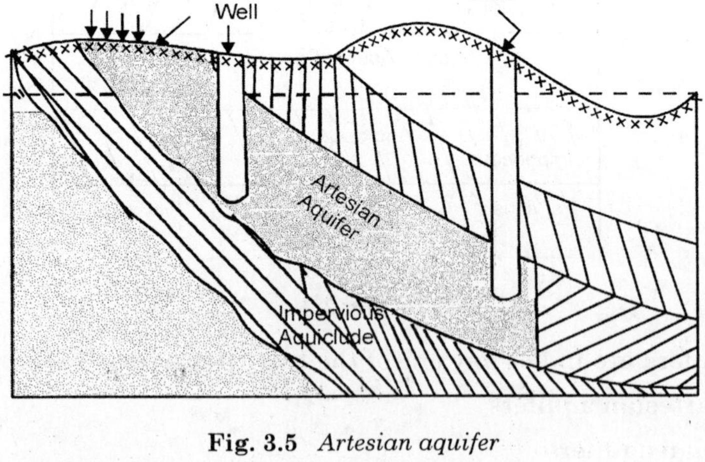

Fig. 3.5 *Artesian aquifer*

out of the surface for reasonable height, automatically, this is called flowing wells.

3.12 PERCHED AQUIFERS

This is a very special type of aquifer which is generally found within unconfined aquifer.

The formation of this aquifer is found when an impervious deposite below a pervious deposite is supported a body of saturated material, then this of saturated material will behave like an aquifer is called perched aquifer. The top surface of this aquifer is called the perched water table.

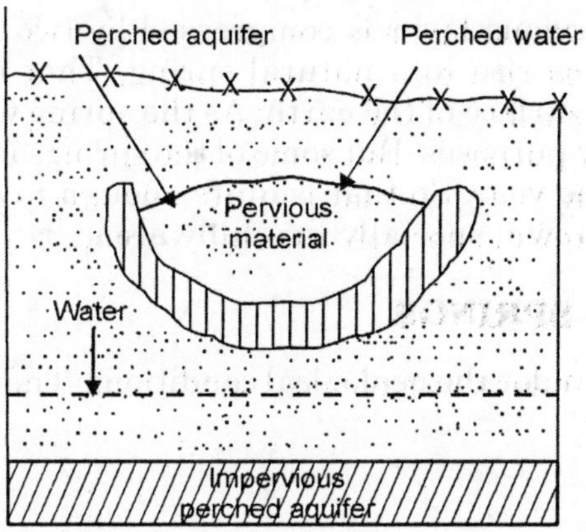

Fig. 3.6

From the following forms the underground water are available.

1. Infiltration galleries
2. Infiltration wells
3. Springs
4. Wells and tube well

1. Infiltration Galleries. Sometimes they are known as horizontal well. This is a shallow horizontal tunnel along the bank of a river bank having water bearing strata. The depth is generally 3 to 6 metres, it varies according to the need. The tunnel is constructed by lateral pipes which will have some slope in it. The pipes are covered by gravels to stop the entry of sand and fine particles in the pipe. Then the water through the pipe collected to a sump well, from there it is pumped out, treated then distributed to consumers.

2. Infiltration Well. This type of wells are generally found in the banks of river. The water is collected into the well from the bottom. The top of the wells are covered having manhole, bottom is kept open.

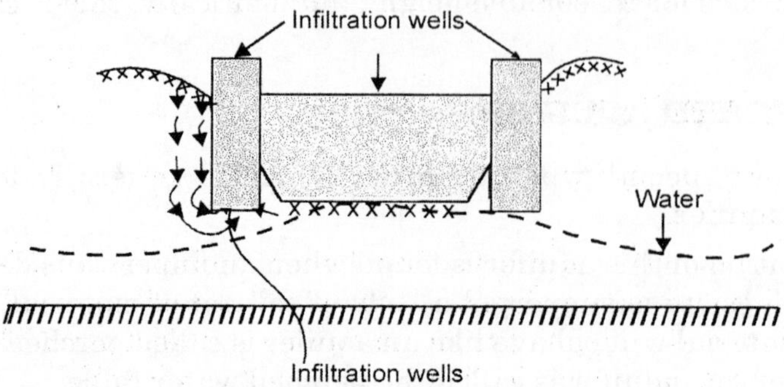

Fig. 3.7

3. Springs. A pervious layer is compressed by two impervious layers at top and bottom, gives rise to a natural spring. That is natural outflow of ground water at the surface of the earth. As the spring water is not enough to use as a water supply purposes. But some of the springs may have good enough water throughout the year. So that is quite enough to serve a water supply purpose for a small town; specially in a hilly area.

3.13 TYPES OF SPRINGS

Springs are formed under the geological conditions. They may have generally three types, such as,

1. Gavity springs.
2. Surface springs.
3. Artesian springs.

3.13.1 Gravity Springs

In an uneven land specially a valley when the water table gets high, the water overflows through the sides of the valley or the depressed side, this

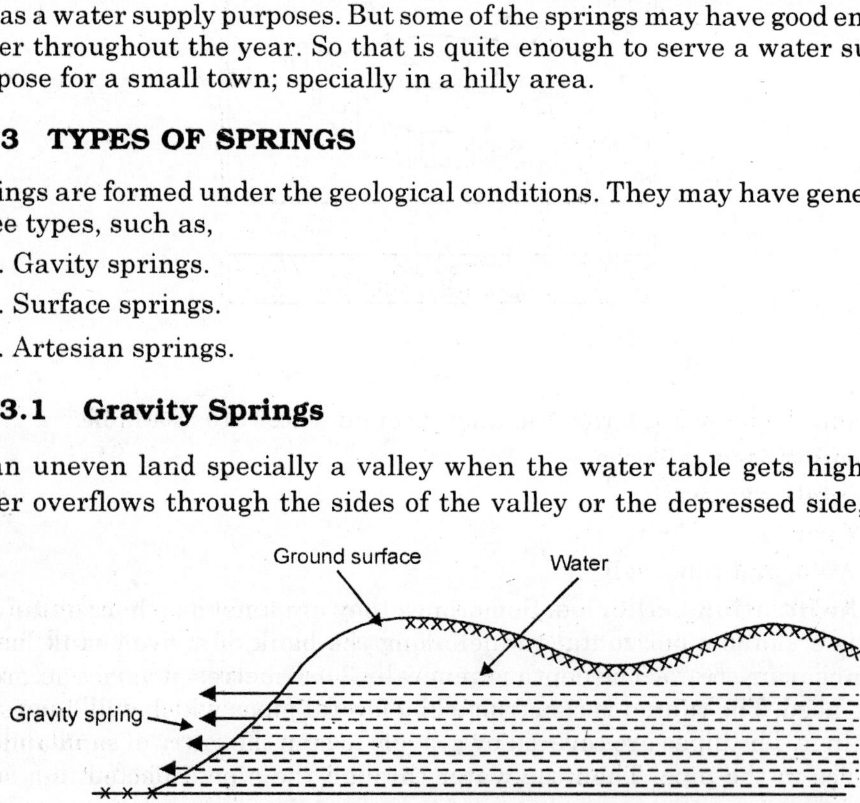

Fig. 3.8 *Gravity spring*

type of spring formed is called the gravity spring. Naturally the flow in the spring vary as the water table vary.

3.13.2 Surface Springs

An impervious strata having a underground storages on top, inclined at an angle underneath the ground, to cause the water table to get high to exposed to the ground surface. This type of stream forms called the surface spring. The water flow in this is not certain at all.

When a storage of water in a pervious layer is under pressure, the water is flowing through the confined aquifer; as shown in the Fig. 3.8, then the spring is formed, such spring is called the artesian spring. This type of spring are

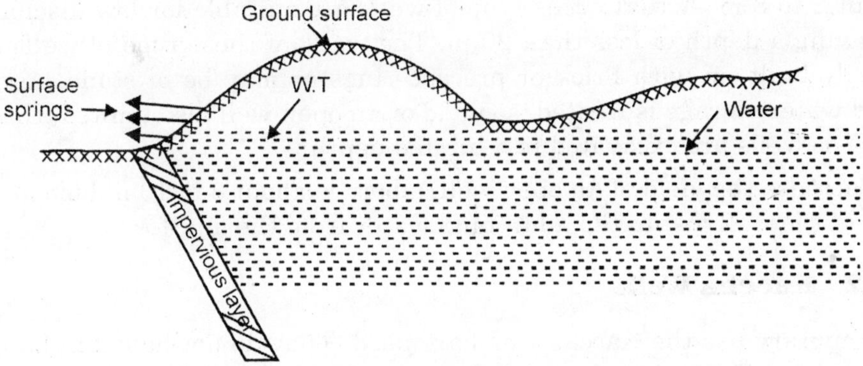

(a) Surface spring

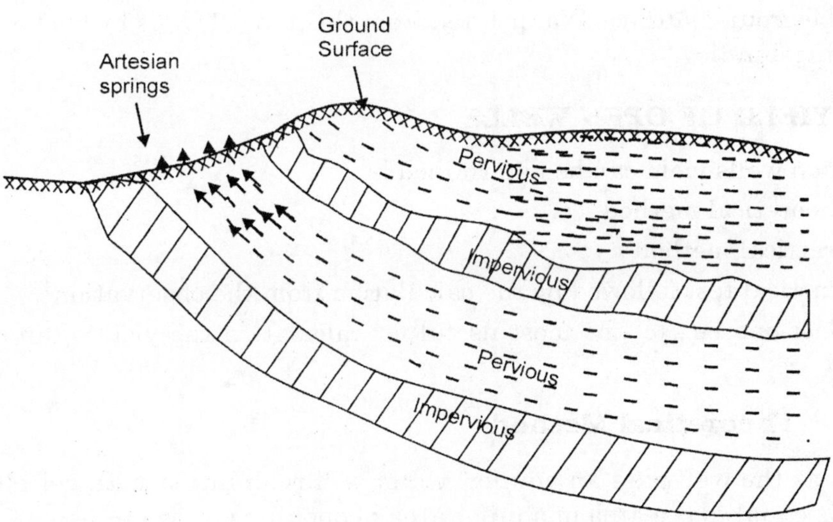

(b) Artesian spring

Fig. 3.9

generally provide a stable quantity of water throughout. Because the water comes out under pressure, so the yields are quite good, they can be used as a possible source of water supply.

3.14 WELLS

The wells are classified into two groups, such as—

1. Dug wells
2. Tube wells

3.14.1 Dug Wells

Open wells and dug wells are the same. Generally they have a big diameter of about 2 to 8 m. Actually this type of wells are suitable for low discharge and having a depth of less than 20 m. The walls of these type of wells are generally built up with brick or precase rings or may be of stone. As the ground water storage is limited, so yield of an open well also limited, that is why these wells are excavated to a limited depth.

To increase the yield of an open well, a bore is about 8 to 10 m hole at the centre of the well just to tap additional water from the aquifer.

3.14.2 Kaccha Wells

For temporary use the Kaccha wells are used. They do not have any lining. In the land soil this Kuccha wells are suitable, the sides are not cave in. Generally they are excavated in soil which have water table of about 3 to 4 m below the ground surface. Water lifted from this type of wells by bucket and rope using a pulley.

3.15 YIELD OF OPEN WELLS

From open wells yield can be determined by

(a) Theoretical method

(b) Practical method

(c) Practical test followed by the calculation from the observation

The last one i.e., (c) the most useful for calculating the yield of an open wells.

3.15.1 Theoretical Method

As soon as the well pass an aquifer, water will rush into it with velocity v. Say A_v is equal to the area of aquifer area of opening into well, then

$$Q = v \,.\, A_v, \quad \text{where } v_v = \eta v_a$$

where v_a = actual flow velocity of ground water

and η = porosity

$$\therefore \quad Q = \eta v_a \,.\, A_s$$

In the above equation, the ground water flow velocity (v_a) found by Slichter's Hazen's formula. As the aquifer area opening to the well found by known dia. of well (d) and the depth (H) as $\pi.d.H$. And η the porosity can be found by soil test. So knowing η, v_a and A the discharge Q can be found.

3.15.2 Practical Test or Yield Test

This can be done by two different tests.

(1) Pumping test.

(2) Recuperating test.

1. Pumping Test. From an open well a pump is installed, and draw water sufficiently so that a heavy drawdown in the water level is observed. Then changing the pumping rate in such a way that the water level in the well remains constant. When the condition of equilibrium is reached the rate of pumping will be equal to the rate of yielding. So the pumping rate will give directly the yield of the well at a definite drawdown. Say Q_1 is the yield, at a definite drawdown S_1, then the yield (Q) at any other drawdown S can be fround as, using Darcy's law

$$Q = KiA$$

$$= K\,(S/L)A$$

$$= \frac{K}{L}.AS$$

or
$$Q = CAS$$

where S = drawdown in the well.

For a known Q_1, at a known drawdown S_1

$$Q_1 = CAS_1$$

where A = Cross section of the well. In case of a cavity the A will become 4/3 times the actual cross-sectional area at the end of the well. The value of S for which the velocity becomes equal to the critical velocity then it will be said that, the critical depression head. But the depression head is generally kept equal to 1/3 times the critical depression head, and this is called the working head.

2. Recuperating Test. Recuperating test is done when it is found very difficult to adjust the rate of pumping. The method of recuperating is described below. First step is taken to drained water from well at a first rate to get sufficient drawdown. Second step is taken to stop pumping. Third step is then allow water to level to start rising. Fourth step is to note the time taken by the water to come back to its normal stage. Then the discharge can be calculated as shown in Fig. 3.10.

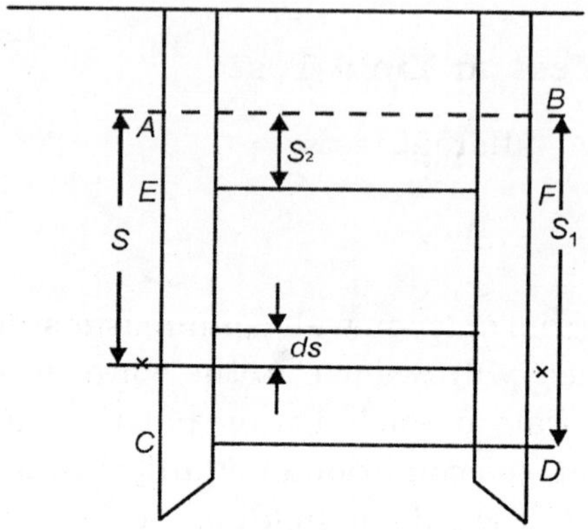

Fig. 3.10

Let AB = Water level before pumping start
CD = Water level when pumping stopped
S_1 = Fall head of water when pumping stopped
EF = Water level at well time T when pumping is stopped.
S_2 = Falling head of water in the well at time T after pumping stopped.

At the time t after pumping was stopped, the position of water level will be at xx. So the depression head found will be S. Let dt be the small increment of time after time t, corresponding to the small depression of fall in level beds. So in time t after pumping stopped the water level recuperates by (S_1– S). After this again it will recuperate ds in time dt.

So the water volume entering in the well in the small interval of time dt = $dV = A.\ dS$ ...(1)

[A = At bottom the cross-sectional area of well].

Again, say Q is the rate of recharge water in the well at time t by falling head S, then is small time interval t under S, then the volume of water entring. The well in small time interval

$$= dV = Q.\ dt$$

$$\text{As } Q \propto S,\ Q = C's \qquad ...(2)$$

[C' = Const. depending on nature of soil, which allow water entry into the well].

$$\therefore \qquad dV = C'S.\,dt \qquad ...(3)$$

Then equating Eqs. (1) and (3)

$$-Ads = C's\,dt$$

(N.B. : –ve sign means the falling head as increases)

$$\text{or} \qquad \frac{C'.dt}{A} = -\left(\frac{ds}{s}\right)$$

Integrating between the limits

$$t = 0 \qquad S = S_1$$
$$t = T \qquad S = S_2$$

We will have,

$$\frac{C'}{A}\int_0^T dt = -\int_{S_1}^{S} \frac{\alpha_s}{S}$$

$$\text{or} \qquad \frac{C'}{A}[6]_0^T = -[\log_e S]_{S_1}^{S_2}$$

$$\text{or} \qquad \frac{C'}{A}(T) = -\log_e \frac{S_2}{S_1}$$

$$= 2.3 \log_{10} \frac{S_1}{S_2}$$

$$\frac{C'}{A} = \frac{2.3}{T}\log_{10}\frac{S_1}{S_2}$$

Values of S_1, S_2, T are known from the test. The value of $\frac{C'}{A}$ can be calculated, which is known as specific capacity of the open wall the unit will be cumecs per m^2 of area under a unit fall of head. Again after knowing the value of $\frac{C'}{A}$, the discharge Q for a well under a constant depression head H can be calculated as shown.

$$Q = C'.S$$

$$\text{or} \qquad Q = \left(\frac{C'}{A}\right)A.S$$

$$\text{or} \qquad Q = \left(\frac{2.3}{T}\log_{10}\frac{S_1}{S_2}\right)A.S.$$

As, A and S are known, so the discharge for any amount of drawdown S can be easily calculated.

3.16 TUBE WELLS

Which is a long pipe or tube is bored or drilled deep into the ground, intercepting one or more water bearing strata called tube wells. Large discharge can be available by getting a larger velocity and at the same time a larger cross sectional area of the water bearing strata. The tube wells provide very good method of getting water supplies.

3.16.1 Deep Tube Wells

These are very deep in nature; may be 70 to 350 m deep can tap more than one aquifers. This is generally done by the state government. Deep tube well can yield about 200 to 220 lit/sec. But on the other hand the standard tube well can yield about 10 to 50 lit/sec. The dia. of this well may be 0.6 m upto 60 m deep and then 0.56 m below 60 m. The strainer dia. is 0.25 m and drawdown may be 10 m.

Apart from this shallow tube wells are generally 20 to 70 m depth can tap only one aquifer; these are only use for private use; These can yield 15 to 20 lit/sec.

A broad classification is drawn on the bases of entry of water. Tube wells are mainly of two types.

1. Cavity type 2. Screen type

1. Cavity Type (Fig. 3.11). In this type of tube wells water is drawn from the bottom of the wells. This type of tube well is very economical, because it

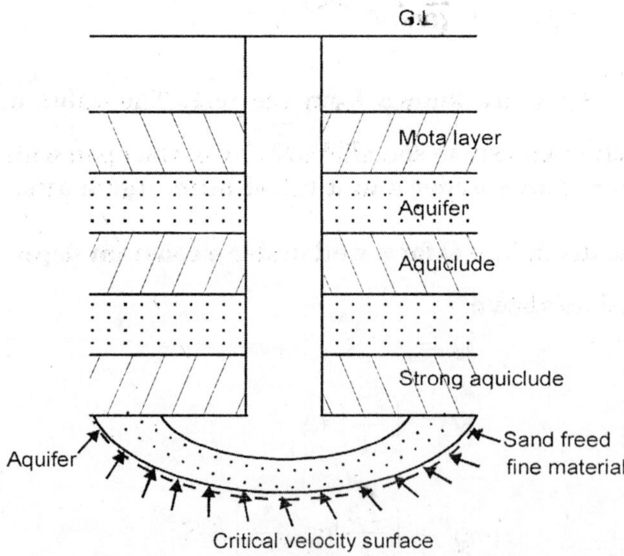

Fig. 3.11 *Cavity type well*

need a plain well pipe in lowered into the bore and send it upto the required strata, through the ground. The same principle is used same as deep open well. The only difference is that open well can tap the first aquifer, just below the layer, but the cavity well can tap the lower or more lower stratum.

The cavity tubewell consists of a pipe bored through the soil and rest on a strong clay layer at the bottom.

So a cavity will form at the bottom, then the water will enters from the aquifer in the well pipe through the cavity. At the beginning of the pumping, fine sand will come out through the pipe, will form a cavity. At the beginning stage the water may have sandy but after some time, clean water will come out.

Generally a cavity need to be formed carefully and slowly by using a centrifugal pump.

2. Screen Type. The term tube well actually, we mean that it is a screen type tube well. By the help of this type tube well number of aquifers can be tapped. This type of tube well is widely accepted in our country to get much more water for water supply. This can also be used for irrigation purposes.

3.16.2 Stranier Tube Well

Through the bore hole plain pipes are placed with the strainer attached to it. Number of strainers can be attached to the plain pipes as the requirement arised. Generally the plain pipes are staying in the non-pervious layer, on

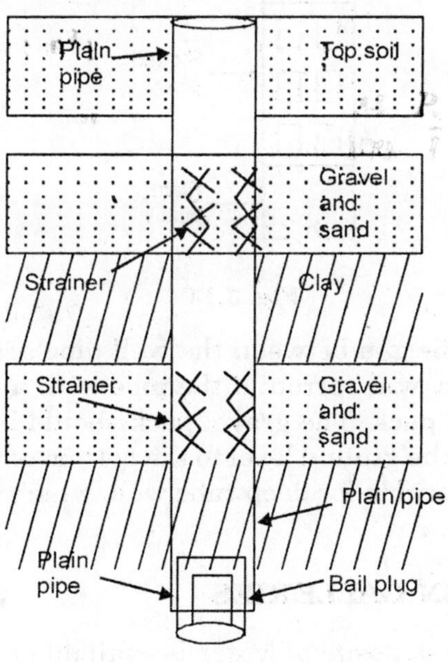

Fig. 3.12 *Strainer tube well*

the other hand the strainers are attached to the plain pipe to stay in the pervious layer. A bail plug is provided at the bottom. Water enters into the well from the side of the stranier and flow through the well as a radial flow.

Strainer is a perforated pipe covered with a fine mesh wrapped round the perforated pipe having a small annular space between the two. But the wire screen prevent the small particles to get into the well, so the water enters the well through the mesh. But particle larger than the size of the mesh is prevented for entering into the well. This will help to get larger velocity of flow.

But it should be kept in mind that strainer type tube well is not suitable in the fine strata of sand. To stop the entry of sand the mesh have to be reduced quite considerably, which may lead to choking of the strainer.

3.16.2 Slotted Tube Well with Gravel Pack

Slotted tube well are the tube well without strainer, the pipe is slotted. The portion of the slotted pipe is placed opposite the water bearing formation the same way as in the strainer in the strainer tube well. After placing the assembly of the slotted pipe with the plain pipe into the bore hole. Then

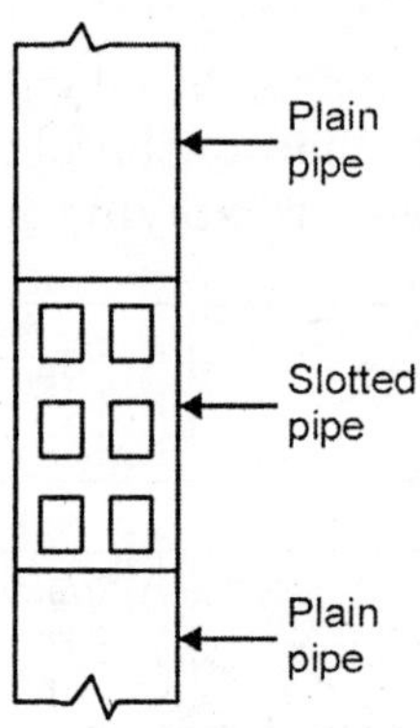

Fig. 3.13

gravel is poured into the gap between the well pipe assembly and the bore hole; in such a way they will sorround the pipe with a thickness say 10–15 mm thickness of gravel pack. The gravel pack should be done entire way of the bore hole, i.e., from the ground level to the bottom of the hole. This type of tube well is highly preferable for deep tube well, which will take water from more than one aquifers.

3.17 INFILTRATION GALLERIES

If it is found that, large deposite of water is available just below the ground level, in that case water supply scheme could be arranged. Generally the long galleries are made below the ground. Practically they may be of 1 m width and depth, and 2 m length but varying from 8 m to 110 m as long, it depends on the extent of the water available. Generally they are found approximately 6 m to 8 m below ground level.

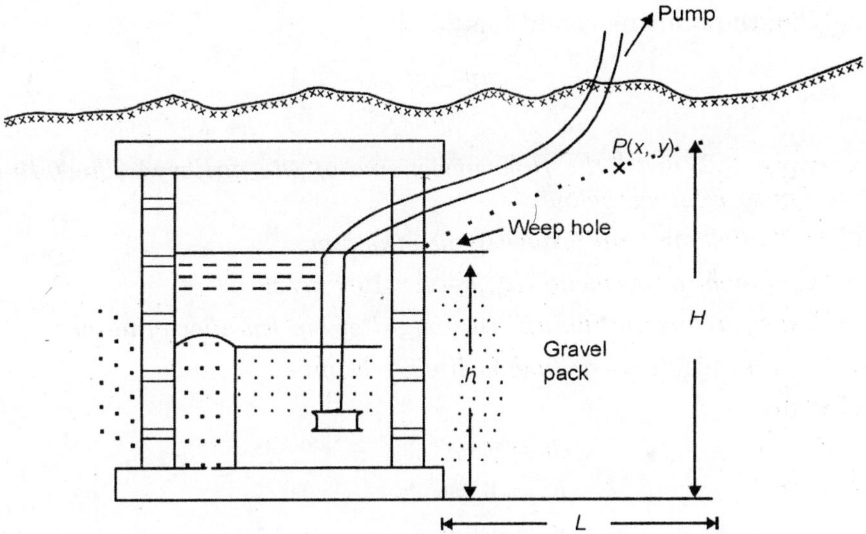

Fig. 3.14 *Infiltration gallery*

Shaft of sufficient length having at intervals could be fed by laterals to tap water from a large area. From the structural point of view they could be made by concrete and bricks box type duct with having openings on side walls and having gravel packed from outside.

This type of construction is quite cheep; but nowadays a different type of system is in use quite often. Simple jointed perforated pipes may be laid underground they can be called infiltration pipes. Only some construction is required when the ground water is near the surface.

But the yield from an infiltration gallery can be calculated on the basis of the knowledge of drawdown conditions then applying Darcy's law. But the drawdown low curve follows Dupit's curve.

From Fig. 3.14.

Static head level = H

Depth of pumping water level = h

Influence zone's distance = L

Then the discharge per unit length of the gallery at any point x is given by

$$Q_x = A \times V = y \times 1 \times K \frac{dy}{dx}$$

$$Q_x = \frac{y^2}{2} \times k \times \frac{1}{x} \text{ (By solving)}$$

Applying the limiting conditions

$$Q = \frac{H^2 - h^2}{2\ell} \times K$$

Example 3.2 *Find the flow per metre of the gallery? The following informations are given below :*

(i) *Depth of clean sand gravel stratum is 5 m.*

(ii) *Co-efficient of permeability k = 1 × 10^-3^ m/s.*

(iii) *Water surface in the infiltration gallery is 1m above the sole.*

(iv) *Its distance to the diffusion ditch = 12 m.*

Solution

$$H = 5 \text{ m}$$

$$h = 1 \text{ m}$$

$$L = 12 \text{ m}$$

$$K = 1 \times 10^{-3} \text{ m/s}$$

Then $$Q = \frac{H^2 - h^2}{2L} \times K = \frac{5^2 \simeq 1^2}{2 \times 12} \times 1 \times 10^{-3}$$

$$= 1 \times 10^{-3} \text{ m}^3\text{/s} = 1 \text{ lit/s}$$

Example 3.3 *The length of the gallery to be found if the draw down is not exceed 4.5 m. The following information is given below.*

(i) *From an infiltration gallery 200 m³/day of water is to be obtained.*

(ii) *At a depth of 6 m from the sub-surface water table the gallery is placed.*

(iii) *The bed will have the permeability of 100 m/day.*

(iv) *The value of L may be taken as 120 m.*

Solution

Let $$Q = \frac{H^2 - h^2}{2L} \times K$$

$$= \frac{6^2 \times 1.5^2}{2 \times 120} \times 100$$

$$= 14 \text{ m}^3/\text{ m/day}$$

∴ The length of gallery will be

$$= \frac{500}{14} = \mathbf{35.8\ m}$$

3.18 WELL DEVELOPMENT

Development of wells means before they put into operation, the pumping rate will be high. Water should be disinfected, and also flushing testing equipping are finished.

At the time, when a new tube well is ready for operation, the temporary equipments are used for construction to get high rate of pumping. This will pumped out the fine sand grains in the close area of the strainer along with water, then creat a type of gap between the strainer and the surrounding. This gap may be filled by the graded gravel. Then be replacing the permanent equipment the normal rate of pumping is obtained.

3.19 STEADY FLOW IN UNCONFINED AQUIFER

At the time of water pumping at a constant rate, a depression cone is formed at that time the draw down will become steady. Then the line of contact of the cone of depression with static water table is called, circle of influence.

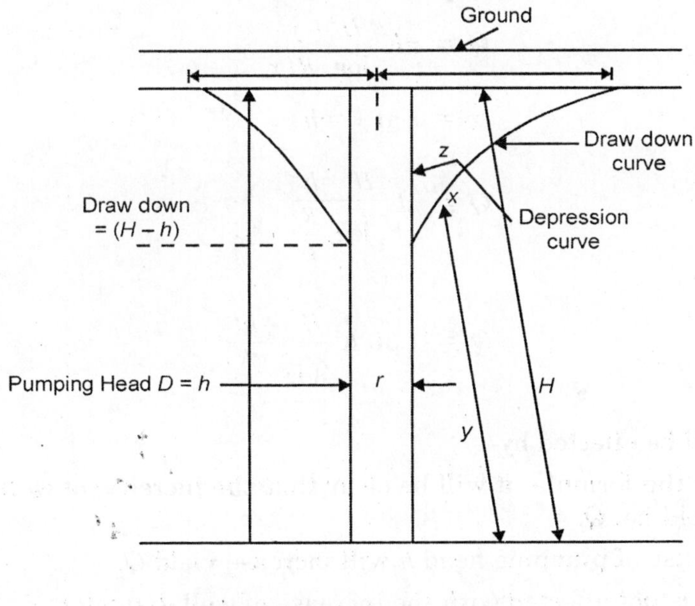

Fig. 3.15 *Unconfined aquifer*

where

H = Well depth in aquifer
h = Pumping depth
R = Radius of circle of influence
r = Well radius
p = Any point on the draw down curve with x–y co-ordinates with center of bottom of well as origin.

As the origin is taken as centre of the bottom of the well, then we will have the velocity.

$$v = KS$$

$$S = \text{draw down curve slope} = \frac{dy}{dx}$$

Then yield = $Q = k\ KS$

where $A = 2\pi xy$

Then $$Q = (2\pi xy)\ K\frac{dy}{dx}$$

or $$\frac{Qdx}{x} = 2\pi\ Kydy$$

Integrating both sides

$$Q = \pi K\ \frac{H^2 - y^2}{\log_e R/x}$$

But $$x = r \text{ at } y = h$$

$$\therefore \quad Q = \pi k\ \frac{H^2 - h^2}{\log_e \frac{R}{r}}$$

or $$Q = 1.36\ K\ \frac{H^2 - h^2}{\log_{10} \frac{R}{r}}$$

Yield will be effected by—

(1) From the formula, it will be clear that the increase of well depth will increase yield i.e. Q.

(2) Decrease of pumping head h will increase yield Q.

(3) Yield is not affected with the increase of well diameter.

3.20 STEADY FLOW IN CONFINED AQUIFER

Flow through the artsian well is discussed here. This is also a case of confined aquifer.

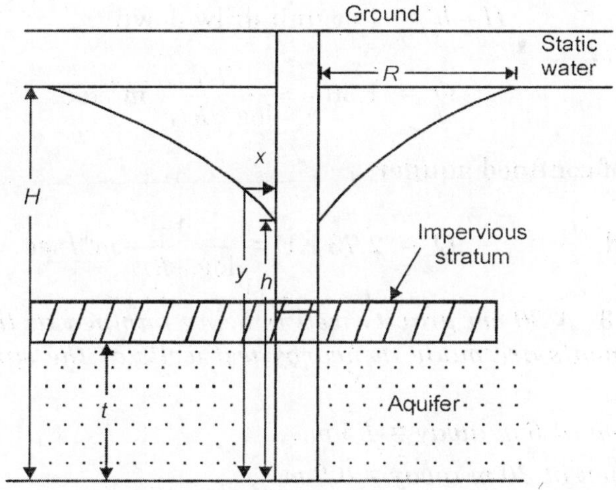

Fig. 3.16 *Confined aquifer*

where

H = Before pumping the water depth.
h = At the time of pumping the depth of water
t = Aquifer thickness below impervious layer
R = Circle of influence radius
r = Well radius
P = Considering the centre at the bottom of well having $x - y$ co-ordinate, any point on the draw down curve.

As before

$$Q = A \times v = (2\pi E)\ x.\ K \frac{dy}{dx}$$

$$\frac{Qdx}{x} = 2\pi tKdy$$

Solving by intregation on both side, and having limits,

$$Q = 2\pi tk \frac{(H-h)}{\log_e R/r} = 2.73 \text{ KE} = \frac{(H-h)}{\log_{10} R/r}$$

N.B. The yield per unit draw down is called the specific capacity of a well. Specific capacity of an unconfined aquifer is as given.

$$Q = 1.36K = \frac{(H+h)(H-h)}{\log_{10} R/r}$$

$$H - h = 1 \text{ or unit draw down}$$

$$\therefore \quad Q = 1.36K = \frac{Hh}{\log_{10} R/r} \text{ m}^3/\text{sec}$$

In the case of confined aquifer

Specific yield, $$Q = 2.73 \text{ KE} = \frac{1}{\log_{10} R/r} \text{m}^3/\text{sec.}$$

Example 3.3 *A 30 cm gravity well is being pumped at the rate of 1350 lpm. Measurements are made in nearby test wells at the same time are as follows.*

(i) Draw down at 6 m away = 4.5 m

(ii) Draw down at 30 m away = 0.9 m

(iii) Distance of ground water table above bottom of the well = 7.5 m.

Determine :

(i) The draw down in the well

(ii) The specific capacity of the well **(C.W.)**

Solution

Discharge

$$Q = 1350 \text{ lpm} = 1.350 \text{ m}^3/\text{min}$$

$$= 0.0225 \text{ m}^3/\text{sec}$$

$$\therefore \quad 0.0225 = K' \times \frac{H^2 - h^2}{\log_{10} R/r}$$

$$\text{or} \quad 0.0225 = K' \times \frac{75^2 - 70.5}{\log_{10} R/r} \quad \text{...(a)}$$

$$\text{and} \quad 0.0225 = K' \times \frac{75^2 - 74.1^2}{\log_{10} R/30} \quad \text{...(b)}$$

By solving (a) and (b)

$$R = 46 \text{ m}$$

$$K' = 0.00003$$

Then
$$0.0225 = 0.0003 \frac{H + h}{\log_{10} r/n} 0.00165$$

$$h = 62 \text{ m}$$
$$75 - 62 = 13.0 \text{ m}$$

$$\text{Sp. capeity is } Q = K' \times \frac{H+h}{\log_{10} R/r}$$
$$= 0.00165 \text{ m}^3/\text{sec.}$$

Example 3.4 *The following informations are given below.*

(i) Thickness of water baring strata = 10 m

(ii) Draw down = 4 m

(iii) Radius of influence circle = 30 m

(iv) Assume permeability constant = 20 m³/day/unit area.

Solution

$$Q = 1.36 K \times \frac{H^2 - h^2}{\log_{10} R/r}$$

$$= 1.36K \times 20 \times \frac{110 - 3.6}{\log_{10} 30/0.4}$$

$$= \mathbf{928 \text{ m}^3/\text{day}}$$

Example 3.5 *Given the following information to calculate the draw down at a discharge of 70 lpm.*

(i) 2 m dia. well having 3 m draw down, when the discharge is 30 lpm.

(ii) When static water depth is 50 m.

Solution

$$30 = 1.36 K \frac{50^2 - 47^2}{\log_{10}(C \times 100)} \quad [\because r = 1 \text{ m}] \quad ...(1)$$

And
$$100 = 1.36 K \frac{50^2 - 35^2}{\log_{10}(C \times 100)} \quad ...(2)$$

Solving these two Eqs. (1) and (2)

$$C = 1.5 \text{ and } K = 0.1255$$

Again
$$70 = 1.36K \frac{50^2 - h^2}{\log_{10}(C + 70)}$$

Substituting the value of CK in the above equation

$$h = 40.87$$

$$\therefore \quad \text{Draw down} = 50 - 40.87 = \mathbf{9.13 \text{ m}}$$

Example 3.6 *At the time of a recuperating test, the water level falls down by pumping by 2.5 metres and is recuperated an amount of 1.5 meters in 60 minutes.*

1. If the well dia. is 2.5 m and the draw down is 40 metres, determine the yield from the well.

2. If the yield of a well is 10 liters/second and a draw down is 2.5 m, determine the diameter of the well.

Solution

Let

$$\text{The initial draw down} = S_1 = 2.5 \text{ m}$$

$$\text{The final draw down} = S_2 = 2.5 - 1.5 = 1.0 \text{ m}$$

$$\text{Time } T = 60 \text{ min.}$$

$$= 60 \times 60 = 3{,}600 \text{ sec.}$$

So

$$\frac{C_1}{A} = \frac{2.3}{T} \log_{10} \frac{S_1}{S_2}$$

$$= \frac{2.3}{3600} \times \log_{10} \frac{2.5}{1.0} = 0.254 \times 10^{-3}$$

1. Then yield from 2.5 m dia well at 40 m draw down, will be

$$Q = \left(\frac{C_1}{A}\right) \times A.S.$$

$$= (0.254 \times 10^{-3}) \times \left(\frac{\pi}{4} \times 2.5^2\right) \times 4$$

$$= 4.8 \times 10^{-3} \text{ cu m/sec}$$

$$= \mathbf{4.8 \text{ lit/sec}}$$

2. Given $Q = 10$ l/s $= 10 \times 10^{-3}$ cu m/sec

or

$$Q = \left(\frac{C_1}{A}\right) \times A.S$$

or

$$10 \times 10^{-3} = (0.254 \times 10^{-3})\, A \times 2.5$$

$\therefore$

$$A = 15.74 \text{ m}^2$$

So the dia. of well will be

$$= \frac{4 \times 1.574}{\pi} = \mathbf{4.48 \text{ m}}$$

Example 3.7 *The following observations made during a pumping test.*

(i) Pumping rate 230 l/s

(ii) Water death before the well 45 m

(iii) Draw down observed in the side walls at the time of pumping

(a) 1.5 m in a well at a distance 30 m

(b) 0.5 m in a well at a distance of 60 m

Find

(i) The influence circle radius

(ii) Co-efficient of permiability.

Solution

At the time of pumping, the water level in the side wells are also lowered, that means that they are inside the influence circle, so they will have the same draw down curve.

$$Q = 1.36K\frac{H^2 - h^2}{\log_{10} R/r}$$

where

$$H = 45 \text{ m } h_1 \ (45 - 1.5)\ 43.5 \text{ m}, h_2 = 44.5 \text{ m}$$
$$r = 0.075 \text{ m}, r_1 = 30, r_2 = 60$$

$$\therefore \quad Q = 1.36\,K\frac{45^2 - 43.5^2}{\log_{10} \text{R}/30} \quad \text{...(a)}$$

and
$$Q = 1.36\,K\frac{45^2 - 43.5^2}{\log_{10} R/60} \quad \text{...(b)}$$

The Eqs. (a) and (b) will be solved by trial and error method. Then

$$R = 96 \text{ m}$$

As
$$Q = 230, \text{ l/s} = 0.23 \text{ m}^3\text{/sec.}$$

$$\therefore \quad 0.23 = 1.36\,K\frac{45^2 - 43.5^2}{\log_{10} 96/30}$$

Therefore the co-efficient of permeability will be

$$K = 50 \times 10^{-4} \text{ m/sec.}$$
$$= 50 \times 10^{-2} \text{ cm/sec.}$$

Example 3.8 *When a well operated under a depression head of 3 m, design an open well in coarse sand to get a yield of 0.004 m³/sec. Assume the value* *of coarse sand i.e.,* $\frac{C'}{A} = 1.00/m^3/hr/m^2/m$ *of depression head.*

Solution

Required discharge from the well

$$= 0.004 \text{ m}^3/\text{sec.}$$

$$= 0.004 \times 60 \times 60K \text{ m}^3/\text{hr.}$$

$$= 14.4 \text{ m}^3/\text{hr.}$$

The depression head = S = 3 m

Now from Eq. $$Q = \left(\frac{C'}{A}\right) A.S.$$

$$14.4 = 1.00 \times A \times 3$$

or $$A = \frac{14.4}{3} = 4.8 \text{ m}^2$$

If the diameter of well taken as d_w then

$$\frac{\pi d_w^2}{A} = 4.8$$

or $$d_w = \sqrt{\frac{4.8 \times 4}{\pi}} = \mathbf{2.48 \text{ say } 2.5 \text{ m}}$$

Example 3.9 *The following inforamations are given to find out the velocities of the ground water flow by using Slichter and Hazen's constant as 400 and 800 respectively.*

1. *The velocity coefficient of water at ground water temperature of 10 °C = 1.*
2. *Effective size of the particles in the aquifer = 0.1 mm.*
3. *Hydraulic gradient = 1 in 80.*

Solution

(a) By Slichter's formula

$$v_a = \frac{K_i D_{10}^2}{\mu} = \frac{400 \times \frac{1}{80} \times (0.1)^2}{1}$$

$$= \frac{400 \times 0.01}{80} = \mathbf{0.05 \text{ m/day.}}$$

(b) By Hazen's formula

$$v_a = \frac{K''.i.D_{10}^2}{60}(1.8\,T + 42)$$

[∵ T is in °C ∵ $K = 273° + C$]

$$= \frac{800 \times \frac{1}{80} \times (0.1)^2}{60}(1.8 \times 10 + 42)$$

[∵ $T = 10°$ C $K = 283°$]

$$= \frac{800 \times \frac{1}{80} \times \frac{1}{100} \times 60}{60} = \mathbf{0.1\ m/day.}$$

Example 3.10 *Datas are gven below. Calculate the velocity of flow in m/ day and discharge in litres per day. Data are :*

(1) Aquifer of very coarse sand 300 m wide and 15 m deep.

(2) Slope of ground water table is l m in 250 m. Also find the standard co-efficient of permeability and co-efficient of transmissibility.

Assume that the ground temperature is [10°C or 283° K] and the coefficient of permeability = 1.0 cm/sec.

Solution

Coefficient of permeability K = 1.0 cm/sec = 864 m/day

Slope = S = 1/250

From the formula $v = K_S$

v = 864 × 1/250 = **3.46 m/day**

Then the discharge will be

Q = 3.64 × 15 × 300

= 15,600 m^3/day = **1.56 × 10^6 lit/day**

Next Part

The co-effcient of permeability K = **864 m/day**

and co-efficient of transmissibility = 864 × 15

= **12.95 × 10^3 m/day**

Example 3.11 *A fully penetrating well of diameter 0.4 is obstracting water from 2.5 m thick confined aquifer. The steady state draw downs at 8 m and 50 m were observed to be 2.5 m and 0.6 m respectively. Compute the steady state discharge from the well. The co-efficient of permeability is 10^{-3} m/s.* **(A.M.I.E.)**

Solution

Using the formula

$$Q = 2.73\, K.t\, \frac{(H-h)}{\log_{10} R/r} \qquad ...(1)$$

or
$$Q = 2.73\, K.t \times \frac{2.5}{\log_{10} R/8} = 2.73\, Kt \frac{0.6}{\log_{10} R/50} \qquad ...(2)$$

Solving the above two Eqs. (1) and (2)

$$K = 90 \text{ m}$$

$$\therefore \qquad Q = 2.73 \times 10^{-3} \times 2.5 \times \frac{2.5}{\log_{10} 90/8}$$

$$= 0.0162 \text{ m}^3\text{/sec.}$$

But again,

$$0.0162 = 2.73 \times 10^{-3} \times 2.5 \times \frac{50-h}{\log_{10} 90/0.2}$$

or
$$(H-h) = 6.29 \text{ m}$$

$\therefore$ The draw down in the well = 6.29 m

N.B. Actually the radius of influence circle is not so easy to get various arbitrary values have been suggested to get R by the different investigators. Among them Slitcher's suggestion is as 150, and Tolman's suggested it will be 300 m. But this seems to be that a more realistic approach will be when discharge Q is correlated with R.

Assuming $R \propto Q$ or $R = CQ$ when C is a constant. Then we will get

$$Q = 1.36K = \frac{H^2 - h^2}{\log_{10} \frac{C}{Q/r}}$$

PROBLEMS

3.1. Differntiate between shallow wells and deep wells. **(P.W.)**

3.2. What are confined and unconfined aquifers ? Explain with the help of neat sketch. Derive an equation for determining the discharge from unconfined aquifer. **(Ranchi IIT)**

3.3. A 2 m diameter well is being pumped at a constant rate of 0.85 m^3/mt. Producing draw downs of 0.80 and 0.25 in two test bores 30 and 60 m away respectively from the well. Depth of water before pumping was 15 m. Find the radius of zone of influence. **(I.I.T.)**

3.4. Sketch and describe briefly the various methods used for drawing water from the tube well. What consideration would determine the choice of a particular method ? **(U.P.S.C.)**

3.5. Differentiate between co-efficient of transmissibility and coefficient of storage. **(A.M.I.E.)**

N.B. The specific yiled of a well is also termed as the coefficient of storage.

3.6. A 2 m diameter well is being pumped at a rate of 6.20 m3/min producing drawdown of 1 m and 0.25 m in two test bore holes 30 m and 60 m away respectively from the well. Depth of aquifer as 15 *m*. Find the drawdown and radius of influence. **(A.M.I.E.)**

3.7. The diameter of a tube well is 300 *m* and the depth of the aquifer is 15 *m*. The radius of circle of influence of the well is 135 m. The coefficient of permeability of soil is 30 $m^3/m^2/d$. Calculate the drawdown of the well when the yield is 200 m^3/d. **(A.M.I.E.)**

3.8. An aquifer of thickness 30 m has the average width of 200 m. The coefficient of permeability of the aquifer is 8 $m^3/day/m^2$. The hydraulic gradient is 2 metres per kilometre. Determine the discharge of the aquifer in cubic metre per hour. **(C.U.)**

3.9. Estimate the velocity of flow in *m*/day and discharge in litres/day through an aquifer of 300 m width and 15 m deep when the slope of ground water table is 1 in 250. Also find the co-efficient of permeability. Assume standard co-efficient of permeability $K = 1$ m^3/s. **(A.M.I.E.)**

3.10. (a) What is meant by 'Porosity' and Permeability' and how do they effect the ground water storage.

(b) Briefly mention the various zones of ground water and point out the importance of the 'zones of saturation' in this connection.

3.11 (a) Write a brief notes on the movements of ground water and define and explain the importance of permeability in this connection.

(b) Define and explain any four of the following :

(i) Capillary fringe

(ii) Specific yield

(iii) Field capacity

(iv) Aquifer and their types

(v) Co-efficient of storage

(vi) Co-efficient of permeability

3.11. Enumerate of various forms in which ground water occurs in nature. Discuss briefly the exploitation of its most important source.

3.12. (a) What is dug well, and what are their types ? How are the dug well constructed ?

(b) Enumerate the method which are used for determining the yield of dug wells. Discuss any one of these methods.

3.13. (a) Differentiate between 'shallow' and 'deep' dug wells. Define and explain the importance of cavity formation in such wells, and compare the possible yields from both these types of dug wells.

(b) A dug well is to be designed in fine sand for the yield of 0.003 cumec when operated under a depression head of 3.5 metres. Assume the value of specific capacity of the aquifer as equal to 0.50 $m^3/hr/m^2$ of area/*m* of depression head.

3.14. Describe briefly the difference methods of sinking tube wells. Draw a neat sketch of a tube well and its various parts. **(Engineering Services)**

3.15. Calculate the discharge from a tube well 50 cm in diameter with a drawdown of 2.5 m in an unconfined water-bearing stratum 15 m thick. Permeability may be taken as 1,600 lit/day and radius of influence as 250 m.

(Engineering Services)

3.16. (a) Give an equation defining Darey's law. What is its limitation ?

(b) What is an artesian well ?

(c) The cost of pumping increases as pipe diameter is reduced but the cost of pipe decreases. How will you make a choice ?

(Engineering Services)

3.17. (a) Describe an artesian well, and how it is formed ?

(b) Design an infiltration gallery, given the following :

Population	= 75,000
Per capita supply	= 150 lit/day
Length of test gallery constructed	= 6 m
Depth of aquifer	= 5 m
Pumping rate	= 600 w.p.m.
Drawdown (corresponding to the above pumping rate)	= 1.5 m

It is desired to maintain a depth of at least 1.0 m above the sole of the aquifers. Draw a typical cross-section of gallery indicating different layers.

(A.M.I.E.)

3.18. (a) Differentiate between confined and unconfined aquifers.

(b) Derive an expression for determining the discharge from an unconfined aquifer.

(c) The diameter of a tube well is 30 cm and the depth of the aquifer is 15 m. The radius of circle of influence of the well 135 m. The co-efficient of permeability of the soil is 30 m/day. Calculate the draw down of the well when the yield of the well is 200 m^2/day. **(A.M.I.E.)**

3.19. What is meant by Tube well ? What are their types ? Describe the most widely used type of tube well with a neat sketch. What is the approximate values of the average yield and depth of such a tube well ?

Chapter 4

Intakes of Water to the Treatment Plant

CHAPTER OBJECTIVE

This chapter, deals the main features of intake of water from the source of supply and take it to the next step. Almost at possible and important ideas has been dealt with very simple language so that this can be easily understood by any one. Some Examples were solved and for further practice, the possible type of problems are given.

From the source of water, the water is taken to the treatment plant safely is called the intake structure. This will depend upon the nature of water sources available. The care should be taken so that the intake structure should have a good life. First of all the location of site should be selected. To locate the site the following points to be considered if possible.

(1) Considering the cost, the site should be as near as possible near the town or city.

(2) The site should in such a place, where the more pure water is available; so the treatment cost is low.

(3) The location of site should be on the upstream side of disposal point.

(4) Care should be taken to see that during the most critical period of the season, it can draw some water.

(5) Care should be taken to see that if necessary, further scope of expansion may be possible.

(6) Particularly in the river intakes, care should be taken to see that it is on the stable zone, i.e., not close to the bank of river. Becuase due to the sliding of bank may create problem for intake water.

(7) Care should be taken that it should not be in the way of navigation channels, if it is so there lies a chance of getting water polluted.

(8) The intake should not be in the area of flood. It should be easy accessible during the flood.

(9) Try to avoid the sharpes bent or curve of a river, if it is not possible, then the intake may be on the concave curve.

4.1 SUBMERGED INTAKES

This is a simple intake system. Consisting with concrete bed laying a withdrawl pipe as shown in Fig. 4.1 This withdrawl pipe takes water into the sump well into the bank, from there the water is lifted by pumps.

Opening of the inlet is covered with a screen, so that is can prevent entries of debris, ice etc. into the withdrawl pipe. The intake opening generally placed about 1.5 to 2.0 m above the bed to avoid entry of silt and other sedimentary materials.

This type of intake is cheaper and do not have any obstruction to nevigation. Generally they are useful for small water supply project. This cannot be used for big project, because, they can not be easily clean and repair.

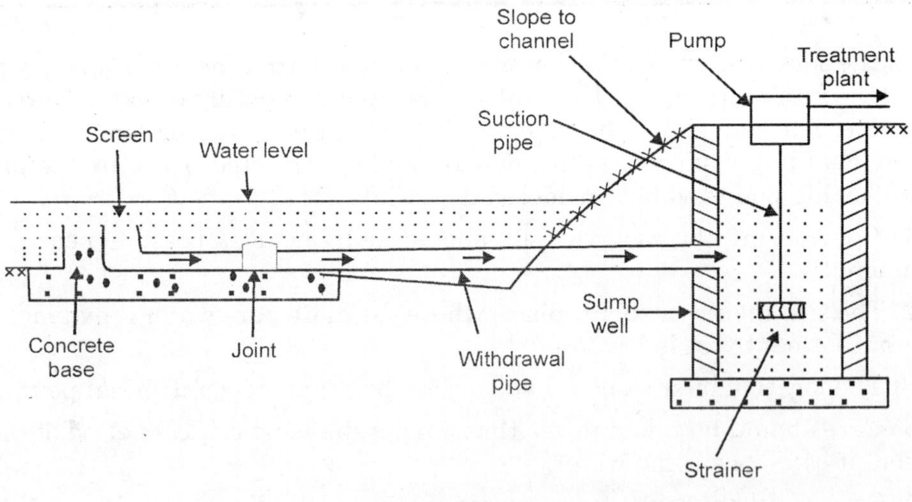

Fig. 4.1 *Submerged intake*

4.2 INTAKE TOWER SYSTEM

Tower intake is generally used for large project, where the water supply is to be high, like cities and big towns. Tower may be situated on the river side or a little away from the source. The fluctuation of water level in the source can be avoided or controlled by the tower. From a big reservoirs tower is used for water supply. Towers are generally rectangular in shape and can be made circular. There will have provision at the top for inspection by some maintenance personals. Two main types of towers are generally used.

1. Intake Tower (Wet Type). It is made of concrete having a inside directly connected from the source. It does not require any pump. Generally it is situated inside the river bed, so that the water pressure will be able to send the intake water to the treatment plant. Lift is not required in the case of reservoir, to take the water directly sump well, but low lift is required in care river. Sometimes the withdrawal pipe maybe on the bed of the river or sometime a tunnel may require to get the supply from river. In case of high treatment plant the water may be required to pump from the intake tower.

2. Intake Tower (Dry Type). This is more or less the same as wet type but the main difference is the intake pipe coming up from the source to the tower, having dry out side. The dry intake tower has a benefit to withdraw water from any level required, that is not possible in wet type. However the tower intake is a heavy structure. So it should be placed to see that the navigation is not interapped. The only difference between the two system is one having wet inside the tower when water is intaking from the source, and the other will be dry at the time of delivery to the treatment plant.

4.3 INTAKE FROM CHANNEL

The channel intake is more or less like the river intake. This structure is built on the side of the bank of the channel. Water enter the sump chamber through the inlet pipe is with a screen to avoid entering some objectionable materials. Since the flow of the cannel is intensified by the structure, the velocity of water in the channel will increase. At the discharge end of the pipe to the well should provide fine screen to avoid entry of the fine particles of dust etc. An outlet valve may be provided at the discharge end to control the outlet flow, and this valve will be operated from top of the control room.

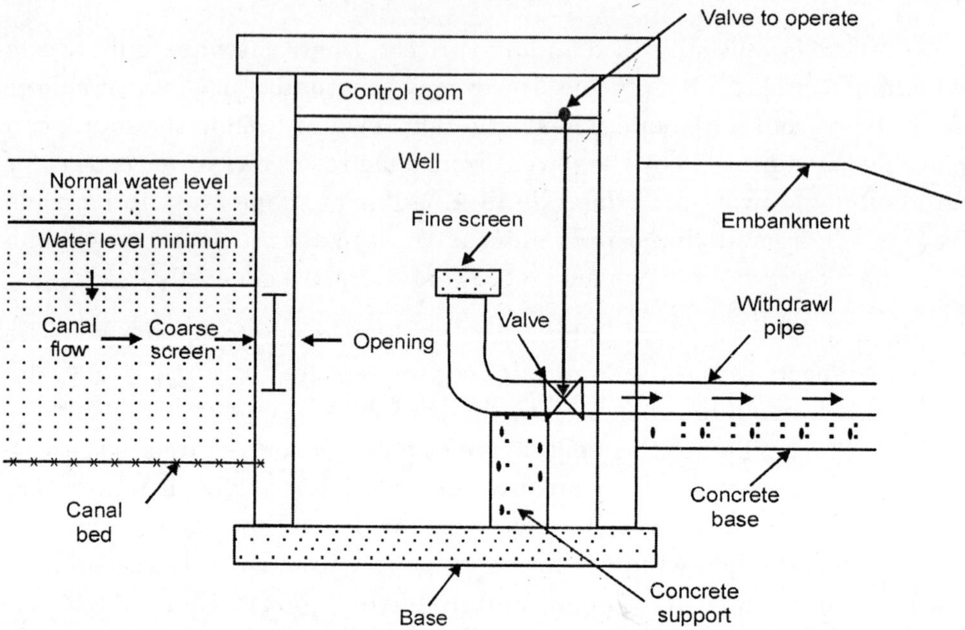

Fig. 4.2 *Intake well in a canal*

Example 4.1 *Design a bell mouth canal intake for a city of 75,000 persons, drawing water from a canal which runs only for 10 hours a day with a depth of 1.5 m. Also calculate the head loss in the intake conduit if the treatment works are 1/4 km away. Draw a neat sketch of the canal intake. Given average consumption per person is150 l/d. Assume the velocity through the screens and bell mouth to be less than16 cm/s and 32 cm/s respectively.*

(Civil Services)

Solution

Required city discharge

$$= 1500 \times 75000 \text{ lit/day} = 11.25 \text{ M lit/day.}$$

As the canal runs daily for 10 hours only, so to get the daily flow within 10 hours.

$\therefore$ Required intake flow per hour.

$$= \frac{11.25}{10} \text{ M lit/hour} = 1.125 \text{ M lit/hour.}$$

$$= \frac{1.125 \times 10^6}{10^3} \text{ m}^3\text{/hr.}$$

$$= \frac{1.125 \times 10^3}{60 \times 60} \text{ m}^3\text{/sec} = 0.3125 \text{ m}^3\text{/sec.}$$

Coarse Screen Design

Choose 20 mm ϕ vertical iron bars at 3 to 4 cm centre to centre.

Coarse Screen Area

$$= \frac{\text{Through this coarse sereen the discharge enterning well}}{\text{Velocity of discharge through screen}}$$

Max. velocity through the screen 0.16 m/sec.

$\therefore$ Minimum area of screen needed

$$= \frac{0.3125}{0.16} \text{ m}^2 = 1.95 \text{ m}^2$$

$$\text{Height of screen} = 1.5 - 0.15 - 0.3$$

$$= 1.05 \text{ m.}$$

N.B. Assuming minimum water head in canal to be 0.3 m below normal water level.

Minimum opening of screen length

$$= \frac{1.95 \text{ m}^2}{1.05 \text{ m}} \text{ m}^2 = 1.86 \text{ m}$$

Assuming the clear opening widths between vertical bars say 30 mm (or 0.03 m) each, then No. of openings $= \frac{1.86}{0.03} = 62$

$\therefore$ No. of bars = 61

Then the length cover by bars of 20 mm ϕ

$$= 61 \times 0.02 \text{ m} = 1.22 \text{ m}$$

So, screen total length = 1.86 + 1.22

$$= 3.08 \text{ ; Say } 3.1 \text{ m}$$

So, coarse screen height 1.05 m and 3.1 m length of rectangular shape will be provided in intake well.

Bell Mouth Entry Design

Bell Mouth Entry Area

$$= \frac{\text{Discharge}}{\text{Velo. through bell mouth}} = \frac{0.3125}{0.32} \text{ m}^2$$

$$= 0.98 \text{ m}^2$$

If d is the dia. of the bell mouth entry, then

$$\frac{\pi}{4}d^2 = 0.98$$

$$\therefore \quad d^2 = \sqrt{\frac{0.98 \times 4}{\pi}} = 1.12 \text{ m say } 1.2 \text{ m}$$

So, 1.2 m ϕ bell mouth will be provided with perforations for fine screen.

Intake Pipe Design

Assuming the flow velocity in pipe is 1.5 m/sec; then

$$\text{The area of pipe} = \frac{\text{Discharge}}{\text{Velocity}} = \frac{0.3215}{1.5} \text{m}^2$$

$$= 0.208 \text{ m}^2$$

$\therefore$ Dia. of (D) pipe will be

$$\frac{\pi}{4}D^2 = 0.208$$

$$D = \sqrt{\frac{0.208 \times 4}{\pi}} = 0.515 \text{m} \simeq 5.5 \text{m}$$

Hence 0.5 m dia pipe will be used.

Velocity of flow through this pipe i.e., 0.5 m dia.

$$= \frac{0.3215}{\pi/4(0.5)^2} = 1.59 \text{ m/sec.}$$

Head loss through intake pipe upto treatment works calculated from the equation as

$$V = 0.85\, C_H.\, R^{0.63}.\, S^{0.54}$$

where C_H = Coeff. of pipe material (130 for cast iron)

R = Hydraulic mean depth of pipe

$= d/4$ (for pipes running full) $= \frac{0.5}{4} = 0.125$ m

V = velocity through the pipe = 1.59 m/sec

S = energy slope line

$$\therefore \quad 1.59 = 0.85 \times 130 \times (0.125)^{0.63} \times S^{0.54}$$

or

$$S^{0.54} = \frac{1.59}{0.85 \times 130 \times (0.125)^{0.63}} = 0.0533$$

$$\therefore \quad S = (0.053)^{\frac{1}{0.54}} = (0.0533)^{1.85} = 4.41 \times 10^{-3}$$

Now $$S = \frac{H_L}{L} = \frac{\text{Head loss}}{\text{Pipe length}}$$

$$H_L = L \times S = \left(\frac{1}{4} \times 1000\right) \times 4.41 \times 10^{-3}$$

$$= \mathbf{1.1025\ m}$$

4.4 SOME IMPORTANT DEFINITIONS

In our country some terms are commonly used to explain clearly.

The maximum actual pressure includes the water hammer is called working pressure.

Due to some abnormal pressure which may occur in the pipe during operation with the maximum pressure, is known as design pressure.

With the help of standard method of testing pipe can stand maximum pressure hydraulilc pressure without leakage, is known as test pressure.

Different types of pressure pipes are used in practice. It depend upon the material used. Such as:

1. Cast iron pipe
2. Steel pipe
3. Reinforced concrete pipe
4. Hume steel pipe
5. Vitrified clay pipe
6. Asbestos cement pipe
7. Other types of pipe

4.5 INTAKE TOWERS

Generally the towers are used as an intake towers where the level of the water intake having the fluctuation of levels. For large projects or on rivers, big reservoirs etc. the towers are used. They are generally placed on the side of the bank. The towers have opening gates controlled by an operator. So that the intake water can be allowed when necessary. There are two types of towers. One is wet and the other is dry type.

4.6 WET INTAKE TOWERS

This will have a vertical circular type made with concrete, filled with water. There will have gate opening on both side. That gate opening will be controlled by an operator. The intake pipe may lie on the bed of the reservoir or on the bed of the river. Sometimes they may be a tunnel below the river bed. Then

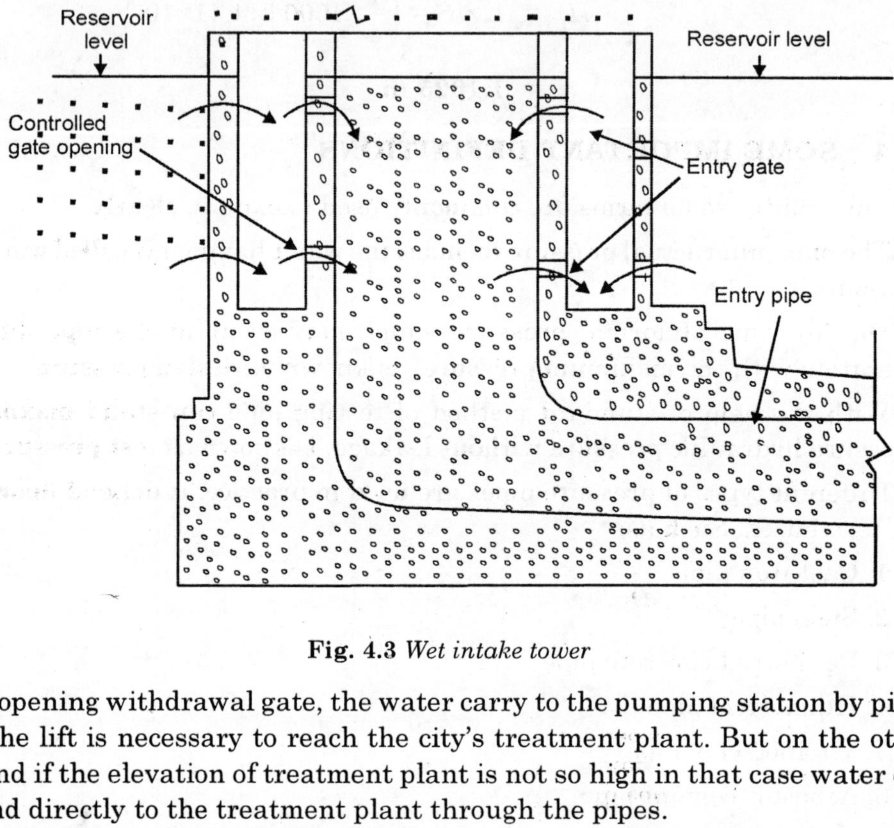

Fig. 4.3 *Wet intake tower*

by opening withdrawal gate, the water carry to the pumping station by pipes if the lift is necessary to reach the city's treatment plant. But on the other hand if the elevation of treatment plant is not so high in that case water can send directly to the treatment plant through the pipes.

4.7 DRY INTAKE TOWERS

In the dry intake towers, the water is directly get into the tower through the pipe. Then that will be taken out by opening the gate to the treatment plant. Otherwise the function of the two types of tower is the same. When the exit gate is closed, the tower remain full of water in the case of dry intake tower.

These intake towers are huge structure they should be properly designed and care should be taken to see that it can stand any type of worst cases, such as hydrostatic pressure, wind force, and earthquake forces.

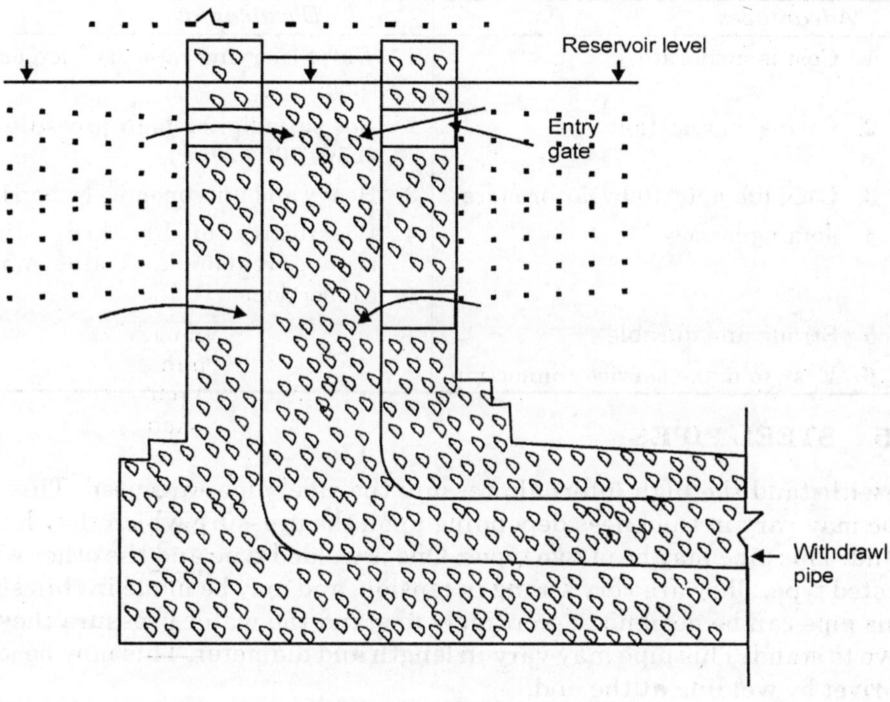

Fig. 4.4 *Dry intake tower*

1. Cost Iron Pipe. It is a very common type of pipe used in water supply system. They can quite satisfactorily stand the water pressure and having good property of corrosion resistance. Generally the manufacture in a length of 3.5 m or sometimes upto 6.0 m length. The cast iron pipe may have two types one is ordinary sand moulding and the other will be centrifugal processes.

As per I.S. : 1536–1976 Types of Cast Iron (Span) pipes are shown in Table 4.1

S.No.	*Category of pipes in the order of increasing for the same diameter*	*Test Pressure in kN/m²*	*Working pressure in kN/m²*
1.	Class LA	3500	1200
2.	Class A	3500	1000
3.	Class B	3500	2000

Advantages	Disadvantages
1. Cost is moderate	1. Water carrying capacity increases in time.
2. Corrosion resistant	2. Not suitable for high pressure (Not above 700 kN/m^2).
3. Long life upto 100 year or more	3. Heavy and uneconomical when large.
4. Joining is easy	4. Possibility of breaking during transportation and also making connections
5. Strong and durable	
6. Easy to make service connection	

4.5 STEEL PIPES

To withstand the high internal pressure the steel pipes are used. This steel pipe may vary in thickness depending upon the pressure which they have to stand. This pipe may be of two types. One is welded type and the other will be riveted type. They are very strong in tension, and may be made in thin shells. This pipe can be designed accordingly to stand the water pressure they will have to stand. This pipe may vary in length and diameter. This may be joined by rivet by welding at the end.

4.6 REINFORCED CONCRETE PIPES

Concrete pour into a mesh made of steel bars and tamped as an ordinary way.

Other way of making a strong R.C.C. pipe is to have a thin steel cylindrical shells both inside and outside and rich cement concrete is used. They are stronger than the other types and can stand large pressure. Generally they are called as cylindrical type. But their patend name in India is hume steel pipes.

As per I.S. : 458–1988 ordinary R.C.C. are manufactured

Category	Dia available	Test pressure in kN/m^2	Places where used
Class–P_1	80–1200 mm	200	Used on gravity mains, design pressure not exceeding 2/3 of test pressure
Class–P_2	80–600 mm	400	Used on pumping mains the design pressure not exceeding 1/2 of the test pressure
Class–P_3	80–400 mm	600	Used on pumping mains the design pressure not exceeding 1/2 of the test pressure

As per I.S. : 1966–1989 steel cylinder R.C.C. pipes

Category	*Test pressure*	*Working pressure*
Class 1	500	In general working pressure to pumping mains is taken as 50% of test pressure, and for gravity mains is taken as 2/3 of the test pressure
Class 2	1000	
Class 3	1500	
Class 4	2000	
Class 5	2500	
Special class	Above 2500 to be specified by purchasers	

R.C.C. Pipes

The advantages and disadvantage of R.C.C. pipes are as follows :

Advantages	*Disadvantages*
1. No corrosion from inside water and outside soil	1. Difficult to repair.
2. Strong and have a long useful life.	2. Very high pressure cannot stand.
3. Resist external compressive load. Not collapse under traffic load.	3. Difficult to carry, they are heavy and bulky in size.
4. Easy construction on site or in factory.	4. Ground water contains acid, alkalis or sulphur, components, corrosion may take place.
5. Expansion co-efficient low. No need of expansion joint at laying.	5. Connection may be difficult to join.
6. Empty pipes are not float in water. They are heavy.	6. Shrinkage cracks and porosity leads to leake.

4.7 HUME STEEL PIPE

A thin steel shell coated inside with cement mortar by centrifugal process. The outside of the shell also coated with cement mortar with a bit thicker thickness, to protect the steel from weather or soil action. This pipe is heavy and difficult to handle. This pipe may have different diameter and size.

4.8 VITRIFIED CLAY PIPES

This pipe is made of clay. They cannot be suitable for water carrying. Because they cannot stand high pressure. They are quite suitable to sewage and drains. Having a smooth efficient surface and free from any corrosion. They are mainly used in sewage disposal. As they are very weak in tension, it is very difficult to have a water tight joints.

4.9 ASBESTOS PIPE

Applying some pressure, asbestos, silica and cement forms a homogeneous material of high strength. This material is then casted to form a pipe. Then the asbestos fibre mixed with cement is used to reinforce it. They are available in different sizes in the local market. They can stand pressure say upto 1400 kN/m^2.

Advantages	*Disadvantges*
1. Highly resistant to corrosion.	1. Cost is high.
2. No need any skilled man for products.	2. Not suitable to carry any petroleum assembled.
3. Light and easy to carry.	3. Pipe strength is very low. They may get damage during carrying.
4. Highly flexible.	
5. Co-efficient of expansion is low. Expansion joints not required.	
6. Hydraulically efficient as the surface is very smooth.	
7. Suitable for small size distribution.	

4.10 OTHER TYPES OF PIPES

Different types of materials can be used to manufacturing pipe. Such as copper, wrought iron, plastics etc.

The cost of copper pipe is very high but the resistant capacity to acidic and alkaline water is high, and can stand high temperature. Easy to built. They are quite good to carry hot water inside the building.

Wrought iron pipes or lighter than cast iron. Workability is easy than that of cast iorn pipe. Finish is neat but cost is high. They are suitable for indoor work. Durability is less and corrode quickly. In the market they are known as galvanised iron pipes.

Plastics pipes are light and corrosion less. Having low strength and low durability. They are quite suitable for domestic supply. They are cheeper. Nowadays they are extensively used in the domestic purpose. P.V.C. pipes are manufactured according to I.S. 4985–1988. They are capable to carry water upto about 100 *m* head of water.

4.11 DISINFECTED OF PIPE LINE BEFORE USE

When the pipe line is ready to supply to the town or city, it should be disinfected before use. The whole pipe line is full of supply water and treated with chlorine in such a way that the resiudal chlorine upto 50 ppm should be there and kept it for 12 hours. After that the water will be allowed to run off from the pipe line. Then the total pipe line is washed with supply water. After that the supply system is ready to supply for public in the town or cities.

4.12 LAYING OF WATER PIPES

Water pipes are used for carrying water from supply source to the city and then to the consumers. Different diameter and type of pipes are used for this purpose .From the supply source the main pipes are of generally 10 cm to 1 m dia cast iron pipes are used. Then for the consumers connection galvanised iron pipes. 2.5 to 3 cm dia is used. When the pipe line is layed through roads and crossing the roads, they should be layed underground. In this case a suitable trench is to be made about 1.0 m depth an 0.5 m wide to lay the pipe. So that the top pressure after filling, created by the road traffic may not do any problem to the supply. Branch lines may be taken from the main line with smaller diameter pipes whichever be the suitable for the purposes. Then from the branch lines, the domestic connection are made. But nowadays it seems that in the domestic connection plastic pipes are getting more and more pupular to the consumers. Because they are cheeper, easy to bend, no corrosion problem.

PROBLEMS

4.1. What is a river intake ? What are the factors which govern the location of an intake structure ?

4.2. Draw a neat sketch showing the intake arrangements which you suggest for drawing water for a city water supply plant being fed direct from river.

4.3. What is an intake structure ? Enumerate the various types of intakes and discuss in details any one of them.

4.4. Write notes on the following :

(a) Simple submerged intake.

(b) Dry intake tower.

(c) Intake well for a channel.

(d) A typical river intake.

4.5. What are the different materials, which are commonly used for water supply pipes ? Discuss their comparative merits and demerits.

4.6. Describe briefly the procedure followed in laying and testing the water supply mains.

4.7. Mention the points which should be taken into consideration in deciding the location and design of an intake for the water supply of a large town, the source being a perennial river. Draw a detailed sketch of the intake and describe fully. **(U.P.S.C.)**

4.8. Sketch and exaplain :

(1) Direction intake from a river. **(Ran)**

(2) Intake tower in a reservoir. **(Ran., Roorkee, AMIE)**

(3) Service connection from a street main. **(AMIE, Ran.)**

Chapter 5

Water Treatment

Water obtained from underground and surface sources; are not free from various types of germs. This type of water should not be used directly. That may cause the epidemic. For that reason water should be treated properly to remove the all possible germs and other unwanted materials; which may be in suspension forms.

Water treatment should be done step by step, which is going to be described here.

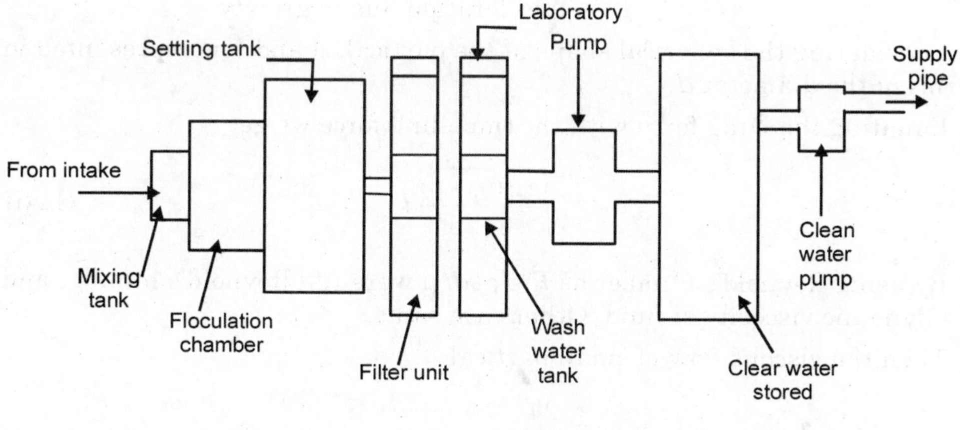

Fig. 5.1 *Treatment plant layout*

A layout diagram of a treatment plant is shown in Fig. 5.1. Water comes from sources, passes through the mixing tank and then after mixing it goes through the floculation channel to the setting tank. From floculation tank it goes to the filter unit, before goes to the next stage a laboratory check should be done. After the laboratory test, water is pumped to the water reservoir, from there the clean water stored into the pumping station ready for supply.

A careful check up should be done in the laboratory by the proper qualified staff regularly to check the physical, chemical and bactereological examination, to make sure that the water is free from these factors.

5.1 SEDIMENTATION THEORY

The theory behind the sedimentation of a suspended particals is as follows :

From Newton's Law Drag force is

$$F_D = C_D A \frac{Pv^2}{2}$$

F_D = Drag force

C_D = Co-efficient of Drag,

A = Area projected by the partical,

v = Settling velocity

ρ = Fluid mass density.

Impelling force,

$$F_1 = (\rho_1 - \rho)\, gV$$

ρ_1 = Mass density of partical

V = Volume of partical

g = Acceleration due to gravity

Considering the spherical shape of the partical, A and V is represented in terms of the diameter d.

Equating the Drag force with the impelling force we get

$$v = \sqrt{\frac{4g(\rho_1 - \rho)}{3C_D\rho} d} \quad \text{... (i)}$$

By using Reynold's number as $R = \rho v d / \mu$ were R = Reynold's number and μ = dynamic viscosity of fluid. Others are same.

Then the viscous flow of small partical

$$C_D = \frac{24}{R}$$

By substituting this in Eq. (i) we get

$$v = \frac{1}{18} \cdot \frac{g}{\mu}(\rho_1 - \rho)d^2 \qquad \text{...(ii)}$$

The above relation is known as Stoke's Law. But velocity is depending upon the temperature. Considering the temperature the settling velocity is given by

$$v = \frac{g}{18}\left(\frac{S_s - 1}{r}\right)d^2 \qquad \text{.... (iii)}$$

where

v = settling velocity (cm/s)
d = partical diameter (cm)
g = gravity (cm/sec^2)
μ = dynamic viscosity of fluid (10^{-2} cm/s)
ρ_1 = partical mass density (gm/cc) or g/ml
ρ = liquid mass density (gm/cc) or g/ml
S_s = Sp. gr. of solid
r = Kinematic viscosity (Centistokes)

Tank for Sedimentation

A horizontal tank is mostly used nowadays in this tank liquid flows through the tank horizontally. In a sedimentation tank there are generally four zones.

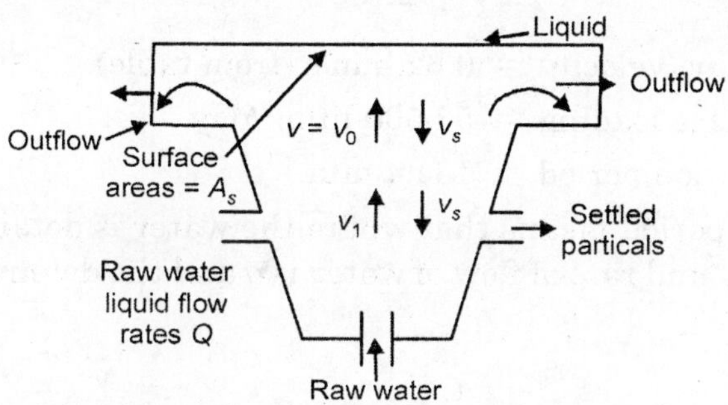

Fig. 5.2 v_s *Settling velocity clarifier*

Let the effective length = l, width of the tank = b and the depth of the Settling = h, and the displacement velocity = oV Also the partical settling velo. = v_1. Then the quantity of water flowing through the tank $q = bhv$. The Partical will be removed if settling time $\frac{h}{v_l} < \frac{l}{v}$

Example 5.1 *Find out the settling velocity, surface loading and the detention period, in a plain sedimentation tank where*

(i) Dia. of particle = 0.02 mm

(ii) Sp. gr. of particle = 2.65

(iii) Settling velocity = 0.62 mm/s

Solution

$$v = \frac{g}{18}\frac{1}{\mu}(\rho_1 - \rho)\, d^2$$

as $$\frac{\rho}{\mu} = \frac{g}{18}(S_s - 1)\frac{d^2}{v}$$ From Eq. (iii)

∴ Setting velocity = 0.62 mm is found from table.

$$\text{The surface loading} = \frac{0.62 \times 24 \times 60 \times 60}{1000}\ \text{m}^3/\text{m}^2/\ \text{day}$$

$$= \frac{0.62 \times 24 \times 60 \times 60}{1000} \times 1000\ \text{lit/m}^2/\text{day}$$

$$= \mathbf{53{,}502 = 53.500\ lit/m^2/day}$$

Assuming the depth of the tank is taken 5 metres, then

$$\text{Detention period} = \frac{5 \times 1000}{0.62 \times 60} = \mathbf{134.4\ min.}$$

$$\text{Settling velocity} = 0.62\ \text{mm/s (from table)}$$

$$\text{Surface loading} = 53{,}500\ \text{lit/m}^2/\text{day}$$

$$\text{Detention period} = 134.4\ \text{min.}$$

The detention period means that where the water is detained in it. If the tank capacity is C and rate of flow of water is q and the detention time period is t.

Then $$t = \frac{C}{q}$$

5.2 COAGULATION

Coagulants are used to settle the very fine particals to deposited at the bottom quickly. The coagulating agents are collect the small and very fine particals and bind them togather to form a heavier particle, so that they can be precipitate at the bottom quicker.

Lots of coagulating agents are generally used, such as, Alum. $Al_2\ (SO_4)^3$, $18H_2O$. Very commonly used, and the other agent is Sodium Aluminate

($Na_2Al_2O_4$), this is an alkaline type of coagulant. Sometimes an iron salt is used, where the water will have the low pH value.

5.3 DOSE OF COAGULATION

Generally the coagulating agents are choosen on the basis of quantity of water. The maximum dose of coagulant should be used by having a test before use. Generally a jar with water having mixed with coagulatent water for 10 to 30 min. By trial repeating the test, when a floc appears at the surface, then it assumed to have the most economical dose of coagulatent should be added. Sometimes it may appear that the dose of coagulant may vary depending the type of water.

5.3.1 Screening

In front of pumps and intake, coarse screen is provided to screen of the materials like debris, trees, animals, branches of trees and bushes etc. Coarse screens are parallel iron roads placed vertically at a slope centre to centre. Nowadays the fine screen are not used, because they need to clean very often, to avoid these the fine particals are separated by sedimentation.

5.3.2 Sedimentation Tanks

Commonly used sedimentation tank is a rectangular type, but circular type of tank is also used. Sometimes the spiral type of tank may use for sedimentation. Whatever may be the type is used, They are generally made of reinforced concrete. Sometimes for quicker sedimentation coagulating agents are also used. The tanks are of generally two types, horizontal type and the other is vertical flow type.

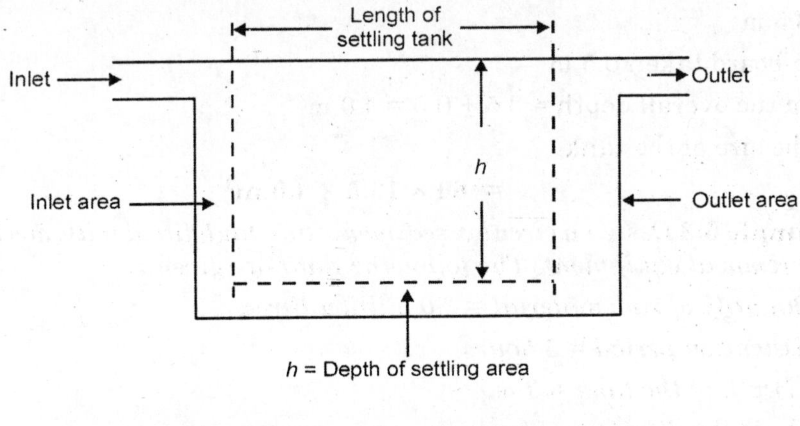

Fig. 5.3 *Sedimentation in a vertical tank*

Example 5.2 *Maximum demand of purified water per day = 10 million litres.*

Design a sedimentation tank for the raw

Supplies considering fitted with mechanical devices to remove sludges. Assume detention period = 5 hours and flow velocity 20 cm/min.

Solution

Water treated per day = 10×10^6 litres

Water treated for 5 hours (Detention period)

$$= \frac{10 \times 10^6}{24} \times 5 \text{ lit.}$$

$$= \frac{10 \times 10^6}{4.8} \text{ lit.} = 2.8 \times 10^6 \text{ lit.}$$

Capacity of tank = 2.8×10^3 cubic metre = 2800 cμ.m

Velocity of flow = 20 cm/min = 0.2 m/min.

Length required = Flow velo. × Detention period

= $0.2 \times (5 \times 60) = 60$ m.

Required Cross Sectional Area

$$= \frac{\text{Tank capacity}}{\text{Length of tank}} = \frac{2800}{60} \text{ m}^2$$

= 46.66, Say 46.7 m^2

Depth of tank assumed = 3.5 m^2, then the width required = $\frac{46.7}{3.5}$ = 13.34 m.

Say 13.5 m

Free-board taken 0.5 m

Then the overall depth = 3.5 + 0.5 = 4.0 m

So the size of the tank

$$= \mathbf{60 \times 13.5 \times 4.0\ m^3}$$

Example 5.3 *Design a circular sedimentation tank fitted with mechanical sludge removal equipment. The following data are given.*

(i) Quantity of raw material = 3.0 million litres.

(ii) Detention period = 5 hours

(iii) Depth of the tank = 3 m.

Find out the diameter of tank.

Solution

Quantity of raw a water = 3.0×10^6 lits. will be treated for 5 hours.

So the capacity of tank

$$= \frac{3 \times 10^6 \times 5}{24} \text{litres}$$

$$= 625 \text{ cm. m}$$

Tank capacity (circular) of depth H and the dia. d is given by (Here $H = 3$ m given)

$$\text{Vol.} = d^2 (0.011d + 0.625 \text{ it})$$

$$625 = d^2 (0.011d + 0.625 \times 3)$$

$$= d^2 (0.011d \times 1.875)$$

This will be solved by trial method, then

$$d = 16.2 \text{ m}$$

∴ The diameter of the tank will be

$$d = 16.2 \text{ m}$$

Note. Function, design and other details of the sedimentation tank is the same as plain sedimentation tank. The assumptions are practically the same.

Nowadays it is possible to use a combined coagulation cum sedimentation tank may be used. This is called a coagulation sedimentation tank. A plain flve chamber is used without having any mechanical devices. Generally the detention period is about to be 15 to 40 minutes, and for the settling tank is about 2 to 4 hours. The floc chambers are having the depth half of the settling tank.

Chapter

6

Filtration

CHAPTER OBJECTIVE

In this chapter the most common type and useful filtering process has been discussed in very short form. The idea is to put the picture as clearly as possible to the learner in a very short and positive form. Though the filtration is one of the most important part of the water supply system, still the most common and practical process are mentioned in this chapter.

After removing the large size of suspended particles by screening and sedimentation, the water still have same fine suspended materials and some bacteria. To get the pure water this should pass through granular materials which is known as filtration. This process will get rid off, colour, odour, and some bacteria presence in it. This process may be classified in two main types, such as

1. Slow sand filter.
2. Rapid sand filter,

And the another process is sometimes used is knows as pressure filter.

1. Slow Sand Filter. This is useful to remove large percentage of impurities and bacteria in comparison to the other process. But slow rate filter yield a very slow rate, require large areas and costly. Considering the disinfection and maximum amount of removal of bacteria, the slow rate filter is not suitable. But in the modern treatment plant the rapid sand filter are commonly used. The coagulation sedimentation tank water is directly sent into the rapid filter and the outcome waters are disinfected to complete killing of germs and the colour is removed.

2. Filter-Materials. Generally the sand either fine or coarse may be used. Sands are supported by gravels; which helps freely to under drain. Sand filter should have the following in properties :

1. Should be clean free from dirt and impurities.
2. Should be uniform in size.
3. Should be hard.
4. Should be not loss weight more than 5%.

When placed in hydrochloric acid for 24 hours.

The uniformity of size is found from the uniformity coefficient i.e. $\frac{D_{60}}{D_{10}}$, this is defined as a ratio of sieve size in mm through which 60% of sample of sand will pass, to the effective size of sand.

The gravel used blow the sand should have hard, dumble, impurities is nil properly rounded in shape and the density of about 1600 kg/m^3.

The slow and filter being used before the rapid sand filter comes in to the filed. Nowadays the slow rate is becoming obsolete. Though they may be used for small plant particularly of warm place, where no cover is required to protect for freezing.

A section of a slow sand filter is shown in Fig. 6.1. The tank–An open water tight rectangular tank made of concrete, The bed of the tank having a slope is about in 100 towards the outlet drain. The depth may have 3 to 4 m and the area of tank from 100 to 2000 sq. m or more.

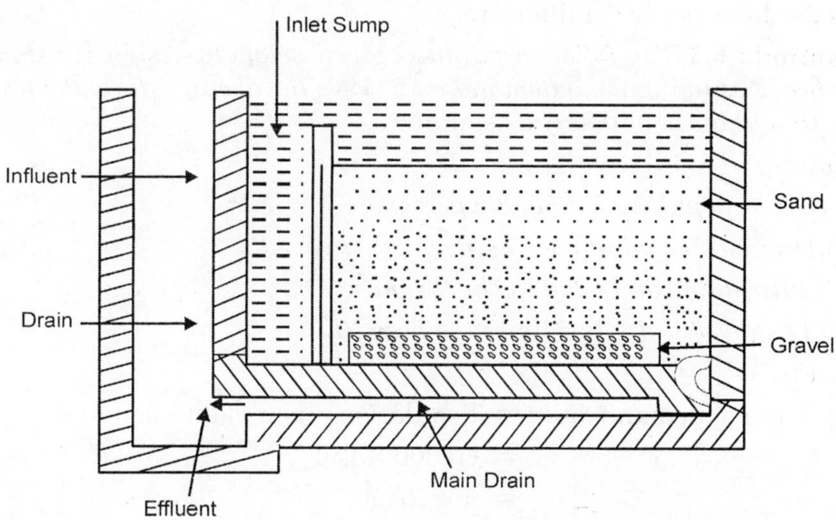

Fig. 6.1

6.1 A SECTION OF A SLOW SAND FILTER

Sand Layer. This is about 100 to 120 cm in depth; placed over the gravel. The effective size (D_{10}) of sand varies from 0.25 to 0.35 mm the uniformity coefficient $\frac{D_{60}}{D_{10}}$ may varies from 2 to 3. The top layer are placed is five and the rests are as follows.

The base materials are gravel; 34 to 80 cm thick of different sizes of gravels. Generally three or four different depths are used each having 15 to 20 cm thick. The low size of gravels are placed at top.

The under-drains are placed at the bottom of the gravel bed, the sub drains connected to a main drains, which lead the filtered water to the outlet well.

6.2 CLEANING OF SLOW SAND FILTER

When the loss of head i.e., filter head is generally limited to a value about 0.7 to 1.3 m. This limiting value is roughly is about 0.7 to 0.85 times the depth of filter sand, then the filter unit must need the cleaning. The cleaning operation is not done by back washing; but done by scraping or removing the top layer i.e., 2 to 3 cm of top sand layer. This washing should be done by clean water. Care should be taken to see the washing is done properly. Cleaning is done when it necessary. If it is found that during this process the sand layer thickness is reduced, then more sand should be added. Normally the interval of the cleaning is needed between 1 to 3 months.

Loading rate of slow sand filter is generally obtained between 110 to 210 litres per hour per m^2 of filter area.

Example 6.1 *The following data is given below to design six slow sand filter bed. Assuming max. demand as 2 times the average daily demand and one unit, out of six will stay as standby :*

Data :

(i) Served population = 60,000 persons

(ii) Demand per capita = 150 lit/head/day

(iii) Filtration rate = 200 lit/hr/sq. m.

(iv) Length of bed = two times the breadth.

Solution

$$\text{Average demand daily} = \text{Population} \times \text{per. capita demand}$$

$$= 60{,}000 \times 150$$

$$= 9 \times 10^6 \text{ lit./day}$$

$$\text{Max. daily demand} = 2 \times 9 \times 10^6 = 18 \times 10^6 \text{ lit./day}$$

$$\text{Rate of filtration daily} = (200 \times 24) \text{ lit./sq. m/day.}$$

$$\text{Taotal surface area} = \frac{\text{Max. daily demand}}{\text{Rate of filtration daily}}$$

$$= \frac{18 \times 106}{200 \times 24} \text{ sq. m}$$

$$= 3750 \text{ m}^2$$

But as per question 5 units are used regularly. So it should provide the necessary filter area required.

∴ For each filter unit area

$$= \frac{1}{5} \times \text{Total area} = \frac{1}{5} \times 3750$$

$$= 750 \text{ m}^2.$$

If L = length and B = breadth of one unit,

Then, $L = 2\,B$ (given)

$$\therefore \quad 2B \times B = 750$$

$$B^2 = 375 \text{ m}^2$$

$$B = 19.36 \text{ Say } \mathbf{2.0 \text{ m}}$$

$$\therefore \quad L = 2 \times 20 = \mathbf{40 \text{ m.}}$$

Hence, the size of each unit will be 4 0 m × 20 m. So it will be better to arrange 3 units on either side.

6.2 RAPID SAND FILTER

In the Rapid Sand Filter, the rate of filtration is much higher than slow rate sand filter and it also require small area than the slow rate. This rapid filter is some times called Mechanical Sand filter.

To satisfy the requirement of demand the rapid filter may classified as (i) having layer size of sand particals, and (ii) Developing pressure over the filtering water, and this may called the pressure filters.

In the rapid sand filter coarse size of sand i.e., 0.5 mm effective size. This type of filter may produce more than 30 times higher rate than that of slow rate. Generally water from coagulation sedimentation tank are passed into these filters, then the fillered water is then treated for disinfetaction after that ready for supply.

Rapid sand filter is generally made of concrete or a brick works, it is an open air tight rectangular tank. Usually the depth may vary from 2 m to 3 m.

To get the uniform filtration the filter bed area should not kept large. Normally the depth is taken to 10 to 8 m^2 per unit.

Roughly estimating the number of units required for a plant is

$$N = 1.22\ \sqrt{Q} \text{ – (by Morrell and Wallaeo)}$$

where

N = No. of filter units

Q = Plant capacity in million litres per day.

6.3 FILTER MEDIA

One of the media is sand and having depth of 75 cm to 85 cm and placed over the bare material gavel. Sand having effective size (D_{10}) varies from 0.3 to 4.5 mm. The uniformity coefficient ranges from 1 to 2. The sand laid in layers, the fine gain sand is on the top and the coarser varieties are at the bottom.

6.4 BASE MATERIAL

In both the filter slow and rapid gavel is used to support the sand. But in rapid filter the gravel also distribute the wash water uniformly. The gravel are about 60 to 90 cm of sizes. Placed at different layers, each layer is about 10 to 15 cm thick. The coarse gravel layer is placed at the bottom layer and finest gravel layer is placed at the top. In the rapid sand filter, the distribution of wash water is most important function, hence the placing of gravel layer is carefully done.

6.5 MANIFOLD LATERAL TYPE

In this type C.I. manifold pipe are placed at the bottom of the filter longitudinally. Laterals are joined perpendicularly in the horizontal plane; are also C.I. pipe of 8 to 20 cm dia; and placed about 15 to 30 cm apart. But in the perforated pipe type system, the lateral drains having holes at the bottom side.

6.5.1 The Wheeler Bottom Type

In this system the floor is casted with concrete and also casted a pyramedial depressions upon with. For collection of filter water, the pressure chambers lies below the floor. On the pyramids source gravels and earthen balls are placed to prevent the downward water flow. This method is cheaper.

6.6 POROUS PLATE TYPE

In this system number of porous plates are provided at the bottom of the tank on top of which the sands are rested. There are some advantages of this

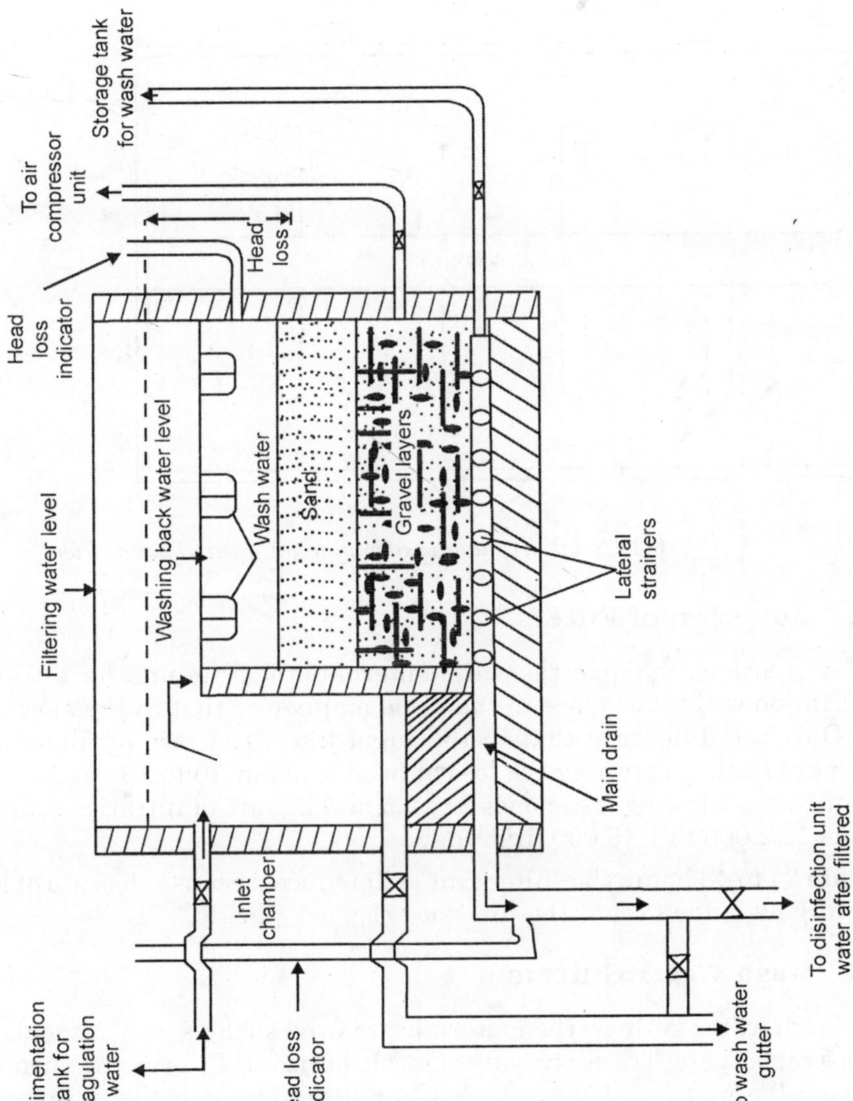

Fig. 6.2 *Section of a rapid sand filter*

system. No need to have any gravel layer. Filtering rate and the flow of wash water is much more uniform. These plates are good resistant to corrosion and easy to clean.

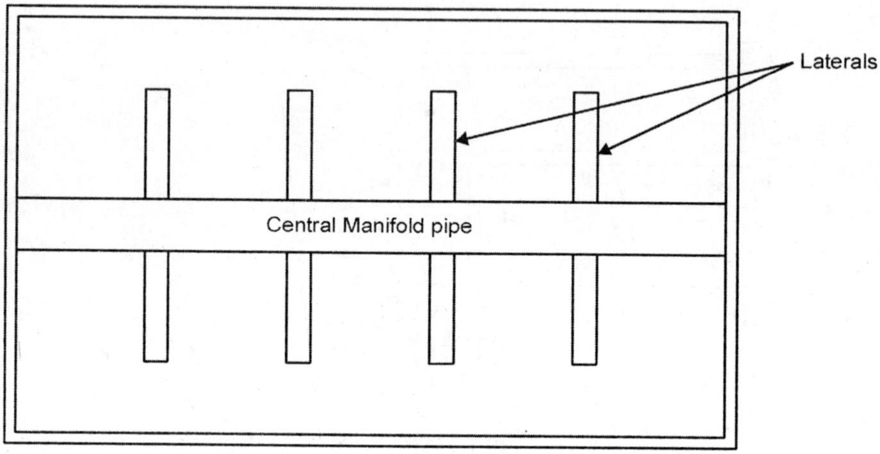

Fig. 6.3 *Under drainage system plan for a rapid sand filter*

6.7.1 Function of Filter

From coagulations chamber the water enter into the filter unit for filtration. The filtration will takes place in the same manner as that in the slow sand filter. Only the difference that in the rapid filter, the rate of filtration is much high. In the beginning the loss of head is about 10 to 18 cm, but with the continuous working it becomes 2 to 2.5 m. The rate of filtration will bear much as 1,12,000 to 1,45,000 lit/m^2/day.

Within 24 to 48 hours the "filter-run" will reduce, because the rate of loading is very high due to that the filter bed clogged quickly.

6.7.2 Wash Water Gutters

In the wash water gutters, the materials are used such as, concrete, plastic, asbestos cement etc. The dirty water travels horizontally over the surface of filter kept 0.5 to 1.0 m before it goes to gutter. Through the main gutter where the dirty water runs as high as 3 m placing the side wall of the filter, if the gutter is placed will be uneconomical, wash water gutter is placed very near to the surface of the sand, so that the good amount of dirty water is not left in the filter after washing. But the top of the wash water gutter should be placed in such a way. The sand surface is not washed in to the gutter. Air and water should not be used simultaneously with such gutter heights.

6.7.3 High Rate Back Wash

For High Rate Back Wash, the sand should expand of its undisturbed volume, for that the water head is provided as in undrains. No other agitation in the filter bed is required when the wash water head pressure is provided. The water head is applied is normally 600 1pm/m^2 only for a period of 10 min. Sometimes this is known as high velocity wash. In this wash system if the flow is increased. Further may carry the sand grain along with the wash water, which is not desirable. For high rate wash, the pressure at the under drainage system should have the pressure head about 6 to 7.5 m, and for washing water requires about 35 to 45 m/hr but the duration for this should be kept minimum 5 to 7 min. The wash water may be supplied by a overhead tank or through a pump.

6.7.4 Top Layer Wash (Surface)

It is expected that the top layer becomes the most dirtiest one and insufficient wash of that layer may lead to the trouble such as, mud balls, cracks, and clogged are formed in the top layer. These shorts of trouble can be avoided by having adequate surface wash, by using mechanical stirrer, may be used jet of water by hydraulic pressure and may use Air Compressor.

(a) Hydraulic Pressure. Two types of washing may have in this system. One is vertically, and the other is horizontally. In the vertical system, pipes are about 25 mm dia, arranged in a way that they should have the minimum intervals about 0.5 to 0.8 m. A nozzle should be provided at bottom end of the pipe. And this nozzle should stay about 10 to 9 cm on top of the surface sand bed.

In the horizontal system pipes are arranged horizontally is such a way that they should have an interval is about 50 cm approximately at a height of 5 cm to 8 cm over the bed surface. The pipes are perforated in such a manner that they have an intervals is about 25 cm at approximately and should provide a non-clogging units, to prevent the entry of filter media.

(b) Compressed Air System. At the time of back washing of filter, the sand grains are generally agitated by water jet or may by compressed air. Sometimes by a mechanical rakes. If the compressed air is going to be used, in that case, air compressor unit with adequate capacity should be installed. Air compression generally able to supply 4 to 5 min. at a rate of 600 to 800 lit/min/m^2 of filter bed. The supply of compressed it may be either through the laterals or by through a separate system.

6.8 OPERATIONAL PROBLEMS IN RAPID SAND FILTER

The main problems of operating a rapid sand filter, are formation of mud balls and the other is cracking of filters.

6.8.1 Mud Balls

Atmospheric mud gathered on the surface of sand. Due to inadequate washing of the filter, these mud sinks in to the bed, then they mixed up with the sand and other impurities to form mud balls. These mud balls will connect to the other mud ball and getting large in size. As they are gaining the sizes, they gets down to the bed and reach the gravel. They will then obstract the upward movement of wash water at the time of cleaning. Around the balls there will be high velocity which will displace the gravel and thereby forming mounds, while on top of the balls they are properly washed and muds are still remain. Generally when mud balls formation starts, they are increasing in numbers, and shortly they cover the total filer bed. To avoid this problem, some methods may be adopted.

By mechanical rakes, the mud balls may be broken and the broken particals will be washed with water. Mud balls may be broken by using water stream.

During back washing for 4 to 5 min. compressed air is used with mannual surface raking will be able to remove mud balls.

By applying 10 cm deep layer of caustic soda on top of the bed and remain for 12 hours for soaking, and the bed is agitated with air wash. After few hours later bed is washed with water until the clean water comes.

6.9 CRACKING OF FILTERS

The fine sand on the top layer of the bed will shrinks, and will cause the shrinkage cracks in the filter bed. Near the wall junction these cracks more prominent. Due to this the pressure on the sand bed will increase, and that will cause the cracks to open more. The mud, floc and the other impurities will penetrate deep down into the filter through these cracks. This will effect washing as well as the efficiency of the filter.

To get rid of this problem, the methods are adopted as used to remove the mud balls.

6.9.1 Rate of Filtration

In a rapid gravity filter, the rate of filtration is very high, generally 50 to 100 lit/min/m^2 of filter area. This high rate of filter save quite a good amount of area as well as filter material.

6.9.2 Performance of Rapid Sand Filter

It can remove 80 to 90% of bacterial load from water. To get bacteria free water, this should be treated into the disinfection unit. Since the water entering into the filter from coagulation sedimentation tanks, generally they are less turbid. The remaining turbidities can be easily removed by the filter and get permissible limits.

The rapid filters are the best and more economical. This should be used for large town and cities for public use. But the treated water from rapid filter is not that much safe as those obtained from slow sand filter. But water from rapid sand filter need to have further treatment before they can be supplied for public use.

Example 6.2 *Design a set of filters, required. Data is given below :*
Population = 60,000
Supply of water = 180 lit/day/capita.
Filters rated to work = 6000 lit/hr/m²
The necessary data can be assumed.

Solution

Ignoring the requirement of wash water,

Assuming that the rate of supply = An average demand.

Also assume that the Max. daily demand is 1.8 times the average daily demand.

Water demand (Max.) per day

$$= \text{Population daily supply rate (Max.)}$$
$$= 60{,}000 \times 1.8\ (180) = 19.44 \times 10^6 \text{ million lit}$$

Demand of water per hour (cleaning loss time neglected)

$$\frac{19.44\times10^6}{24}\text{lit/hr} = 810\times10^3$$

$$\text{Filtration rate} = 6000 \text{ lit/hr/m}^2$$

$$\therefore \quad \text{Filter bed area} = \frac{\text{Demand of water}}{\text{Filtration rate}}$$

$$\frac{810\times10^3}{5000}\text{m}^2 = 162 \text{ m}^2$$

As per question two units are to be designed.

$$\text{The area of each unit} = \frac{162}{2} = 81.0\,\text{m}^2$$

Assuming $L = 1.5\,B$, where L = Length, B = Breadth

$$\therefore \quad 1.5 \times B^2 = 81.0 \text{ m}^2$$
$$B^2 = 54 \text{ m}^2,$$
$$\therefore \quad B = 7.34 \simeq 8 \text{ m.}$$

Hence decide to take 8 m width and 12 m length. So, two unit of 12 m × 8 m is required.

Table 6.1 *Difference between slow sand filter and rapid sand filter.*

S. No.	*Cases*	*Slow sand filters*	*Rapid sand filter*
1.	Requirements for pre-treatment	Water from plain sedimentation tank of raw water may not have any treatment to enter into the slow sand filter. No need to have any coagulation.	Floculation, coagulation and sedimentation must be done before entering the rapid sand filter.
2.	Material used as base	The sand is supported by gavel at the base having 3 to 65 mm in size and 35 to 80 cm thick.	The sand is supported by gravel at the base having 3 to 40 mm size and the depth will be about 55 to 85 cm. The water will uniformly spread over the sand on top.
3.	The size of filter sand and its distribution	Generally the range of the filter sand lies between 0.2 and 0.4 mm and having the uniformity coefficient between 1.8 to 3.0.	Filter sand effective size ranges from 0.35 to 5.5 having uniformity coefficient as 1.2 to 1.8.
		Generally the uniform distribution of the grain size throughout the depth of the filter only the top 10 to 12 cm may have the fine quality.	Having fine sand laid on top followed by the coarsest size at the bottom.

Contd.

Contd.

S. No.	*Cases*	*Slow sand filters*	*Rapid sand filter*
4.	System of under-drainage	To receive filter water laid in order	To receive filter water laid in order, but for washing at a very high rate water is allowed to pass for back washing.
		Covered drain or open jointed pipes are used.	For lateral discharge perfoated pipes are used to discharge into the main or at the bottom diffuse plate may be used.
5.	Filtration rate	Filtration rate is very slow, such as 100 to 200 lit/h of filter area.	Filtration rate is very high such as 3000 to 6000 lit/hr/m^2 of filter area.
6.	Unit size	Large size require sach as. 30 m × 60 m. May vary 200 m^2 or more.	Small size required i.e. 5 m × 8 m. This may vary 20 to 100 m^2.
7.	Costing	Considering land and materials, the cost is high. But the operational and maintenance costs are low.	Initial cost is low. But the operational and maintenance costs are considerably high. However this is economical.
8.	Cost of depreciation	Low High	
9.	Efficiency	Removing bacteria it is very efficient. But removing colour it is not much efficient.	Removing bacteria it is not very efficient but on the other hand removing colour is it very efficient.
10.	Adjustibility	Not adjustable to meet the variations of demand.	Good adjustibility can meet the reasonable variations of demand.
11.	Post treatment required	More or less pure water is obtained, but to make sure it may be disinfected.	Disinfection must be done.

Contd.

Contd.

S. No.	*Cases*	*Slow sand filters*	*Rapid sand filter*
12.	Construction	Simple	complicated because the underdrain must be properly designed.
13.	Supervision	No skilled supervision is needed.	Skilled supervision is essential.
14.	Cleaning	Cleaning is a laborious process.	Easy and short method.
15.	Wash water needed	Small amount of wash water needed.	Large amount of wash water needed.
16.	Time of cleaning	Cleaning may required at an intervals of 1 to 3 months time.	Frequent cleaning at about an interval of 1 to 3 days time is required.

6.10 PRESSURE FILTERS

It is like a small rapid sand filter. Situated in a closed vessel, the water to be treated passed through this vessel under pressure, which is higher than atmospheric pressure. So it should be noticed that the filters located in air tight vessels. The raw water is generally pumped into the vessel by means of a pump. The pressure normally developed may 25 to 65 metre head of water.

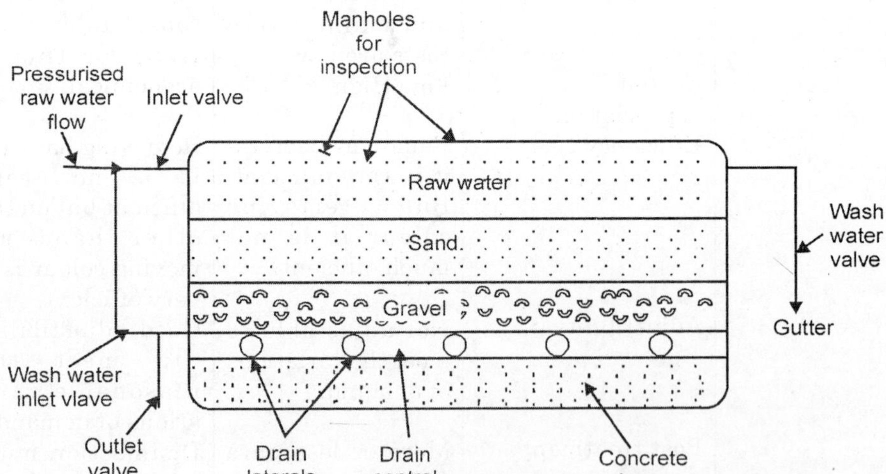

Fig. 6.3 *Pressure filter*

The pressure filter can have rate of filtration 2 to 5 times higher than a rapid sand filter, lower rate of filtration may be used for domestic, industrial supplies. And the high rate is used for swimming pool recirculating supplies.

Considering the water quality this is not efficient than the rapid sand filter; in removing bacteria and turbidities. The quality of filter water is very poor considering for consumption of public supply.

Example 6.3 *A filter unit is 4.5 m by 9.0 m. After filtering 10,000 cubic meter per day in 24 hours period, the filter is back washed at a rate of 10 lit/ sq.m/sec. for 15 min. Compute the average filtration rate, quantity, and percentage of treated water used in washing, and the rate of wash water flow in each trough. Assume 4 troughs.*

(Engg. Services 1980 : Civil Services 1985)

Solution

$$\text{Filter Area} = 4.5 \times 9.0 = 40.5 \text{ m}^2$$

In 24 hrs quantity of filtered water = 10,000 m^3/day

$$\text{Filter Area} = \frac{\text{Filtered water}}{\text{Filtration rate}}$$

Filtration rate in lit/hr/m^2

$$\frac{10{,}000 \times 1000}{24 \times 40.5} = 10{,}288 \text{ l/hr/m}^2$$

Considering $\frac{1}{2}$ hr clearing lost, the average filteration rate

$$= \frac{10{,}288}{24.5} \times 24 = \mathbf{10.078 \ l/hr/m^2}$$

For cleaning water used @ 10 l/m^2/ sec. for 15 min

$$= 10 \times (4.5 \times 9.0) \times (15 \times 60) \text{ lit.}$$

$$= 364500 \text{ lit.}$$

$$= 364.5 \text{ m}^3$$

Expressing the quantity of wash water as percentage of total water filtered,

$$= \frac{364.5 \times 100}{10{,}000} = \mathbf{3.645\%}$$

Discharge of wash water through one trough

$$= \frac{\text{Discharge of total wash water through the filter}}{\text{Troughs number}}$$

$$\frac{10\,\text{l/m}^2/\text{sec}\times(4.5\times9)\text{m}^2}{4} = \textbf{101.25 l/sec.}$$

PROBLEMS

6.1. Explain the theory of filtration water. **(Ranchi)**

6.2. What are the three classes into which the filters classified ? Describe each of them. **(Ranchi)**

6.3. Compare and contrast by points a slow and filter with a rapid of gravity filter. **(U.P.S.C., AMIE, C.U.)**

6.4. Discuss the detail construction and working of a rapid gravity filter. How does it differ from slow sand filter ?

(AMIE, U.P.S.C., Roorkee. B.H.U.)

6.5. Design a slow sand filter,

(a) For a town of population 40,000 and maximum rate of supply 200 l.p.h.d. (45 g/hd) **(U.P.S.C.)**

(b) For a population to be served =50,000 and maximum rate of supply 200 l.p.h.d. (45 of phd) **(U.P.S.C.)**

6.6. Illustrate your answers with detailed sketches of the filters.

(a) Design a 13.6 m.l.d. (3 m.g.d.) water treatment plant with rapid sand filter. **(Ranchi)**

(b) Design a rapid sand filter unit for a population of 30,000. Give a detail sketch with procedure and amount of back-washing water required. **(Ranchi)**

6.7. Design the following rapid gravity filter for the three under-mentioned towns :

(i) Number of each filter beds.

(ii) Dimentioned in each filter beds.

(iii) Composition of filter media.

(iv) Approximate quantity of wash water for bed for wash.

(v) Spacing and position of wash water gutters.

(vi) Height of tip of the gutter above the unexpanded sand. Structural depth of the gutter may be taken as 90 cm.

(a) For a town with population 45.5 m.l.d. (16 m.g.d.) of water. **(C.W.)**

(b) For a town with population 600,000 and rate of supply 270 l.p.c.d (60 g.p.h.d) **(C.W.)**

(c) For a town with population 2,00,000 and rate of supply 230 l.p.c.d.) (50 g.a.c.d) **(U.P.S.C.)**

6.8. Write notes on :

(a) Air binding of filter **(U.P.S.C.)**

(b) Mud ball formation **(U.P.S.C.)**

(c) Filtering film **(U.P.S.C.)**

(d) Negative head **(Bha.)**

(e) Back washing **(Bha.)**

6.9. Mention the factors that control the working efficiency of rapid gravity filter. **(Roorkee)**

6.10. Answer with respect to rapid sand filters.

(a) What is the maximum permissible head loss?

(b) What is the filtration rate?

(c) Function of rate of flow controller.

(d) Effective size and uniformity co-efficient **(A.M.I.E.)**

6.11. Describe with are next sketch the working of a pressure filter, what are the relative advantages and disadvantages of this type over those of the gravity filter. **(Engg. Services)**

6.12. Design a rapid sand filter unit for a population of 100,000 to be served by a 200 lit/capita/day water supply. Assume the following data :

Rate of filtration = 3×10^5 m^3/ha/day

Amount of wash water = 5% of the filtered water/day

Filter dimensions of each unit= 17.5 m 10 m

Filter needs back washing once in 24 hours. **(Engg. Services)**

Chapter

7

Disinfection of Water

CHAPTER OBJECTIVE

Main and common disinfectants are described here. It is the disinfection of water is very important for public health before use, so for large scale of water supply like water supply system. Chlorine is most common and useful. But for villages, and urban area, where the public water system is not available, in that case, the commonly used lime bleaching powder, and potassium permanganate. All these items are discussed here.

After filtration, the water is not ready to supply for public utility. Because the water after filtration will have some bacteria and other impurities. These bacteria may cause injurious effect to the human bodies, or may lead to epidemic. To make sure the using for public drinking or use water is free from bacteria and quite safe. So after filtration they need some treatment to kill the bacteria, this is called disinfection of water or sterilisation of water. After filtration water goes to the disinfection plant before public supply.

Different methods are adopted for disinfection by using some chemicals. But it has been found that the chlorine is the most useful and the best chemical for disinfection. Chlorine is used broadly to sterilized the water. The method of use chlorine for disinfection is called the chlorination of water. It is also true that the slow sand filter cannot give guarantee of cent per cent removal of bacteria. But the excess number of harmless bacteria is also not expected to have drinking water. The presence of pathogenic bacteria in water is objectionable and harmful for public health. So disinfection must be done to kill or remove all these objectionable elements before final supply to the public.

A disinfectant must have the following qualities :

(i) They should be cheap and reliable.

(ii) Expected to kill bacterias in water.

(iii) Should not make water toxic or harmful to the consumers.

(iv) By application of dose, they should not have objectionable tests and odour.

(v) Within the time for disinfection they should be able to kill all the bacterias.

(vi) Their lasting period in the disinfection chamber should be fairly good, so that they do not allow any further growth of bacteria.

(vii) They should have such property, that any part of the distribution system, their strength can be easily determined.

(viii) They should be easy to handle, easy to store and easy to carry.

Chlorine is universally accepted as a water disinfectant, because it can satisfy the above qualities.

But there are different methods may be applied for disinfection of water. They are as follows :

(1) Water boiling

(2) Ozone of disinfectant

(3) Ultra-violet ray may be used

(4) Excess lime

(5) Potassium permanganate

(6) Use of chlorine.

The above methods are discussed below briefly.

1. Water Boiling. By boiling water the bacteria present may be killed effectively. This is very simple method but effective one. This method can be used for domestic purposes. Only for a house particular house. But not for a huge supply or public water supply. However, this can kill the existing bacteria but can not prevent further contamination. So this method is not suitable for public water supply, but may be good for an individual house hold purposes.

2. Ozone as Disinfectant. Ozone is a gas having faint blue colour with a pungent smell. This is very good for disinfection. Ozone has three oxygen atoms. When passing through the water with the presence of electric current in a closed chamber a chemical reaction take place. A nascent oxygen is produced. This nascent oxygen is a powerful oxidising agent and good enough to remove, the organic matters as well as kill the bacteria from water. Generally 2 to 3 ppm of ozone is consumed for about 10 minutes. Ozone and chlorine are very good competitive disinfectants. But chlorine is used universally as a disinfectant except few countries like France and Russia and in India cities like Chandigarh.

3. Ultra-violet Ray may be used. This is an invisible light ray having wavelength of 1000 to 4000 mμ. This ray found in sun ray. They are very effective killing all types of bacteria. So they will give a perfect sterilized water. For this treatment the water should pass several time round the bulbs producing this ray. The depth of water should not be move than 10 cm or so. This method is very costly and need a technical person to look after. This method is not good for treating large quantities of water for public supply. But this method may be suitable for small quantities of water, like hospital, college, school, surgical use or for drinking purposes in a house. In these case the costing may be bearable. In a developed country this method is used for treating swimming pool water.

4. Excess Lime. Lime treatment for disinfection is an old method which is not used in nowadays for disinfecting the water for public in general. But in rural villages this method is still applied. Because this is cheaper and easy to apply.

When lime is added to water, it raises the pH value of water and makes it alkaline. At thus stage some bacteria may be present, so can not get rid of all the bacterias. Further addition 06 lime, the pH value of water rises more when reaches up to 99.9, i.e., extreme alkalinity will remove almost all the bacteria. But this method cannot protect the water for a long time from recontamination. Specially in village wells they may be suitable for a short period. To get better drinking water, the process of adding excess lime should be applied time to time.

5. Use of Potassium Permanganate ($KMnO_4$). This method of application of Potassium Permaganate is still popular in villagers for disinfecting their well water. It need small amount and cheep at the same time easy to apply. In villages there will have generally lesser amount of bacteria, which will be killed and at the same time it produces the better test of water by killing the organic matter.

In well water treatment, small amount of Potassium Permaganate mixed. With well water in a bucket full of water, then this bucket water will thoroughly mixed with well water and leave if for some times say at least for 8 to 10 hours before use. Generally in villages they do this before going to bed i.e., at about say 8.0 P.M., and use this water in the morning at about 6.0 A.M. Though Potassium Permanganate is cheap, handy and quite useful, but it cannot remove all the bacterias, about 98% may be removed. But the cholera bacteria will be removed 100% by this method. This method may be recommended for rural areas.

6. Use of Chlorine. Chlorine is almost universally accepted as a disinfecting agent for public water supplies. This is very cheap, easy to handle, easy to measure, it is also capable to provide disinfecting effects for a long time. If also can protect further recontamination of water in the distribution system. If the dose of chlorine is bit high, in that case the drinking water may have bitter and bad taste.

7.1 CHLORINE AS DISINFECTION

As the chlorine is added to water, it produces hypo-chlorous acid (HOCl) and hydrochloric acid (HCl). Then in the next stage HOCl i.e. hypochorous acid is then ionize to hypochlorite ions (OCl^-). Reaction is

$$Cl_2 + H_2O = HOCl + HCl$$

$$HOCl = H^+OCl^-$$

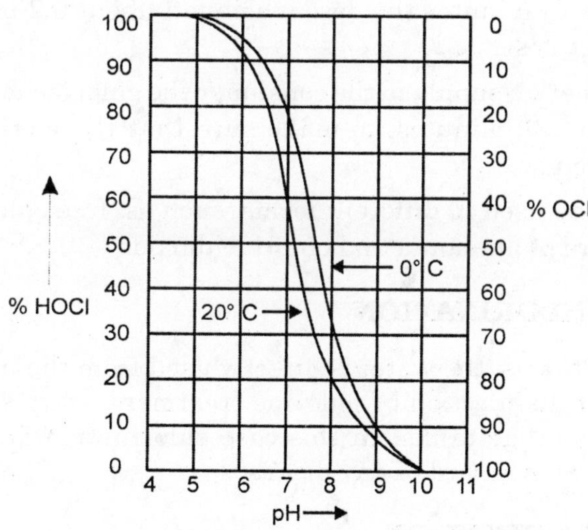

Fig. 7.1 *pH effect on HOCL and OCl⁻ in water*

Considering the pH value of water, below pH of 5 chlorine will remain a elemental Chlorine. But on the other hand above pH of 5 the water will have a mixture of hypochlorous acid and hypochlorite ions. But above pH value of

10 the water will have the hypochlorite ions only. In Fig. 7.1 the HOCl and OCl^- variation with pH value is shown. It is known that HOCl is the most powerful agent for disinfection.

7.2 CHLORINE DOSES

Depending upon the quality of water. The dose of chlorine should be adjusted. Because the chlorine will react first with inorganic impurities and then the residual chlorine will react with organic impurities. Reaction of chlorine with organic matter.

After the reaction there will have free chlorine. This free chlorine will cause germicidal action on bacteria and pathogens. This free chlorine will immediately kill the pathogens first and then bacteria. But the form of combined chlorine will have a long term germicidal effect on water.

Generally the most of the water is disinfected satisfactorily if the free chlorine is residual is about 0.2 mg/l, and stay about 10 minutes after the chlorine is added. If the dose is a bit high it will cause bad test, but on the other hand if the dose is small, they may not be reliable.

Generally the maximum dose of chlorine in water is decided by having an experiment after 10 minutes the dose which left about 0.2 mg/l is then that will be the correct dose.

To make sure before supply to the consumer the chlorine mixed with water should be at least 30 minutes, to make sure that the water is completely disinfected before use.

Chlorine may be used in different forms, such as, free chlorine, bleaching powder, a mixture of ammonia and chlorine dioxide.

7.3 PLAIN CHLORINATION

In some cities or towns, the water is directly used from the intake by having the treatment for disinfection by chlorine treatment. In this case they need to have a large storage tanks. In this case chlorination is very important factor against the safeguard of any diseases.

7.4 PRE-CHLORINATION

This is nothing but the addition of chlorine before coagulation. This is generally used in most polluted water. Because by adding chlorine before coagulation. This will reduce the tests and odours and also rescuing the algae present and also reduces the other organisms. So the effect of that it will keep the filter sand and increase the running time of filter.

7.5 ADDITION OF CHLORINE AFTER FILTRATION

Addition of chlorine does not depend on prechlorination; it is always applied to the addition of chlorine after filtration. Depending on the water charter dosage will vary. Generally it varies from 0.25 to 0.5 mg/lit. A residual of chlorine will obtain 0.2 mg/lit. to get the disinfecting effect throughout the distribution system, greater residual chlorine may be needed.

7.6 BREAK POINT CHLORINATION

This means that how much chlorine should be added to water. Actually, it means that the dose of chlorine added to water. For chlorination, beyond which any further addition of chlorine will appear as free residual chlorine.

When chlorine will add to water, it reacts with ammonia and other organic matter. They will form a combination of chlorine. Actually it is found that the addition of further chlorine doses, the residual chlorine will increases. Say upto a point *A* and suddenly drops to point *B*. But after the point *B*, it increases steadly to a straight line. The point *B* is called the "Break Point". Further addition of chlorine beyond *B* is known as "Break Point Chlorination."

This process is goes on in the water. But when chlorine is not a certain stage, in the water it will not start reacting. But after some stage. When the chlorine dosage and ammonia and the organic impurities reached a certain

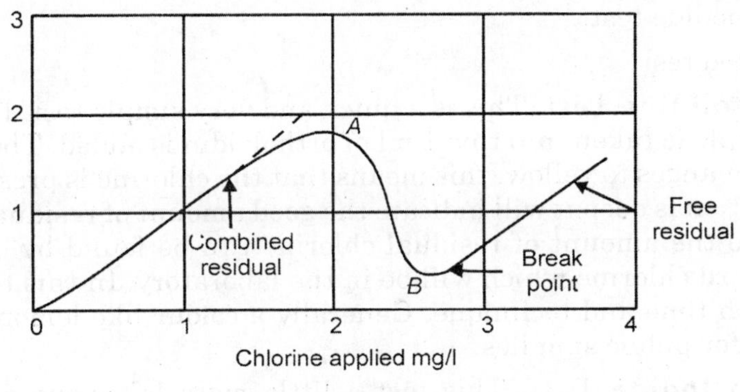

Fig. 7.2 *Break-point chlorination*

proportion, then the chlorine will start to react and combine with them. Still further increasing the chlorine dosage the chlorine drops from *A* to *B* inspite of increasing chlorine.

OA is known as combined residual. *A* to *B* is the oxidation of chloramines, and from *B* onward is the residual chlorine.

7.7 SUPER CHLORINATION

If the high dose of chlorine is added to the water, to get the free residual chlorine 2 mg/lit beyond the breaking point only at the time when epidemic is spread or any water borne diseases breaks out in a locality, this process may be applied. But not applied in normal time. Dechlorination is applied to avoid the strong odour and a different test.

7.8 DECHLORINATION

The dechlorination, is nothing but to get rid of tests and odour due to chlorine from water.

This dechlorination can be done by using some chemicals, these chemicals commonly known as dechlorination agents and as follows :

(1) Activated carbon, sodium thiosulphate, sodium sulphite, sodium bisulphate, ammonia etc.

Dechlorination may also be done by aeration method.

7.9 TEST FOR RESIDUAL CHLORINE

Sometime it is required to examine that after fulfilling the required contact period of chlorine with water, how much residual chlorine is left in the water. This can be done by some tests.

(1) Orthotolidine test

(2) Starch-iodide test

(3) Chloroted test

(1) Orthotolidine Test. This is a quick and very simple test. In this test a 100 ml sample is taken into this 1 ml of ortholoidin is added. The colour of the solution changes to yellow, this means that the chlorine is present. If the colour of solution is deep it will indicate the good amount of residual chlorine is present. So the amount of residual chlorine can be found by comparing with a standard chlorine which will be in the laboratory. In this test it does not take much time and technique. Generally a colour like lemon yellow is good enough for public supplies.

(2) Starch Iodide Test. This test a little more laborious and costly. Normally in this test water sample of one litre is taken and to this water 10 ml of potassium iodide is mixed thoroughly. After mixing properly 5 ml of starch solution is added to it. It will give a blue colour. This the colour will be removed by titration. The titration will be done blue colour water sample against sodium thiosulphate solution of strength N/100. Using the simple titration equation. The amount of chlorine can be found.

In the original water sample of water, quantity of 0.355 chlorine in mg/lit

To remove the blue colour number of ml of thiosulphate required.

(3) Chlorotex Test. This test is a simple one, in this test BDW chlorotex reagent is used, which will give a colour matching.

Test – 50 ml of water taken in a glass jar, 5 ml of chlorotext reagent in added to it. Colour will change immediately this means chlorine is present. The developed colour will be matched to found out the deposite of chlorine.

White milky colour shows, presence of chlorine is Nil.

Lightly milky colour shows, presence of chlorine	—	0.1 mg/lit
Pink milky colour shows, presence of chlorine	—	0.2 mg/lit
Red milky colour shows, presence of chlorine	—	0.5 mg/lit
Purple milky colour shows, presence of chlorine	—	0.6 mg/lit
Violet milky colour shows, presence of chlorine	—	0.8 mg/lit
Blue milky colour shows, presence of chlorine	—	1.0 or more

7.10 BLEACHING POWDER

Bleaching powder, may be used for disinfection in a small treatment plants, domestic wells and village ponds when a epidemic is break out in the surrounding areas. Water containing good amount of organic material, it is quite suitable for disinfection treatment. Practically bleaching power containing 30 to 33% of chlorine available, as the time goes on its power decreases. This is a poor disinfectant giving rise or pH value of water and having low chlorine available.

Example 7.1 *By using bleaching powder a water is going to be disinfected. The powder containing 2.4 mg/lit of chlorine.*

Datas are given below

1. 6×10^6 liters of water need daily.

2. 25% of chlorine available in bleaching powder.

Find the bleaching powder needed per day.

Solution

Demand of chlorine total = 2.4 × (0.25 as chlorine residual)

∴ Water required (per litre) 2.5 mg chlorine

6×10^6 water need = $6 \times 10^6 \times 2.5$ mg chlorine

= 6×2.5 mg of chlorine

Chlorine available is 25%

Chlorine available for 1 kg of chlorine in 4 kg of powder

∴ 6×2.5 kg chlorine found from $4 \times 6 \times 2.5$

= 60 kg of powder

∴ Powder needed per day = **60 kg/per day.**

Example 7.2 *Chlorine usage in the treatment of 20,000 cubic meter per day is 8 kg/day. The residual of the 10 min. contact is 0.20 mg/lit. Calculate the dosage in milligrams per litre and chlorine demand of water.*

(Engineering Services)

Solution

Treated water city = 20,000 m^3

= 20,000 × 10^3 lit

= 20 × 10^6 lits = 20 ml.

Required chlorine per day = 8 kg (given)

= 8 M. mg.

$$\text{Per litre of water chlorine used} = \frac{8\,\text{M. mg}}{20\,\text{M.C}} = 0.4\,\text{mg/l}$$

Then the dosage of chlorine =0.4 mg/l

The residual chlorine left (given in question)

= 0.2 mg/l

Hence the chlorine demand of water

= **0.4 – 0.2 = 0.2 mg/l**

N.B. Actual chlorine dosage means i.e., which will be reacted with water is the same as chlorine demand.

PROBLEMS

7.1. Explain the theory of disinfection by chlorine. **(A.M.I.E.)**

It is required to treat 3.5 m.l.d. of water with 0.3 mg/l of chlorine. If the disinfection is in the form of bleach that contain 30% of available chlorine. How many kg of bleach are needed to treat the daily flow of water.

(A.M.I.E.)

7.2. What is meant by :

(a) Plain chlorination **(W.A.S.C., AMIE : Pl)**

(b) Prechlorination **(W.P.S.C.)**

(c) Disinfection of water **(C.W, W.P.S.C.)**

(d) Chlorine process **(W.P.S.C.)**

(e) Super chlorination **(W.P.S.C.)**

(f) Residual chlorine **(Roorkee; C.W. Ranchi)**

(g) Break point chlorination **(Bha., AHIE, Pb)**

(h) Marginal chlorination **(Rachi S)**

7.3. What are the various methods adopted for sterilisation of water. Discuss the advantages and disadvantages. **(Ranchi, W.P.S.E.)**

7.4. Why chlorine is widely used in disinfectant in municipal water treatment? Discuss the method of determination of chlorine dose for a water supply. **(C.W. Roorkee)**

7.5. Discuss the effect of pH on efficiencies of chlorination. **(AMIE)**

7.6. Compare ultra-violet-ray and chlorine as sterilising agents for drinking water. What effects temperature and pH value of water have on the sterilization activity of chlorine. **(Roorkee)**

7.7. Mention the various methods used for sterilisation of water and describe the one most commonly used in India. **(U.P.S.C.)**

Chapter **8**

Joints of Pipes and Valves

CHAPTER OBJECTIVE

Pipe lines are jointed by using some joints.
They are a special type of joints with specific trade name. The most common and vital type of joints are discussed here.
Valves are quite useful for a water supply system. Special types of valves are used for special occasion. The most common and important valves are generally used in water supply system are discussed here.

8.1 SOCKET AND SPIGOT JOINT

This type of joint are widely used in our country. The term socket is some time known as bell. At the time of using, one end of the pipe is enlarged, and the other end is normal. The enlarged end is called socket and the normal

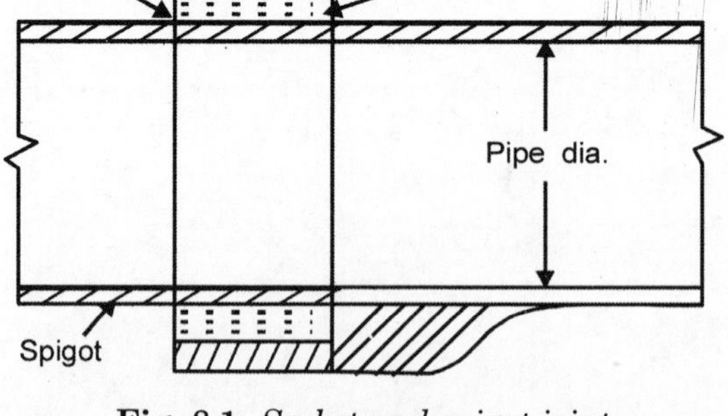

Fig. 8.1 *Socket and spigot joint*

end is called the spigot end. Some jutes are wrapped into the spigot end, and then inserted into the socket end. There will have some gap. That gap may be filled by sand and cement mortar, to make it water tight. Generally this type of pipe are quite flexible and quite comfortably laid on the flat ground.

8.2 FLANGED JOINT

Cast iron pipes or steel pipes are cabted with a flanges at both the ends. They are joined to gather by nuts and bolts or may be riveted where necessary. These joints are generally used in a pumping station or a filter plant.

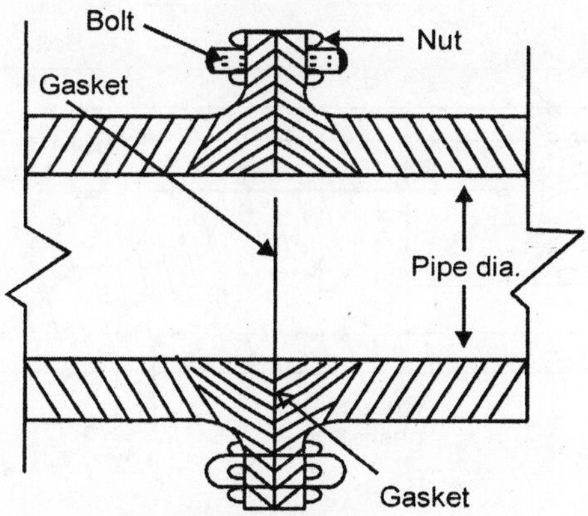

Fig. 8.2 *Flanged joint*

At the time of joining two flanges are bought to gather, and having a rubber washer (which is commonly known as gasket) just to make sure that the joint is tight enough and no water leakage is allowed. Then by nuts and bolts they are fixed.

They are not suitable for any detection or any variations. They are strong enough, but an expensive one.

8.3 DRESSER COUPLING

When both the ends of a cast iron pipes are to be joined in that case this type of dresser coupling is used.

Special type of metallic collar is fitted and tightened over the abutting ends. An iron ring and a gasket are slipped over each other of the ends of the

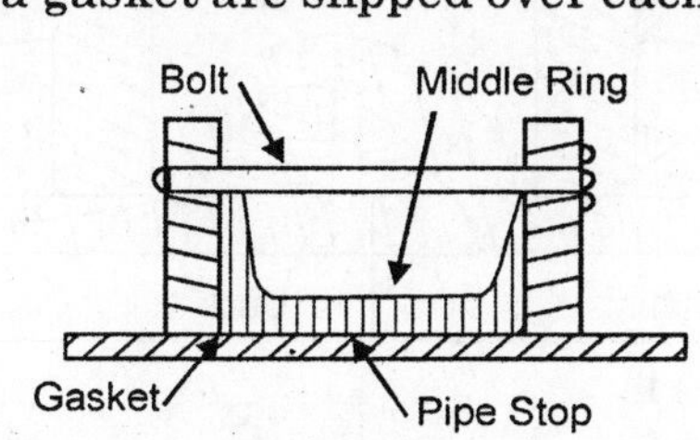

Fig. 8.3

pipe and between the gaskets an iron sleeve is inserted with the help of bolts. This joint is strong and quite rigid. They can stand vibrations. So they are generally used when a pipe line is passing over a bridge or under the bridge.

8.4 EXPANSION JOINT

Due to the temperature variations of stresses are produced. To overcome this stresses at some intervals of a pipe of line these expansion joints are used. Here

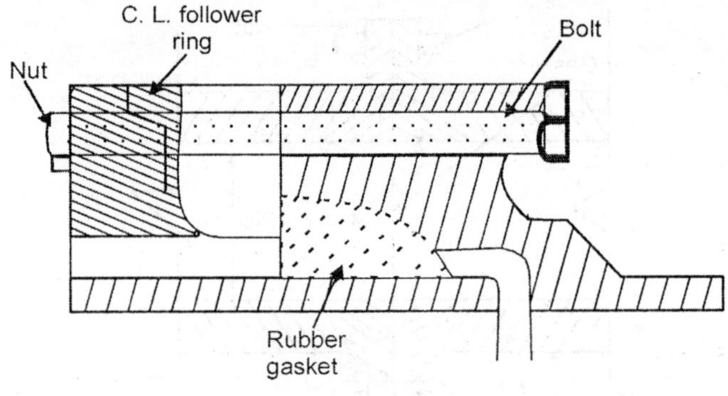

Fig. 8.4 *Expansion joint*

the socket end is tightly connected to can annular ring which can slide. Over the spigot end. At the time of making joint a small space is remain between the spigot end and socket end. A gasket is filled in the spigot end, then it is tighten by the bolts and nuts.

8.5 FLEXIBLE JOINT

To avoid unwanted stresses due to bending occur. This is due to settlement. In this case the socket is a spherical in one end the spigot will be at the bed. To

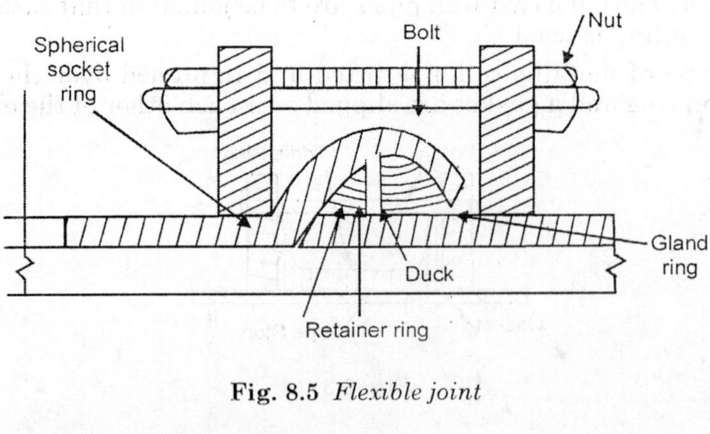

Fig. 8.5 *Flexible joint*

keep the spherical rubber gasket, a retainer ring is placed over the bed, and a gland is placed over it. Then this will be tighten by means of nuts and bolts.

8.6 VALVES

Sometimes a typical situation may occur in a pipe line to get rid off this situation, require to have some valves. At different types of situations, different type of valves are used.

8.7 SLUICE VALVES

These are sometimes known as gate valves. This is generally use to control the flow in a pipe line. When a question of repair is needed, this valve close the supply of water beyond the valve in the pipe line.

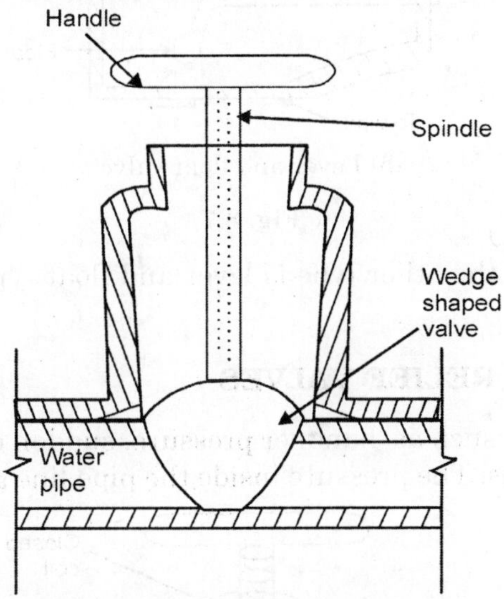

Fig. 8.6 *Sluice valve*

This is made of cast iron having a brass or stainless steel mounting at its end fitted with a pipe. A circular wedge shaped disc attached to the end of a threaded spindle passing through a gland. This is connected to a handle by means of a threaded spindle. This can be raised or lowered by manually to stop the flow or allow to have flow when raised. This also can be used to control or regulate the flow.

8.8 AIR RELIEF VALVE

To avoid the interference of flow in a long pipe, this type of air relief valve is used. Because air will store in the high point of the line, and will create trouble to have smooth flow. By the help of this type of valve air will go out to the atmosphere from the pipe. When air fill the float chamber, the opening will be cleared, air will escape until the water rises and fill the chamber. The float will be lifted and closed the opening. These valves are of two types one

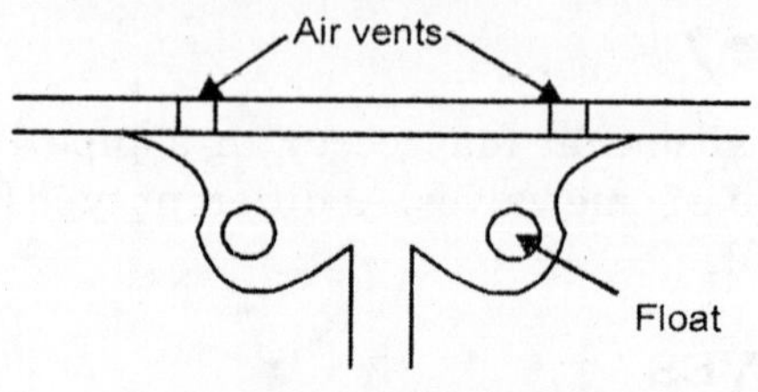

(a) Floating Ball valve

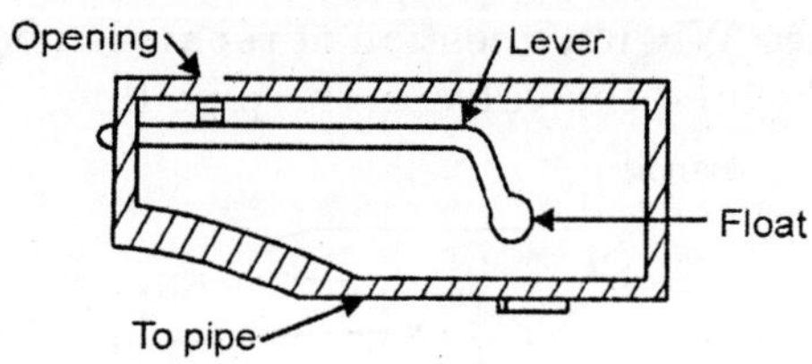

(b) Lever and float valve

Fig. 8.7

is floating balls and the other type is lever and float type. The function will be the same.

8.9 PRESSURE RELIEF VALVES

Due to some reasons such as 'hammer pressure, sudden closer of sluice valve, or other same reasons. The pressure inside the pipe line get high enough then

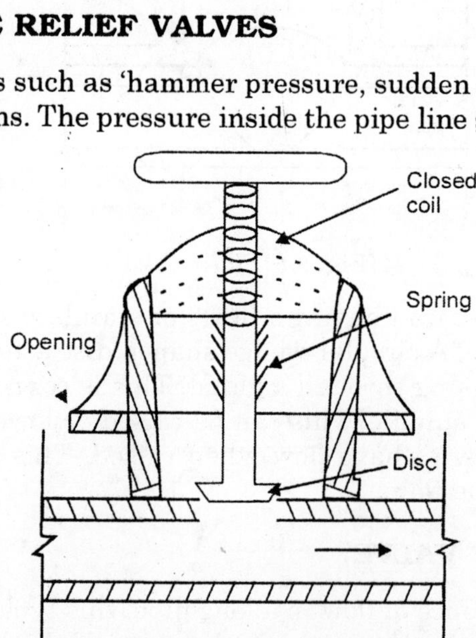

Fig. 8.8 *Lever and float valve*

the design maximum pressure of the pipe line. If the pressure is high than that of the limit of the maximum pressure, there would be a danger to get burst. To avoid this a pressure relief valve is used. This is connected in the pipe line. There will have an opening between them, which will be normally closed by a disc attached by a spring. At the time of pressure increases the disc is pressed back and making an opening to the outlet. As soon as the excess pressure is released, the spring push back the disc to come back to close the opening.

8.10 CHECK VALVES

This valve allows water to flow in one direction. This is sometimes called the non-return valves. Generally this valve is placed in the delivery side. This will prevent water to fall back from rising means. This is a disc covering the inside area of the pipe. The disc is open only in one direction. When water will try to flow in the opposite direction the disc closed the opening and the flow will be stopped.

8.11 GLOBE VALVES

It consists of a flat disc in parallel to the direction of flow, and its seat is also parallel. Head loss will be high, due to the change in direction of flow through this valve. Generally this valve is used in building distribution system. They are in small sizes and made of brass.

PROBLEMS

8.1. What are mechanical joints? Sketch and describe any one of them. **(Ran)**

8.2. Sketch and describe

(a) Pipe joint–two types **(Ran)**

(b) Lead joints in C.E. pipe **(A.M.I.E.)**

(c) Wipe joint **(Ran)**

(d) Spigot and socket joint **(W.P.S.E, Pat)**

8.3. At what point in a pipe system you will provide a single acting and double acting air valve and why? **(W.P.S.E.)**

8.4. Sketch and explain :

(a) Sluice valve **(W.P.S.E)**

(b) Air relief valve **(Ran, Pat; Roorkee, W.P.S.E.)**

(c) Check valve or non-return valve **(Ran, Bha.)**

(d) Globe valve **(Ran)**

(e) Hydrant **(Ran)**

Chapter

9

Softening of Water

CHAPTER OBJECTIVE

Possible ways and means are discussed in this chapter to make the water soft. For domestic purpose the hard water is not practically suitable for use. The most common and handy methods for softening of water is also discussed here.

For public uses hard water does not practically make any effect to health. But on the other hand it has a serious effect in industries. Water from ground water source may contain very high amount of sulphates, carbonates, calcium, chlorine etc. in a soluble state. These soluble materials means the water hard. Due to those there will have same trouble like :

(a) Extra soap needs to get rid of hardness. About 25 mg/lit will require per lit or water that will produce washing charge high.

(b) Due to the hardness the precipitate soap may damage the fabrics of clothes.

(c) Scales will formed in the boiler will cause the fuel expenses at the same time it may be in danger of overheating of boiler.

(d) Sometimes the pipe line will chocked and clogged making trouble of water flow.

Water may be formed as hard when the water having above 400 mg/lit of total hardness and when water having 100 mg/lit of total hardness is called soft water.

To avoid the above problems, the water should be make soft.

9.1 WATER SOFTENING PROCESS IN MUNICIPAL SUPPLY

Generally they use the following methods.

1. Lime-soda
2. Ion exchange

In the lime soda process the hardness will be removed by using lime and soda ash. There will have some chemical reactions as follows :

1. $Ca\,(HCO_3)_2 + (Ca(OH)_2 \rightarrow 2\,CaCO_3 + 2H_2O$
2. $MgCl_2 + Ca(OH)_2 + Na_2CO_3 \rightarrow Mg(OH)_2 + CaCO_3 + 2NaCl$
3. $MgSO_4 + Ca(OH)_2 + NaCO_3 \rightarrow Mg(OH)_2 + CaCO_3 + Na_2SO_4$

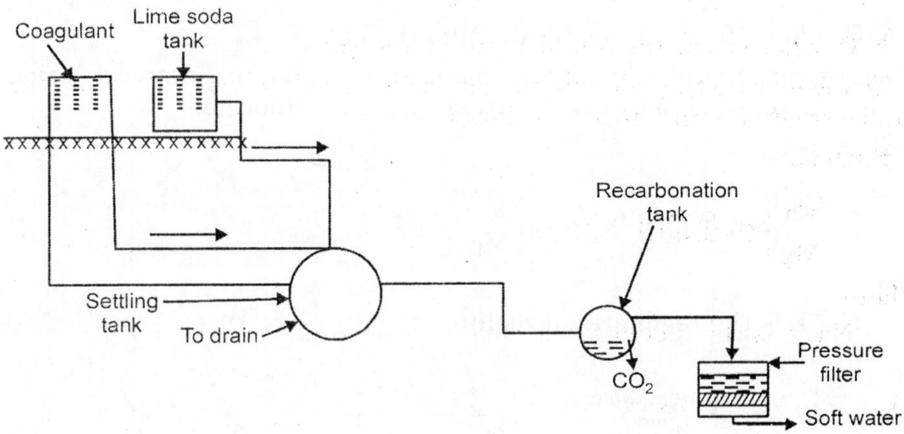

Fig. 9.1 *Softening plant (Lime Soda)*

For water coagulation feeding and mixing machine is used and a recommended detention provided in about 40 minutes.

In the settling tank coagulated and sedimented materials are collected, in this case generally more longer detention time is allowed :

Water super saturated with calcium carbonate leaving softening lime behind goes to the filter sand through the pipe lines. In between them re-carbonation may be necessary.

Sometimes alum may be used in the settling tank to help the sedimentation of calcium carbonate and carbon dioxide.

In industries the pressure filter is used but in softening process in municipality, sand filter will be quite useful.

9.2 ION EXCHANGE (BASE – EXCHANGE)

To get rid of hardness from water base exchange or ion exchange method is commonly used in many municipality or public water supply. Zeolite is the most commonly used having the formula as $Na_2OAl_2O_3 = SiO_2\, y.\, H_2O$. The usual value is about 2 or more and *y* having ranging amounts. Natural zeolite is known as green send because of its colour. But the artificial zeolite is white in colour known as permufil. It has got wonderful quality to exchange their cations. During the softening operation zeolite replaced the magnesium and calcium ions from water. The reactions are follows :

$$\text{Naz} + \text{Ca}\begin{bmatrix} \text{CH(O}_3)_2 \\ \text{SO}_4 \\ \text{Cl}_2 \end{bmatrix} \rightarrow \text{Na}_2\begin{bmatrix} \text{CH(O}_3)_2 & \text{Ca} \\ \text{SO}_4 & + \;\; \text{Mg} \\ \text{Cl}_2 & \end{bmatrix}\text{Z}$$

N.B. where Z means complex zeolite radical.

To regenerate the calcium and magnesium zeolite into active solution this can be treated with 5 to 10% solution of sodium chloride.

Such as :

$$\left.\begin{matrix} \text{Ca} \\ \text{Mg} \end{matrix}\right] \text{Z} + 2\text{NaCl} \rightarrow \text{Na}_2\text{Z} + \left.\begin{matrix} \text{Ca} \\ \text{Mg} \end{matrix}\right] \text{Cl}_2$$

where

Na_2 Z is the regenerated zeolite.

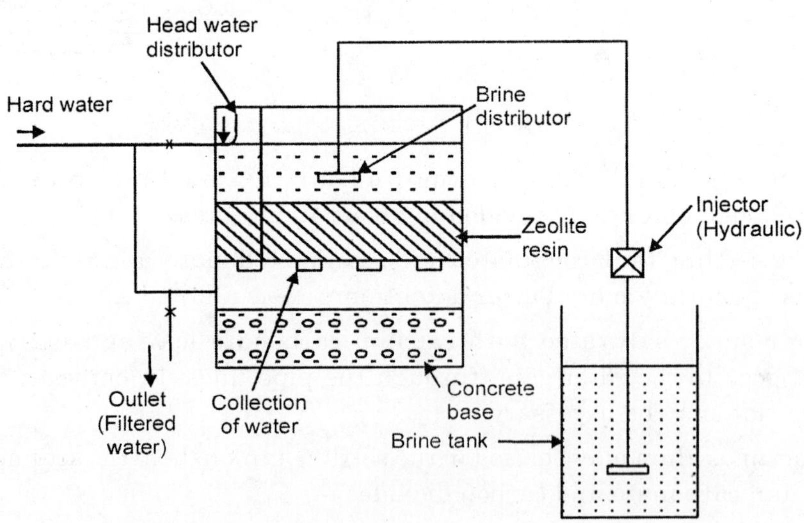

Fig. 9.2 *Sectional view of zeolite water softener*

Zeolite softener may have two types gravity filter or pressure filter. The rate of filter water through the pipe may be upto 3.50 lit per m^2 per minute. In this processes the water may have no hardness at all. Truely speaking this type of water is not desirable for public supply. To get the desired public supply, some of unsoftened water is mixed with this water.

For zeolite process

Advantages	*Disadvantages*
1. Zero hardness may be obtained. This is useful for some industry. 2. Operation is simple and the plant is compact. 3. This process is cheaper. 4. This can be used automatically. 5. No sludge formation. 6. No difficulty of treating water of varying quality. 7. From manganese. Ferrous iron is removed.	1. Not suitable for high turbid water. 2. Sodium carbonate is present in the water. 3. Costly for the treatment of water having iron and manganese. 4. Zero hardness may not required many times as this type of water will be corrosive to the distribution system. 5. Dissolved mineral contents may slightly increase.

Example 9.1 *What will be the cost of resin required for treating 10×10^6 lit/day of water, with regeneration after 24 hours. Given :*

(i) Water of 500 mg/lit of $CaCO_3$ hardness per m^3

(ii) Costing Rs. 8,000 per m^3

(iii) Using 25% extra bed.

(iv) Capacity of resin 70 kg/m^3

Solution

500 mg per one lit hardness :

$\therefore$ Hardness in 10×10^6 litre $= 500 \times 10 \times 10^6$ mg $= 5000$ kg

70 kg hardness removed by m^3 resin.

$\therefore$ 500 kg hardness removed

$$= \frac{5000}{70} = 71.1 \text{ m}^3 \text{ of resin.}$$

$$\simeq 72 \text{ m}^3 \text{ of resin}$$

25% extra bed providing $= 1.25 \times 72 = 90$ m^3

Resin cost = 8,000 per m^3

$\therefore$ Total cost = Rs. $8,000 \times 90 = 72,0000.00$

= 7,20,000.00

9.3 AERATION

In this method water is made in contact with air, so that it can absorb oxygen and can remove carbon dioxide. From the treated water it may kill some bacteria. This may be done in the way as stated below:

(a) Spray Nozzles. Through special nozzles, water is sprinkled in to the atmosphere in such a way that water breaks into droplets, in this way the dissolved gases will be removed at the same time carbon dioxide also.

But to obtain this efficiently need to have a pressure head of water is about 10 to 15 m.

(b) Freefall Aerator. Water is allowed to fall freely from a certain height is about 2 m over the series of steps with a fall of 0.2 m each approximately. The easiest type of structure is called aerator. This type of aerator is commonly used because it can serve large quantities of water, need small area and easy to clean and also having 10 m head and long life.

(c) Air Diffusion. To mixed through the compressed air is bubbled through the bottom of the tank. In the bottom of the tank there will have perforated pipes. Water will mixed thoroughly with water, due to the upward movement of air.

9.4 AERATION OVER TRAYS

Water is sprayed over onto the number of trays. Trays containing coarse materials, like coke, stone or ceramic balls ranging from 45 to 140 mm dia. These materials are spread over each trays.

Generally there will have 5 to 10 number of trays with a spacing of 0.2 to 1 m^2 per m^3/day/flow. In this case water will make to fall through the edges of the trays and collected at the bottom.

In the aeration process there will have some disadvantages and at the same time there will have advantages. They are as follows :

Advantages	*Disadvantages*
1. Removal of iron it is quite effective.	1. Removal of taste and odour is not so effective. 2. Due to more dissolved oxygen. Corrosiveness is increased. 3. High water head required.

9.5 ACTIVATED CARBON

Activated carbon is made from paper-mill waste, sawdust and other similar materials. This has a powerful property of absorbing gases, liquids, and fine impure particles. It also can attract impurities. This activated carbon is available in powder form and as well as granular form. The trade name is Dacro, Nuchar etc.

The granular from is generally used in gravity filter and pressure filter. The powder form is used before and after the coagulation, before filtration. Generally, some powder is added in the mixing tank and remaining portion is added just before the entry of the filter tank. This method is called split method, this is done in two stages. Sometimes this activated carbon may be applied in a slurry form. But the application of powder form may create some trouble of blowing dust due to wind. The usual dose of this is 6 to 20 mg/l.

The activated carbon have the advantages. It helps the coagulation when added in powder form. Organic matters are removed at the same time chlorine demand reduces. Removes the odours, tests, odours, but the over-dose is not harmful.

9.6 WATER-STABILISATION

Judging the pH value of water, the stabilisation of water is selected pH value of water is found by a simple test, called the Lavibond Comparator test. This test is simple and quick. Put value found 7 means the water is neutral neither acidic nor Alkaline. Over 7 the water will be alkaline and below 7 means acidic.

When water is acidic that will be corrosive in nature. So they may create corrosion in the pipes. In the Alkaline state that will protect against corrosion. From acidic state to get high pH value to make it alkaline state, add calcium carbonate to make a saturated.

9.7 DESALINATION

Water may contain two types of salt i.e., common salt and the other is sodium chloride in dissolved form. The taste of this water will be salty. This water is not suitable for drinking. The process of removing these dissolved salt from water is known as desalination, after that the water will be free from salt, then this water will be called fresh water. Some methods are adopted to get fresh water.

9.8 EVAPORATORS

The countries very near the sea, they do not have any source of fresh water, then they have to use sea water which is salty. To get the fresh water they have to use distillation of sea water. Which is a very expensive one. The running cost is quite high.

9.9 SOLAR STILLS

Through the shallow compartment covered with glass or plastic, salt water runs through them. Vapour collected on the glass cover which is salt free can be collected in buckets. This method is only effective where the sun light is available throughout the year.

9.10 METHOD OF FREEZING

Ice crystals are formed by lowering the temperature of salt water. By washing the crystals of ice is formed with fresh water successively to get rid of salt from salt water. Fresh water is obtained.

9.11 FLUORIDATION

It has been found that presence of fluorine in drinking water is essentialy for health, particularly for tooth, and bones. It also found that the maximum amount of fluorine i.e., 0.5 to 1 mg/l will give the best result. Excess of that amount may cause the adverse effect to the human health. If the presence of fluorine is less that amount, the add sodium fluoride (NaF) is very common; to achieve the maximum amount. If the excess amount of fluorine is formed, then that should be out from the water by other process. Excess fluoride in water will have the effect of teeth enamel.

9.12 WATER DEFLUORIDATION

To get rid of excess amount of fluorine in drinking water. There should have a plant for removal of this. But that will have very costly. So lots of various methods have been developed. Such as Dolomite lime, magnesium sulphate etc. Some time Bone chars may be used, but this will be an expensive method. Bone char has a good tendency to absorb fluorine. In any case if the fluorine concentration is found excess, then any one method should be adopted to get rid of the excess Fluoride from drinking water. So the water will have a concentration of fluoride in the safer zone.

9.13 PRESENCE OF ARSENIC

Presence of Arsenic in West Bengal is a great problem at the moment. The ground water contain more than 0.04 mg/l causes problems to the human bodies. Lots of research work is going on to get Arsenic free water for the public.

It has been found that water getting of by tube well below the depth of 18 m to 110 m depth are contain excess of Arsenic. But below this depth i.e., less than 18 m, the water will be quite safe. Due to the effect of Arsenic presence bad type of skin lesions can be visible. But the children will have the very bad effect of this. At present this a real great problem in West Bengal, specially in the villages.

9.14 REMOVABLE OF ARSENIC

The methods are as follows :

1. Activated Alumina (AA)
2. Reverse osmosis
3. Coagulation precipitation

9.15 BY ACTIVATED ALUMINA (AA)

In this process the water having Arsenic passed over the bed of insoluble Activated alumina, they may in granular form. The bud containing Actived Alumina absorb the Arsenic from percolating water. The Alumina gets saturated with Arsenic can be used again by back washing with water having approximately 1% solution of caustic soda. This regeneration method will help to get back almost of the bed capacity. The waste water should be handled carefully because it will have high concentration of Arsenic.

9.16 REVERSE OSMOSIS

This is a very costly method. Needs a regular replacement of chemicals. As in this method, Arsenic removed with other dissolved salts because the membranes are not selected to removal of iron of only a particular metal or mineral. The maintenance cost is high and needs pretreatment of water. On the other hand it require small space and the efficiency is high.

9.17 COAGULATION PRECIPITATION

In this method Arsenic can be removed effectively by coagulation then by sedimentation and filtration will separate, the solid particles then the clear water can be treated. The filter bed need to be clean frequently, as the Arsenic gets deposited on filter bed. This should be back wash with water having caustic soda. This is safe for household and community. The chemicals are easy available. Capital cost is low. But it will have a problem of disposal of sludge which is highly toxic in nature. It needs a trained operators.

9.18 DOMESTICALLY TREATING WATER

In cities and towns, water available from Cooperation and Municipalities are quite safe, free from mud, bacteria and other viruses.

Still sometimes the treated water stored in overhead tanks and sometimes the water get through the pumps. Due to the leakage in the pump line ground water gets into the supply line. This way the drinking water gets contaminated with some bacteria, and viruses. That is why to get safe water for househould purposes and offices nowadays they uses their own treatment arrangement like filter, ultra violet ray treatment, etc. The branded name in the market is Aquaguards, pure it etc. Sometimes boiling is done and chlorine tablets is used.

PROBLEMS

9.1. Explain the sedmentation process used in a water treatment plant. Draw a neat sketch of a sedimentation tank in which coagulant is used. Compute the dimensions of continuous flow rectangular settling tank for a population of 20,000 persons with a daily per capita water allowance of 120 litres. Assume detention period to be 6 hours. **(Engineering Services)**

9.2. Which are the discrete particles removed in an ideal sedimentation tank under the following conditions:

(a) horizontal flow ; and

(b) vertical flow **(Engineering Services)**

9.3. Distinguish between slow sand and rapid sand filters with reference to :

(i) Rate of filtration,

(ii) Filter media of sand,

(iii) Period of cleaning; and

(iv) Loss of head. **(Engineering Services)**

9.4. Discuss the relative merits of Rapid Gravity filters and pressure filter indicating the conditions favourable for the choice of each.

Describe the working any one of them with the aid of neat sketches. **(Engineering Services)**

9.5. Design rapid sand filtration units for a population of 100,000 to be served by a 200 lit/capita/day water supply. Assume the following :

Rate of filtration = 3×10^5 m^3/ha/day

Amount of wash water = 5% of filtered water/day

Filter dimensions of each unit = 17.5 m × 10 m.

The filter needs back washing once in 24 hours. Assume any other data not given. **(Engineering Services)**

9.6. Describe the procedure of lime-soda process for water softening. **(A.M.I.E.)**

9.7. What is meant by hardness of water ? How is it expressed ? Describe the method of soften by zeolite process. **(U.P.S.C. ; AMIE)**

9.8. Write briefly on carbonate and non-corbonates hardness in water and their removal. **(Roorkee)**

9.9. Write notes on :

(a) Zeolite water softening **(Ran.)**

(b) Hardness of water **(Ran. S)**

(c) Softening **(Roorkee)**

(d) Taste and odour control **(A.M.I.E.)**

(e) Ion exchange process **(A.M.I.E.)**

(f) Recarbonation following softening.

9.10. Why softening of water is necessary ?

9.11. (a) Diffferentiate between temporary and permanent hardness.

(b) Mention any three methods of softening water. Describe colite process of softening wter in detail. **(Engineering Services)**

9.12. Write short notes on any four of the following :

(i) Chlorination,

(ii) Water Aeration,

(iii) Fluoridation

(iv) Desalination

(v) Chemical Coagulation

(vi) Break point method of chlorine

(vii) Carbonate hardness

(viii) Non carbonate hardness.

Chapter 10

Water Distribution System

CHAPTER OBJECTIVE

This distribution system method is an important one. Because if the system is failed to discharge supply water, there will be a great problem to be faced by the consumers from this point of view this is very important.

In this chapter the whole system is discussed in a manner, so that any one can have a clear idea. What is the distribution system and also know how to do it perfectly. To make this chapter clear some Examples are worked out to support the theory. It also try to show that how the theory works in practice. The portion which is most important to know is discussed here only.

The distribution system is that system, the treated water should be made possible to reach the consumer houses. This is possible to do this by having a well planned notwork of pipe lines, having different sizes of pipes. There may need to have a pump to maintain the pressure head through the distribution pipes. The supply to the consumer may be for 24 hours or may have two to three times a day. The water may directly pumped to the system of pipe network or water can be stored in an elevated reservoir and then fed into the distribution pipes.

10.1 FOR A GOOD DISTRIBUTION SYSTEM

The following points should be maintained.

(1) Able to supply water with reasonably good pressure at each and every corner of the city.

(2) Reasonably good amount of water to fullfill the demand.

(3) Least capital cost for construction of the supply system should be well look after.

(4) Easy to repair and operate.

(5) Try to prevent the future pollution of water.

(6) Due to leakage loss of water should be prevented.

(7) Care should be taken to see that bursting of pipe should not interfere the system.

10.2 LAYING OF PIPE LINES ON BOTH SIDES OF A STREET

On both sides of a street, houses and flats are existed. To supply water, the pipe lines will have to lay as shown in the Fig. 10.1. The sizes of the distribution pipe will be selected depending upon the amount of flow will be needed for the smallest lateral pipes are about 8 to 10 cm dia. and for fire supply hydraint, it will be 15 to 20 cm in dia. The cross connection in this case will have to be at a suitable distance. The distribution system will have different type of network system, depending upon the local condition and the position of roads.

The systems are :

1. Tee system.
2. Reticulation system.
3. Circular system.
4. Radial system.

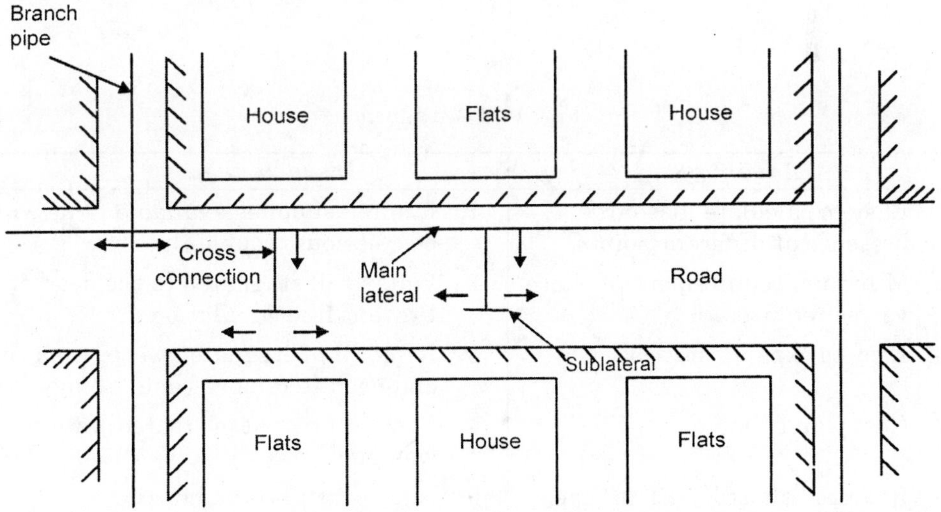

Fig. 10.1 *Pipe lines on both side of street*

10.2.1 Tee System

The Tee system is shown in the Fig. 10.2. In this system a main pipe connected with submain pipes, generally the connection of sub-main with main will be at right angles. Then from sub-main there will have branch line. From this branch lines the domestic connections are given to the individual consumers.

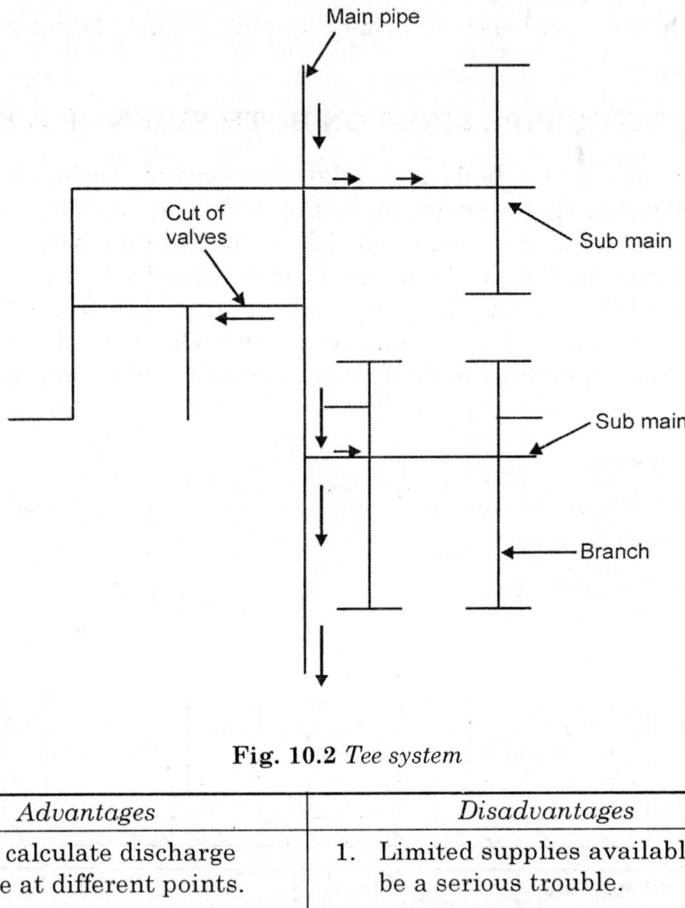

Fig. 10.2 *Tee system*

Advantages	*Disadvantages*
1. Easy to calculate discharge pressure at different points.	1. Limited supplies available. This will be a serious trouble.
2. Minimum requirement of sluice valves required.	2. Water will stagnated to the dead end. Prevent free circulation.
3. Easy to expand the system.	3. Removal of stagnated water from the end need to remove time to time.
4. Cheap and simple.	4. Greater amount of treated water wastage.
5. Required short length of pipe.	5. Need careful attendance.
6. Easy to lay pipes.	6. Repairing of main and sub-main lines will cause trouble to the consumers.

10.3 RETICULATION SYSTEM

In this system, the main sub-main and the branches are inter connected to each other in such a way that they will form a complete network system. This system is quite good and helpful in a well planned city or town.

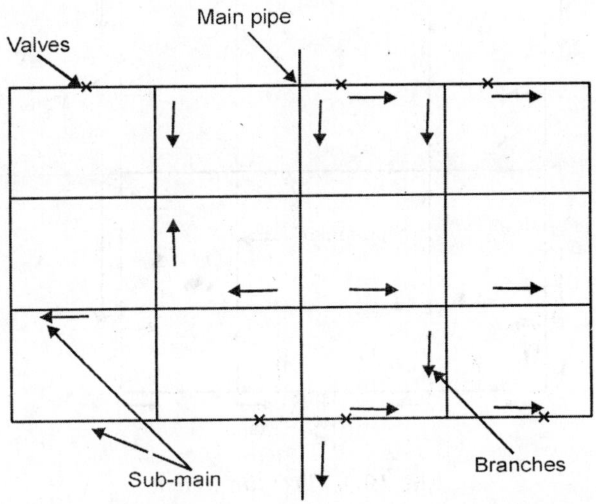

Fig. 10.3

Advantages	*Disadvantages*
1. At different places water reaches with different pipes.	1. Require more pipe lengths.
2. Loss of friction and pipe size reduces.	2. Large number of valves required.
3. Repairing effect the small area.	3. Construction cost high.
4. Due to continuous supply. Pollution is practically nil.	4. Design is complecated and difficult.
5. For fire fighting water can be available from nearer to that end by closing the valve.	5. Cost of design is high. Need an expert for that.

10.4 CIRCULAR SYSTEM

In this system the main pipes are placed at the periphery of the served area, and the distribution pipes are placed as shown in Fig. 10.4. They will form a rectangular block. This type of supply system is used in a town or cities. The water pressure at the various points will be good enough to serve the population. This is a costly system and need more length of pipe lines.

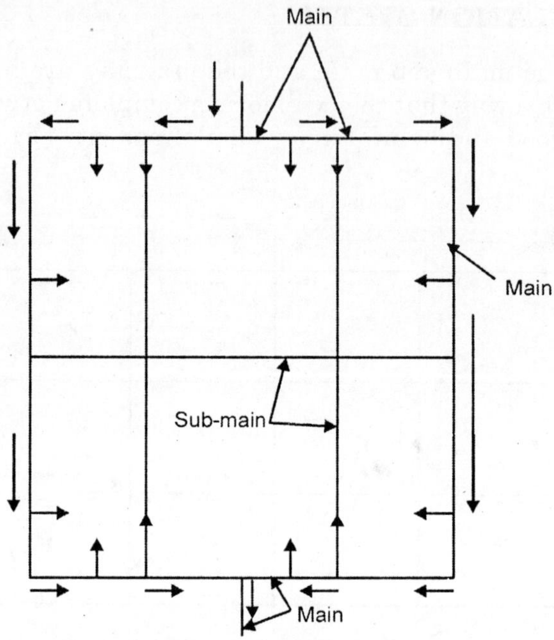

Fig. 10.4 *Circular system*

10.5 RADIAL SYSTEM

In this system water is taken out from the main line and pumped into the reservoirs and then the water will distributed radially through the pipe lines

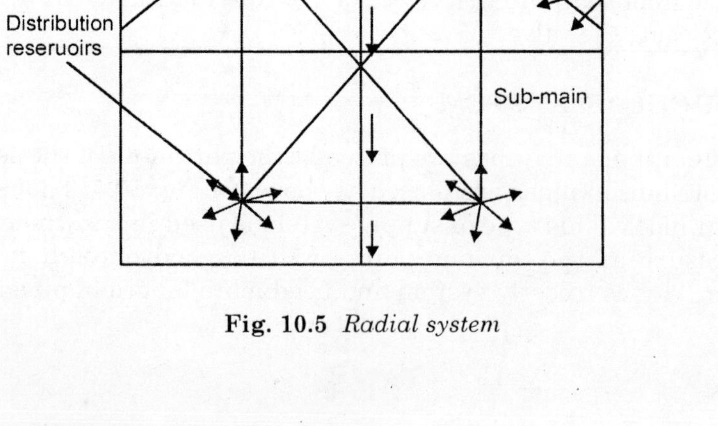

Fig. 10.5 *Radial system*

into the different parts. In this method high pressure of water is available and the distribution water is quite efficient. But lots of pipe involved in this system to form a good pipe network.

10.6 DISTRIBUTION METHODS

To get the adequate water supply to the consumer's end, the following methods are adopted; Depending upon the condition and topography of the area to be served.

1. Gravitational method
2. Pumping method
3. Combination of both

1. Gravitational Method. If the source of supply is in high level from the distributed part, so the sufficient pressure head can be maintained in the main pipe lines. This is the most economical and reliable method.

2. Pumping Method. In the pumping method, water is pumped out and directly supplied to the supply network system for the public use. There may be required high lift pump. This method is cheap by a regular attendance is required. But when there is not electric current, the whole system will be in trouble. In this case no storage tank is required.

To avoid the electric problem, this pumping method can be modified. The water can be stored in a storage tank by the pump at high altitude. Then supply of water will be done from the storage tank.

In both cases the pressure in the supply line is fairly uniform and the rate of pumping also be uniform. Hence this method is found economical enough.

3. Combination of Both the System. In this combined system, the water is pumped to the supply main directly and reservoir simultaneously. This method is that is why called a combined method. In this case generally the

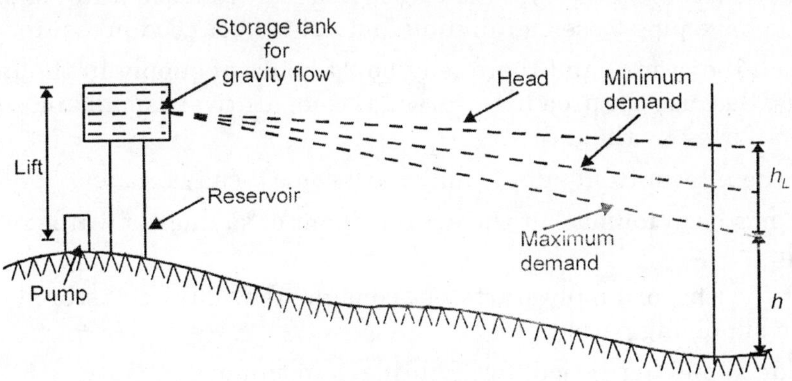

Fig. 10.6 *Radial system*

pump works at a constant rate. Sometimes the excess water is stored in the reservoir to meet up the demand for dry period; and for fire service, also stand by for any emergency like breakdown. It is quite a reliable method of water supply.

Advantages

1. At failure of pump and power, this method is quite effective.
2. Efficient and effective and also cheap.
3. Pumping rate is uniform. Supervision need less. May be 8 to 12 hours a day.
4. It is quite effective for fire fighting.

10.7 DISTRIBUTION SYSTEM

In this system water pressure play an importnat part. When the water enters in the distribution main and reach the last end of the consumer's place. The pressure drop will be high. It will may be inconvenient to the consumer. This is due to the pipe friction as well as some leakage in the pipe system. To avoid this there will have some different storage tanks with pump to supply water with reasonable good pressure to the consumer. The storage tank will be so designed that it will quite sufficient to cover the surrounding area.

This way of having different storage tank may cut down the cost quit conveniently. At the same time maintenance cost will be quite low. The greater the design pressure, the costlier will be the system.

10.8 SUPPLY SYSTEM

In our country most of the cities and towns will have the intermittent supply. Peak period of the day will be in the morning and in the evening. So they will supply water say in the morning from 6 A.M. to 10 A.M. and in the evening from 2.00 P.M. to 6.00 P.M. This will save water from the water source. This will also have pipe losses minimum, and able to get good pressure.

But on the other hand there may be 24 hours of supply in the industrial campus. Because the area to be served is comparatively small and so the cost can be afforted by the industry.

But the intermittent supply may have some disadvantages.

1. It has been found that the arrangement of savings of water is not good enough.
2. At the time of supply starts, the consumers will have to stored water for the non supplying period.
3. Non supplying period, fire fighting is in trouble.
4. Lots of valves and gate way required and a continuous checking is needed.

So the maintenance cost is getting high. This can be improved by making a different zones, and the different zones will supply water at different times of the day.

10.9 TYPES OF RESERVOIRS DISTRIBUTION

There are mainly two types.

1. Surface reservoirs.
2. Elevated reservoirs.

1. Surface Reservoirs. This type of reservoir is provided in a high point of the city. They may be a circular type or a rectangular one. Consturcted at the ground level or may at below ground level.

Water stored in the reservoir and then directly sent from there to the distribution system. In some cases, water need to have a pump to send into the distribution system from the reservoir. It depends upon the position of the reservoir is placed. Whatever it may be the case, the direct distribution from the reservoir to the distribution system is due to gravity, but when pumping is required, then the system is called combination of pump and gravity system.

2. Elevated Reservoir. Tanks of circular, rectangular or may be eliptical which one is suitable by the designer. These tanks are elevated at a high level and water pumped from the ground reservoir or from filter units. The elevated tanks are placed on top of a tower of pillars generally made of R.C.C. Then water from the tank is sent to the distribution system to the consumers end.

10.10 CAPACITY OF STORAGE

The capacity of storage of a town is generally found on the basis of (a) balancing storage, (b) Break down storage

(a) Balancing Storage. When the fluctuating demand with a constant rate of supply from the treatment plant is the main function to meet this. Water to be stored in the reservoir for balancing the variable demand against the constant supply is called the balancing storage.

To find out this capacity by availability a mass curve method can be adopted. In this method ploting of accumulated supply or demand against time. Mass curve of supply is drawn first, then the demand curve is superimposed into it. Then the balancing storage can be determined by the maximum ordinates between the two curves.

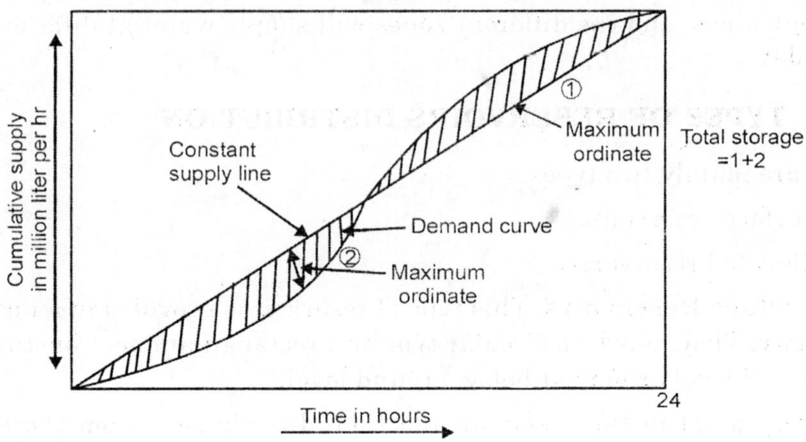

Fig. 10.7 *Mass diagram for 24 hours*

(b) Break Down Storage. Due to breakdown of pump or failrue of power supply the amount of storage during this period is called the break down storage. This is very difficult to ascertain. To avoid the uneconomical design of the tank generally the capacity for 2 to 4 hours of pumping is provided to avoid this type of problem.

10.11 FIRE RESERVE

Considering the Indian condition, the water reserved is taken for a period of 2 hours pumping capacity storage. But generally about 5 litres per capita is stored for fire fighting.

Example 10.1 *A town with a population of 1 million has a continuous water supply. Average supply in 270 lcpd the water being supplied by direct pumping. The total supply of 270 lpcd is phased as follows :*

Time	*lpc*
5 AM to 11 AM	*90*
11 AM to 3 PM	*54*
3 PM to 9 PM	*81*
9 PM to 12 mid-night	*27*
12 mid-night to 5 AM	*18*

Water is supplied from the treatment plant at a uniform rate of 11.25 million litres per hour, for all the 24 hours. Find out the capacity of the reservoir required for distributon of water. Assume no loss or drawal from the trunk main. **(AMIE)**

Solution

Total daily supply = Supply rate × Population

$= 270 \times 10^6$ lit = 270 M lt.

Converted cumulative demand shown in Table 10.1.

Table 10.1

Time	*Consumption per capita in lpc*	*M lit consumption for 1 million population*	*Demand in in M lit cumulative*
(1)	(2)	(3)	(4)
5 AM to 11 AM	90	90	90
11 AM to 3 PM	54	54	144
3 PM to 9 PM	81	81	225
9 PM to 12 (mid-night)	27	27	252
12 (mid-night) to 5 AM	18	18	270

The graph is not drawn to scale. Just to explain a sketch is shown. If the graph drawn in scale this will be obtained as shown in Fig. 10.8.

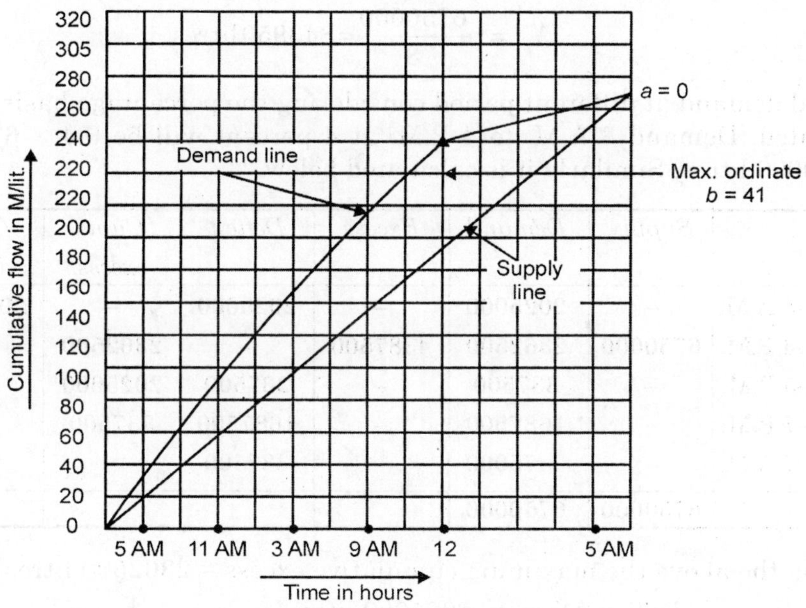

Fig. 10.8

The supply water obtained at a rate of 11.25 M lit. per hour. The line of supply in the mass dia. will be a straight line and the slope will be 11.25 M lit/hour *a* and *b*. The two maximum ordinates are in the demand line and supply line taken from the graph.

a = 0 million lit.

b = 41 million lit.

Total storage required = $a + b$ = **41 million litres.**

Example 10.2 *For the water supply of a small town with population of 50,000, it is proposed to construct a distributing reservoir. The pattern of draw off is as under.*

7 A.M. – 8 A.M............30 percent of day's supply

8 A.M. – 5 P.M............40 percent of day's supply

5 P.M. – 7 P.M............25 percent of day's supply

7 P.M. – 7 A.M............5 percent of day's supply

The pumping is to be done at a constant rate of 8 hours per day from 8 A.M. to 4 P.M. Assume water requirement of 135 lpcd. **(B.I.T.)**

Solution

Total consumption of water = 50,000 × 135

For 8 hours pumping rate (8 A.M. to 4 P.M.)

$$= \frac{6750000}{8} = 843950 \text{ litres}$$

Total demand at different period considering the percentage basis can be calculated. Demand 8 A.M. to 4 P.M. at 4 percent will be 0.4 × 6750000 = 2700000 litres. Similarly it is calculated below.

Time	*Supply*	*Demand*	*Excess*	*Deficit*	*Cum. excess*	*Cum. deficit*
7 A.M.–8 A.M.	–	2025000	–	2025000	–	2025000
8 A.M.–4 P.M.	6750000	2362500	4387500	–	2362500	–
4 P.M.–5 P.M.	–	337500	–	337500	2025000	–
5 P.M.–7 P.M.	–	1687500	–	1687500	337500	–
7 P.M.–7 A.M.	–	3375000	–	337500	–	–
Total	6750000	6750000				

From the above the maximum cumulative excess = 2362500 litres

Deficit Max. = 2025000 litres.

∴ Storage balance = 2362500 + 2025000 = 4387500 litres

Providing break down storage for 1 hour

Storage for break down = 843750 litres

Reserve for fire 2 lit/head = 2 × 50,000

= 100,000 litres

So the total capacity = 438,7500 + 843750 + 1,00,000

= **5,331,250 litres.**

It will be better to have uniform pressure distribution in the whole city. The number of reservoir having capacity of 4.85 lacs litre will be 11 numbers.

Example 10.3 *The peak days water consumption of a country of 50,000 persons is 200 l.p.c.d. The demand variation being as follows : 6 A.M. to 9 A.M. 35% of total ; 9 A.M. to 12 noon 10% of total, 12 Noon to 3 P.M. 15% of total, 3 P.M. to 6 P.M. 15% of total, 6 P.M. to 9 P.M. 25% of total. Find analytically the service storage capacity for the city, inflow is at uniform rate over 24 hours.* **(Roorkee)**

Solution

Hr.	*Supply*	*Demand*	*Excess*	*Deficit*	*Cum. Excess*	*Cum. Deficit*
12 mid night to 6 A.M.	2,400,000		24,00,000	–	2,400,000	
6 A.M. to 9 A.M.	1,200,000	3,500,000	–	2,300,000	1,00,000	–
9 A.M. to 12 noon	1,200,000	1,000,000	200,000	3,00,000	3,00,000	–
12 Noon to 3 P.M.	1,200,000	1,500,000	–	3,00,000	–	–
3 P.M. to 6 P.M.	1,200,000	1,500,000	–		–	–
6 P.M. to 9 P.M.	1,200,000	2,500,000	–	1,300,000	–	3,00,000
9 P.M. to 12 noon	1,200,000		1,200,000	–	4,00,000	1,600,00

= 200 × 50,000 = 10,000, 000 l/d

$$\text{Pumping rate} = \frac{10,00,000}{24} = 416,667 \text{ lit/hr}$$

Balancing storage = Max. cum. excess + max. cum. deficit

= 2,400,000 + 1,600,000

= 4,000,000 litres

Extra 15% provided = 0.15 × 400,000

For brake down and fire reserve = 6,000,000

∴ Total capacity = 4,000,000 + 600,000

= **46,00,000 litres**

10.12 DISTRIBUTION NET WORK DESIGN

10.12.1 Sections Choise Method

This is a rough method to select the size and shape and the pipe diameter in the network. To get the more accurately a cross checking should be carried out. The way to follow this the steps are as following.

If we look for the similarity, we will be able to see that in case of a hydraulic flow, the flow to a junction will be equal to the total flow out of that point. Same as electric circuit for a flow into a point will be equal to the flow out from that point.

Again it is found that the total head loss in a clossed circuit is equal to zero. So as in the case of electric circuit is equal to zero. So as in the case of electric circuit, the total voltage drop in a closed circuit is equal to zero.

But in the case of hydraulic flow, the head loss is proportional to the discharge to the power $\propto$ ($\propto = 1.85$). As again in the electric circuit, the voltage is directly proportional to the current.

Example 10.4 *Two 150 mm pipes of equal diameters form a loop in a distribution network. Calculate the diameter of an equivalent of same length to represent the loop (There are no intermediate draw offs).* **(CUPSC)**

Solution

As $$h_s = \frac{flv^2}{2gd} \propto \frac{Q^2}{d^5} \text{ (Standard formula)}$$

Say in pipe 1

$$Q = \text{distance}$$

$$\text{Head loss } h_f = K = \frac{Q_1^2}{(15)^5}$$

Like in pipe 2,

$$\text{Head loss } h_f = K\frac{Q_1^2}{(15)^5}$$

Total discharge in two pipes = $Q_1 + Q_1 = 2Q_1$

As per question, if the loop is replaced by an equivalent pipe dia. D, there will have the same head loss.

$$\therefore \quad h_f = K\frac{Q_1^2}{(15)^5} = K\frac{(2Q_1)^2}{D^5}$$

(1) The total network is divided into some sections.

(2) 1st section will be at right angle to the flow direction.

(3) On the basis of population and other factors, the supplied water beyond that section is calculated.

(4) Across the each section the capacity of distribution system is determined keeping in mind at this stage of analysis the velocity range should be within 0.75 to 1.25 m/s and the hydraulic gradient at 1 to 3 unit per 1000.

(5) The capacity of the existing pipe will be determined on the basis of available or expected hydraulic gradient.

(6) If the capacity is not found adequate, then add some more pipes to get that or change the size of the pipe chosen before.

(7) Flow velocity should be checked.

To explain this an simple example is given as follows.

Section 1–1

Pipe dia. 60 cm exixting say the hydraulic gradient taken 20 cm per 100 m. Then pipe capacity = 275 lit/sec and flow velocity = 0.9 m/s.

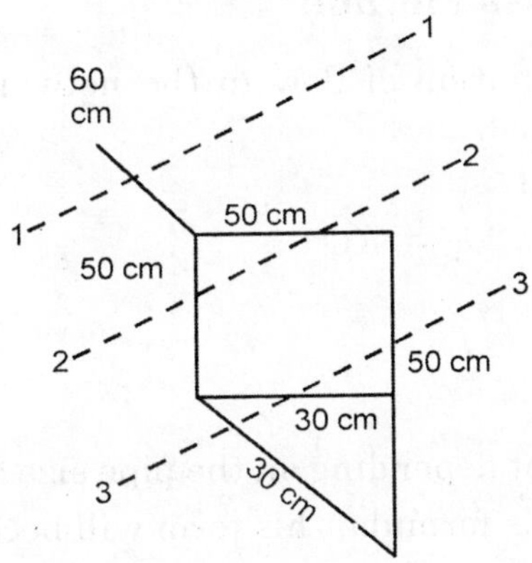

Fig. 10.9

Section 2 –2

Each pipe capacity = 136 lit/sec.

Velocity of flow = 0.80 m/sc.

Section 3–3

Through pipes the flow are 138, 69, and 69 lit/sec. From the sketch, the hydrualic radient will be 18 cm per 100 m, 130 cm per 100 m, and 130 cm per 100 m. And the velocity will be, 0.8 m/sec, 1.50 m/sec. and 1.50 m/sec.

10.12.2 Method of Electrical Analogy

This method is nothing but a comparison between the hydraulic and the electric circuts. The job can be done more faster by using computer analysis.

$$\therefore \quad D^5 = 4x\,(15)^5$$

and $$D = 19.8 \text{ cm } \underline{\Omega}\ 20 \text{ cm}$$

To check this result, this problem can be done by applying Hazen's nomogram. Say each pipe discharges is 20/ ps.

Accrdingly head loss = 150 cm/100 m

So discharge through = Pipe (20 + 20)

= 401.p.s. head loss 150 cm/100 m.

Equivalent × diameter = 21 cm.

From Table 6.4,

2.05 No. of 15 cm dia. pipes= 20 cm dia.

So this will be the same as found before.

10.12.3 Hardy–Cross Method

In this method a distribution of flow in the network is assumed and the resulting head is balanced.

This is expressed

$$h = KQ^x$$

where

h = head loss

Q = the flow

K = a constant depending on the pipe size and internal conditons.

Using Hazen–Williams formula, this term will become $h = KQ^{1.85}$

Then according to Hardy cross analysis on any pipe in a circuit will become

$$Q = Q_1 + \Delta$$

where

Q = acutal flow of water

Q_1 = amount assumed

Δ = flow correction

Then the total equation become

$$KQ^x = K(Q_1 + \Delta)^2\,x = K\left[Q_1^x + xQ^{x-1}\,\Delta + ...\right]$$

The other remaining terms in this expression is neglected, because Δ is very small with respect to Q. Then for a pipe circuit

$$\Sigma\, kQ^x = 0$$

Then from the above $\Sigma\, kQ^x = \Sigma kQ_1^x - \Sigma xKQ^{x-1}\,\Delta = 0$

Hence it can be write

$$\Delta = \frac{\Sigma KQ^x}{\Sigma KQ_1^{\,x-1}}$$

To use the method of the following steps are to be taken.

1. Assume any amount of flow with its direction.
2. Head loss in each pipe is calculated from equation or from nomogram.
3. Calculate the total head loss around each circuit $\Sigma h = \Sigma KQ_1^x$, but the attention should be paid to the sign.
4. Calculate the sum ΣKQ_1^{x-1} for the same circuit without regard to sign. It should be noted that heads which make up ΣKQ_1^x can be obtained from the nomogram of Hazen–Williams formula for $C = 100$. It also noticed that h or KQ_1^x of the pipe will make up the circuit, then dividing that by Q_1 the value becomes as KQ_1^{x-1} in that case the calculation will be easier.

Example 10.4 *In the network shown in Fig. 10.10, water enters 5000 l/m and from each of the joints B, C, D, E an amount 500 l/m, leaves the net given the length AB = BC = CF = AD = DE = EF = BE = 500 metres and their respective diameters are given against the section. Calculate the flow through each section by Hardy cross method. A nomograph is herewith enclosed.*

(C.U.)

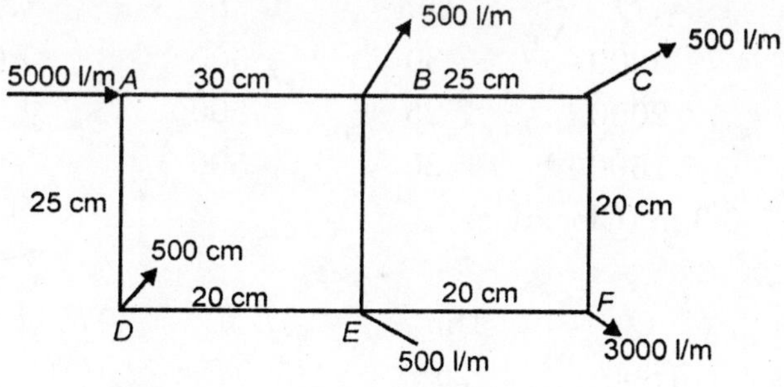

Fig. 10.10

Solution

Considering loop *ABED* a tabular form of results are given assuming discharge *Q* in *AB* and *AD*.

Fine	*Length*	*Discharge* Q	*Dia*	h_f	$\frac{h_f}{Q}$	*Remarks*
AB	500	3000	30	1.5	0.0005	
BE	500	500	20	0.5	0.001	
			Total =	2.0		Clockwise
AD	500	2000	25	1.25	0.0006	
DE	500	1500	20	2.00	0.0013	
			Total =	3.25		Anticlockwise

$$\Sigma\frac{h_f}{Q} = 0.0034$$

$$\text{Correction } \Delta Q = \frac{3.25-2.0}{1.85\times0.0034} = 198.72 \simeq 200 \text{l/m}$$

∴ Flow corrected

$$AB = 3200 \text{ l/m}$$
$$BE = 700 \text{ l/m}$$
$$AD = 2800 \text{ l/m}$$
$$DE = 1300 \text{ l/m}$$

Considering the above flow, the flow will be in

$$BC = 2000 \text{ l/m}$$
$$CF = 1500 \text{ l/m}$$
$$EF = 1500 \text{ l/m}$$

Head loss is considered in the circuit *ACFD*.

Line	Q	Dia	L	h_f
AB	3200	30	500	1.6
BC	2000	25	500	1.7
CF	1500	20	500	3.0

Total h_f = 6.3 m (Clockwise)

Same way

Line	Q	Dia	L	h_f
AD	1800	25	500	1.3
DE	1300	20	500	2.0
EF	1500	20	500	3.0

Total h_f = 6.3 (anticlockwise)

Because the loss of heads are the same, so, the flows will be correct.

Example 10.5 *Calculate the diameter of cast iron pipe required for the distribution system of a part of the town with a population of 1,000 persons. Make reasonable assumptions regarding the rate of supply, terminal pressure etc.* **(Patna)**

Solution

Assuming flow per capita

$$= 1501.\text{p.c.d.} = 150{,}000\ \text{l.p.d.}$$

So total flow $= 0.042\ \text{m}^3/\text{sec.}$

Considering 0.8 m/s velocity maintained, then d is give by

$$d = \sqrt{4/\pi \times \frac{0.042}{0.8}} = 0.258\ \text{m} \simeq 25\ \text{cm}$$

Loss of head

Friction 0.01 is assumed

$$\therefore \quad h_f = \frac{4fv^2}{2gd} = \frac{4\times0.01\times100\times0.8\times0.8}{2\times9.81\times25}$$

N.B. Length of pipe is considered as 100 m

$$= 0.512\ \text{m}$$

So, at 100 m length of head loss seems to be quite reasonable.

From the above calculation the pressure head in the pipe appears to be high enough, if the losses is increased still at the consumers end 14 m of water head will be good enough to have good supply.

Example 10.6 *A town is receiving its water supply from an impounding reservoir through 90 cm gravity main with a fall of 79.2 m and 80.5 km. Due to growth of population it is proposed to increase the supply by 50 percent taying a part length of additonal 150 cm pipe and new line being jointed cross connections. Find out the length of the new pipelline required (f = .01).*

(Pb change into metric unit)

Solution

From the formula we have

$$h_f = \frac{4f\,l\,v^2}{2gd}$$

$$79.2 = \frac{4\times0.4\times80.5\times1000\times v^2}{2\times9.81\times0.90}$$

$$\upsilon = 0.47$$

Now, $$Q = \frac{\pi}{4} \times (0.9)^2 \times 0.47 = 0.30 \text{ m}^3/\text{sec}$$

Considering the new length of the pipe be l

Then new discharge will be = q = 0.45 m^3/sec

So, $$79.2 - \frac{4 \times 0.01(80.5 - l) \times 1000 \times \left[0.45/\left(\pi/_4 \times 0.9\right)^2\right]^2}{2 \times 9.81 \times 0.90}$$

$$+ \frac{4 \times 0.01 \times \left\{\dfrac{0.45 - 0.30}{\pi/4 \times 1.5^2}\right\}^2 \times l \times 1000}{2 \times 9.81 \times 1.50}$$

By solving the above calculation will be found

$$\boldsymbol{l = 11.7 \text{ km}}$$

Example 10.7 *Determine the distribution of flow in the pipe network shown in fig 10.11. The head loss, he may be assumed as KQn. The flow is turbulent and pipes are rough. The value of K for each pipe is indicated in the figure. Use Hardy–Cross method.* **(Engg. Services)**

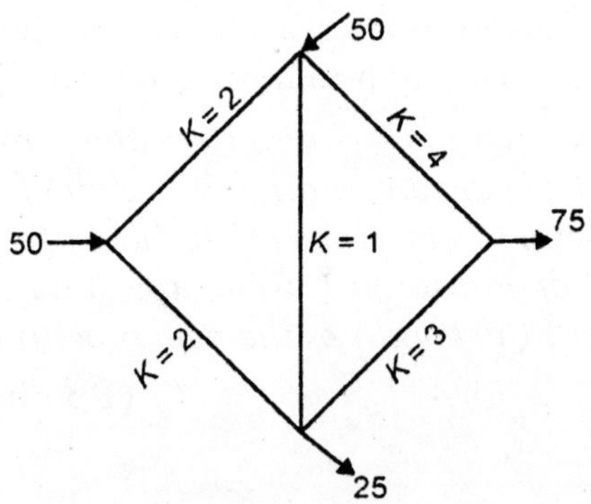

Fig. 10.11

Solution

Considering the law of continuity at each points, i.e., assuming the input is equal to the output at each point. In Fig. 10.12 the assumed flow are shown.

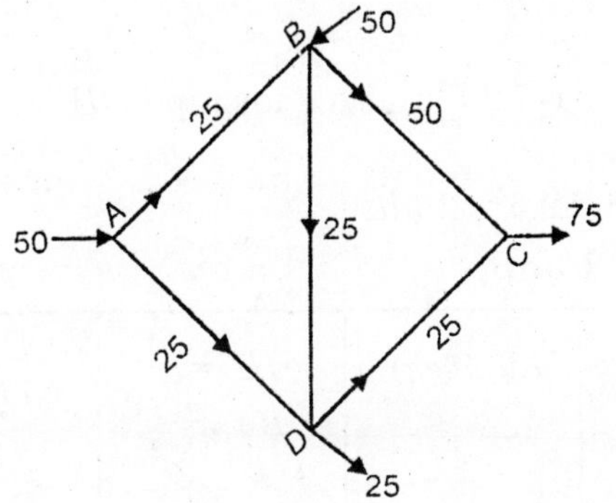

Fig. 10.12

So head loss = KQ^n, and assuming $n = 2$. From Darcy's formula of rough pipes and turbulent flow.

$$H_L = KQ^2$$

The values of K for different pipes are given in the questions. These values can be used for the loops *ABDA* and loop *BCDB*. Applying Hary cross method these loops analysed as follows.

Pipe	*Assumed flols*= Q_a *l/s*	*K (given)*	$H_L - KQ_a^{\ 2}$	$\frac{H_L}{Q_a}$	*Corrected Q after 1st correction* $= Q_{a_1} = Q_a + \Delta_1$ *l/s*
1	*2*	*3*	*4*	*5*	*6*
For loop ABDA–Loop (1)					
AB	25	2	1250	50	(+) 22.5
BD	25	1	625	25	(+) 35.0
Common pipe					
DA	– 25	2	(–) 1250	50	(–) 27.5
		Σ	+ 625	125	

$$\Delta_1 = \frac{(-)\Sigma H_L}{x \cdot \Sigma\left(\frac{H_L}{Q_a}\right)}$$

$$\Delta_1 = \frac{625}{2 \times 125} = (-)2.5$$

N.B. Flow correction in common pipe BD

$$= 25 + \Delta_1 + \Delta'_1$$

where Δ_1 for pipe BD = (–) 2.5 and Δ'_1 for pipe DB (–) 12.5 and hence Δ'_1 for BD will be (+) 12.5

So $Q_{ED} = 25 - 2.5 + 12.5 = 35$ l/s

For loop BCDB – Loop

Pipe	*Assumed flows Qa L/s*	*K (given)*	$HL = KQ_a^2$	$\frac{H_L}{Q_a}$	*Corrected Q after 1st correction* $= Qa_1 = Qa + \Delta_1$ *l/s*
1	2	3	4	5	6
BC	50	4	10000	200	37.5
CD	–25	3	(–) 1875	75	(–) 37.5
DB (Common pipe)	–25	1	(–) 625	25	(–) 35.0
		Σ	7500	300	

$$\Delta_1 \text{ in 2nd loop} = (-)\frac{7500}{2\times300} = (-)12.5$$

After 1st Correction (Flow)

Pipe	*Discharge corrected after 1st correction l/s*
For Loop *ABDA*	25
AB	35.0
BD (Comon pipe)	(–) 27.5
DA	(–) 37.5
For loop *BCDB*	
BC	37.5
CD (Common pipe)	(–) 37.5
DB	(–) 35.0

These results of discharges are used for 2nd correction.

Second correstion

Pipe	Assumed flows (Qa_1) l/s	K	$H_L = K Q_a^2$	$\frac{H_L}{Q_a}$	Corrected flow for 2nd correction $Q_{a2} = Q_{a1} + \Delta_2$
1	2	3	4	5	6
For Loop *ABDA*-(Loop-I)					
AB	22.5	2	1012.5	50	19.8
BD (Common pipe)	35.0	1	1225.0	25	32.6
DA	(–) 27.5	2	(–) 1512.5	50	(–) 30.2
		Σ	725	135	

$$\Delta_2 = \frac{(-)725}{2 \times 135} = (-)2.7$$

Pipe	Assumed flows (Qa_1) l/s	K	$H_L = K Q_a^2$	$\frac{H_L}{Q_a}$	Corrected flow for 2nd correction $Q_{a2} = Q_{a1} + \Delta_2$
For Loop *BCDB*-Loop-II					
BC	37.5	4	5625	150	37.2
CD	(–) 37.5	3	(–) 4218.75	112.5	(–) 37.8
DB (Common pipe)	(–) 35.0	1	(–) 1225	35	32.6*
		Σ	181.25	297.5	

$$\Delta_2' = (-)\frac{181.25}{2 \times 297.5} = (-)0.3$$

* Flow corrected in common pipe *BD*

$$= 35 - 2.7 - (-0.3) = 35 - 2.7 + 0.3$$

$$= 32.6 \text{ l/s}$$

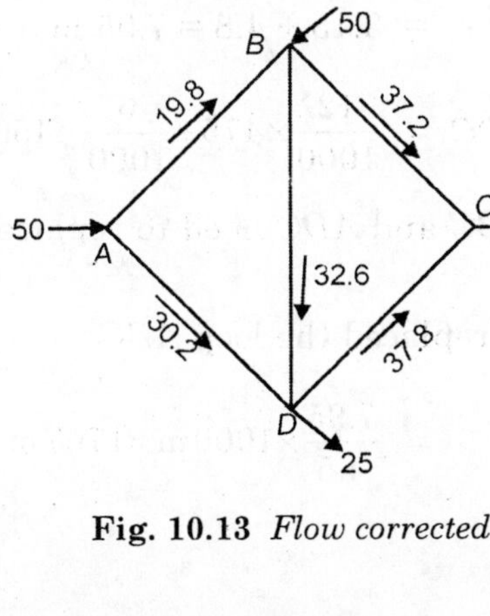

Fig. 10.13 *Flow corrected*

Flow corrected after second correction is plotted in the net work (Fig. 10.13) to check the balance of flow at each point to justify the answer.

10.13 METHOD OF EQUIVALENT PIPE

Sometimes when a pipe network causing problems i.e., flow is interepted. In that case without changing the network a single pipe with different diameter may be used to replace the portion of the disturbed loop of the network. This system of replacement is called the method of equivalent pipe system. The equivalent pipe is calculated in such a manner, that the flow of water in the replaced loop will be same and head loss will also be same.

This can be explained by an illustration (Fig. 10.14).

Size and length of pipes are shown in Fig 10.14.

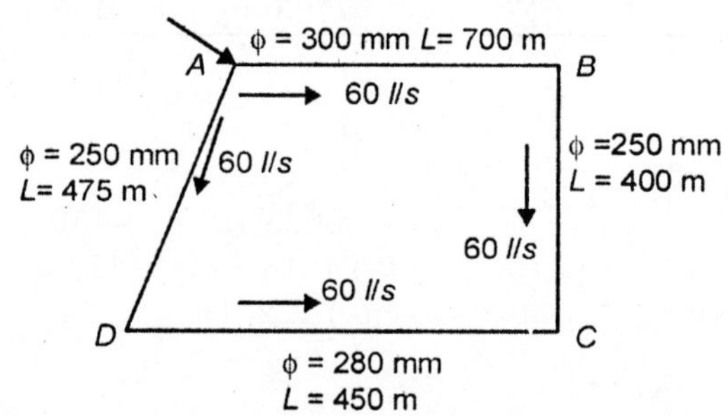

Fig. 10.14

ABCD is a pipe circuit and assumed 60 l/s of flow flowing through the way *ABC* and *ADC*. From William–Hazen's nomogram the loss of head in the circuit will be

For $$ABC = \frac{4.5}{1000}\times 700 + \frac{12}{1000}\times 4000$$

$$= 3.15 + 4.8 = 7.95 \text{ m}$$

For $$ADC = \frac{12}{1000}\times 475 + \frac{6}{1000}\times 450 = 5.7 + 2.7 = 8.4 \text{ m}$$

Now these two loops *ABC* and *ADC* need to replace by two 300 m length pipe

Then, 300 mm dia pipe replaced the loop *ABC*

$$= \frac{7.95}{4.5}\times 1000 \text{ m} = 1767 \text{ m}$$

Again, 300 m dia pipe replaced the loop *ADC*.

$$= \frac{8.4}{4.5} \times 1000\,\text{m} = 1865\,\text{m}$$

If the flow is properly divided in that case the head loss in both routes will be same. If the head loss in route *ADC* is 7.95 m, then the head loss per 1000 m will be

$$= \frac{7.95}{1865} \times 1000 = 4.27\,\text{m}$$

In 300 mm dia. pipe loss of head i.e., 4.27 m per 1000 m will occur if the discharge is 56 l/s (i.e., from William–Hazen's nomogram).

So the circuit *ABC* and *ADC* may be changed by an equivalent pipe carrying a total discharge of 60 + 56 =116 l/s.

Now pipe diameter 300 mm having discharge of 116 l/s, the head loss rate per 1000 m is from nomogram is found to be as 15 m. But to get head loss of 7.95 m in the equivalent pipe, the pipe length should be

$$= \frac{1000}{15} \times 7.95\,\text{m} = 530\,\text{m}$$

So, the toal Loop *ABCD* can be changed into a single pipe of 300 mm dia. and length should be 530 m, without changing the discharge.

10.14 OTHER METHODS

Various types of methods are used nowadays they are, sectional method, circular method, the Pitometer distribution studies method etc. The above methods are often used when necessary.

PROBLEMS

10.1. State the factors you would take into consideration and the procedure you would follow in designing a distribution system for the water supply of a city. **(Engineering Services)**

10.2. Explain the Hardy Cross method used for pipe network analysis in water distribution system. **(A.M.I.E.)**

10.3. Write short notes on the following :

(a) Distribution of water to consumers.

(b) Causes of water wastes, its detection and preventions.

(Engg. Services)

10.4. A pipe network consists of the following pipes.

Pipe	Length (m)	Diameter (cm)	Friction factor
AB	400	30	0.04
BC	600	30	0.010
AD	500	40	0.012
DC	500	25	0.011

In flow at *A* 1.0 m^3/sec, while outflows at *B*,*C* and D are 0.3, 0.5 and 0.2 m^3/sec. respectively. Find the flow in each pipe taking only one trial. The pressure at *A* is 100 m of water. **(A.M.I.E.)**

10.5. (a) Discuss briefly the procedure commonly adopted for planning and designing of a distribution system in a water supply scheme. What precautions you would take to make the design economical.

(b) Two 150 mm dia. pipes of equal lengths form a loop in a distribution network. Calculate the diameter of an equivalent pipe of the same length to represent the loop (there are no intermediate draw offs).

(Engg. Services)

10.6. A water transmission main to carry a flow of 0.05 m^3/s through a length of 2.5 km, where maximum allowable head loss is 8 cm. Calculate the size of the main and the velocity of flow, assuming suitable other data. **(C.U.)**

10.7. A town is receiving its water supply from a reservoir through 90 cm diameter gravity main with a fall of 100 m in 100 km. Due to the growth of population it is proposed to increase the supply by 50% laying a port of length an additional 150 cm dia. pipe and the new like being cross connected. Find out the length of the new pipe. (Assume friction factor = 0.01). **(A.M.I.E.)**

10.8. Describe the various layouts or distribution network in a water supply system and state their advantages, and disadvantages.

(W.P.S.C., Ran. Bha.)

10.9. Explain the method of determining the capacity of a service reservoir for storatge. **(Ran. UPSC, CW, IT)**

10.10. State the principles of designing the pipe line. **(Ran.)**

10.11. Discuss briefly the procedure commonly adopted for planning and designing of a dstribution system in a water supply scheme. What precautions you would take to make the design perfect ?

10.12. Write short notes on the followings :

(i) Distribution of water to consumers.

(ii) Hardy Cross method.

(iii) Different types of distribution networks.

(iv) Electrical analogy method.

(v) Distribution mains.

(vi) Method of sections.

(vii) Equivalent pipe method.

Chapter

11

Basic Definitions and Design of Sewer System

11.1 INTRODUCTION

To maintain the standard of the health of the public, the disposal of all types of waste is essential, in cities, towns and where the population is thick. If the production of waste is large an immediate disposal is an utmost necessity, otherwise habitation will be simply impossible on account of nuisance created by the waste.

These waste includes household waste solid and liquid including human extract, industrial waste, waste from public places, such as market, public latrine and waste from institutions like hospitals, commercial buildings etc.

To prevent the communication of deceases, the immediate and careful disposal of the waste is extremely needed, as the waste may contain big lot of diseases germs.

If the raw waste is discharged into a stream which is used for many other purposes it may prove dangerous by the break out of epidemic scum may be formed on the surface which is unsightly and unhealthy.

The materials in the wastes are highly decomposable organic substance which during putrification give rise to abnoxas gases. These are the reasons for which the treatment and disposal of waste is essential.

11.2 METHODS OF COLLECTION

11.2.1 Old Method

Old method of collection of human extract consisted of accumulation of night soil previces and cesepoles, etc. then removal by mannual labour, transportation in animal drawn vehicles to the point of ultimate disposal. This method is still in use in many towns and municipalities where the modern method of sewage has not been adopted. This method of collection is called dry method.

11.2.2 Modern Method

The modern method of collection which is known as waste-carriage method consists of mixing of waste with a considerable amount of water and collection by means of pipe system.

11.3 DEFINITIONS

Sewage. Combination of various types of liquid wastes conducted away from residence, buildings and institutions, industrial establishment.

Sullage. Discharge of kitchens, bathrooms wash basins etc. excluding discharge from hospitals, operation theater and slaughter houses.

Refused. Common term applied to all waste except human extract, this can be divided into three categories: (1) Garbage, (2) Rubbish, (3) Ashes.

1. *Garbage.* Waste food materials, waste from market etc. excluding dead animals.
2. *Rubbish.* Dry, non-putrescible other than ashes such as, paper, broken furniture, sweepings, brickbats and metal pieces etc.
3. *Ashes.* Dry non-putrescible residue of fuel after burning.

Nisht Soil. Human extract.

11.3.1 Historical Background of Sewage System

From ancient times the collection of storm water and drainage is done in their own way. As the science is developing in all respects, so as these methods are changed in 18th century from the year 1800. From the year 1800 the collections were treated and the systematic treatment of waste water followed in the late 1800 and the early in the year 1900. Latest part of the 19th century germ theory has been developed by Pasteur, that is the beginning of a new era in sanitation.

By the year 1900 nuisance and health condition brought about an increasing factor, then the thinking started to have a more effective means of waste water treatment is needed. New ideas have been developed for big cities and large town for the disposal of waste water.

11.4 CLASSIFICATION OF SEWERS

Sewers may be classified with respect to their use are as follows:

1. *Sanitary sewers.* A sewer which carries sewage from residence,

commercial buildings and institutions and to which storm, surface and ground water are not internally admitted also refer to as separate sanitary sewer or separate sewer.

2. *Storm sewer. A* sewer which carries storm water and surface water, street wash and other wash water as drainage, but excludes other sewage and industrial waste. This sewer is also called 'Storm drain'.

3. *Combine sewer.* A sewer receiving both surface run off and sewerage is called the 'Combined Sewer'.

4. *Storm over flew sewer.* A sewer used to carry the excess of storm flow from a main or intercepting sewer to an independent outlet generally in connection with a combined sewer system having limited main or intercepting sewer capacity.

5. *Relive or auxiliary sewer.* A sewer built to carry the flows in excess of the capacity of an existing sewer.

Sewer may be classified according to their positions as follows:

1. *Building sewer.* The extension from a building drain to the public sewer or other places of disposal, also called the house sewer or house connection.

2. *Lateral sewer.* A sewer which discharges into a branch or other sewer and has only building sewer tributary to it.

3. *Branch sewer.* A sewer which receives sewage from a relatively small area and discharges into a main sewer.

4. *Sub-main sewer.* An arbitrary term used for relatively large branch sewer.

5. *Main sewer.* A sewer to which one or more branch sewers are tributary and which serves a large territory, also called trunk sewer.

6. *Intercepting sewer.* A sewer which receives dry weather flow from a number of transverse sewers or outlets and frequently additional predetermined quantities of storm water (if from a combined system) and conduct such water to a point for treatment or disposal.

7. *Outfall sewer.* A sewer which receives sewage from a collecting system and carries it to a point of final discharge.

Inverd. The lowest point of the interior surface of a sewer is called the 'inverd'.

It is often necessary to classify a system as a whole as constructed to individual parts of the system and in such cases the following classification may be used:

1. *Separate system.* A sewer system comprised exclusively a sanitary only sewage and to which storm water, surface water and ground water are not intentionally admitted also refer to as sanitary system or separate sanitary system.

2. *Storm sewer system.* A system composed only of sewer carrying storm water, surface water, street wash and other wash waters or drainage and from which sewage and industrial waste are excluded.

3. *Combined sewer system.* A system of sewers receiving both surface run off and sewage.

The above mentioned classifications are only for the sewers only, not for the sewages. The classification of sewage has been described separately.

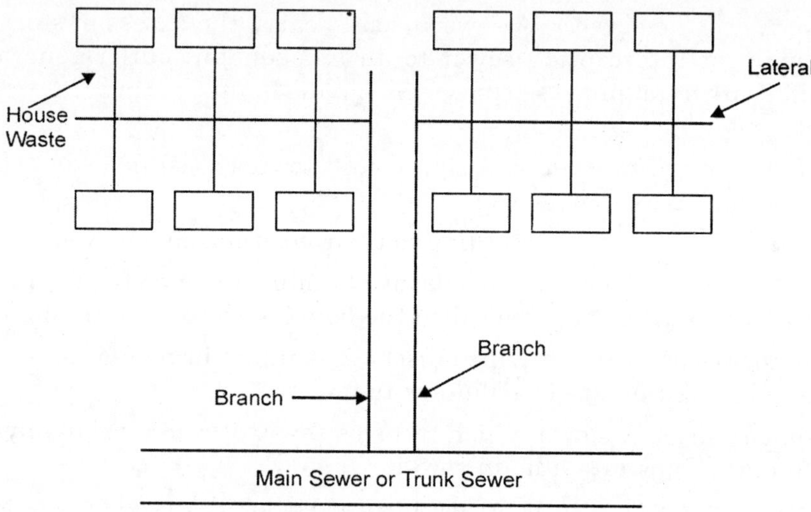

Fig. 11.1. *Connections of lateral branch and main.*

11.5 CLASSIFICATION OF SEWAGE BASED ON ORIGIN

1. *Sanitary sewage.* Sewage derives from dwelling, business building, public institution etc. also called 'Domestic sewage'.

2. *Storm sewage.* Surface run off caused by rainfall which is conveyed through sewer are called 'Storm sewage'.

3. *Industrial sewage.* Liquid wastes from industry such as dying, paper making etc. are called 'Industrial sewage'.

4. *Combined sewage.* Combination of sanitary and storm sewage with or without the industrial waste is called the 'Combined sewage'.

Sewage also can be classified according to condition or stage of putrification such as fresh sewage when much time is not passed since the production of sewage so that putrification has not started and it contains still some dissolved oxygen.

State Sewage. When the sewage has a little or no oxygen in it but still free from putrification.

Septic Sewage. A sewage which is putrifacing under anaerobic condition that is no oxygen condition.

Sewage Treatment. It means treatment by different methods, in order to remove or stabilise the objectionable substance to make it inoffensive.

Sewage Disposal. It means to dispose of the sewage with or without treatment in such a manner as it may not harm to anybody.

Quantity of Sewage. Quantity of sanitary sewage — sanitary sewage and industrial sewage is made of used water, supplied by the water supply. So actually the quantity of sanitary sewage and industrial waste is proportional to the water supplied by public supply.

But all water supply does not reach the sewage, for example, water lost through leakage in water supply mains water used for gardening, street washing, manufacturing processes etc. So there is some reduction from the water supply.

Again there are some addition also such as water from private supply infiltration of subsoil water though leaky joints in the sewer.

Quantity of Sewage. 70 to 130 per cent of the water consumed in different cities. But most of the designers prepare to adopt average sewer flow, equal to the average rate of water consumption.

Fluctuation in Sewage Flow. Knowledge of rate of flow including maximum and minimum rates is of importance for the design of sewers pumping plants etc. The size of the sewers should be such that it can carry maximum flow and the slope should be such, that there is no deposition during minimum flow (Fig. 11.2).

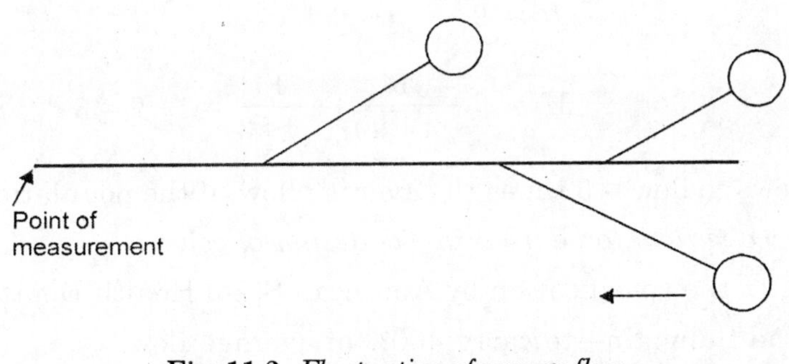

Fig. 11.2. *Fluctuation of sewage flow.*

Fluctuation in sewage flow follows similar fluctuation in the rate of water uses, but the deviation from average in sewage flow is smaller, because of greater capacity of sewer which tend to equalise the flow.

Again such flow fluctuation lag behind, similar fluctuation of water uses, because some time will be taken by the sewage to rich the point of measurement, again corresponding to a peak in water consumption there may be several peak in sewage flow, because of different distances to be travelled by sewage from different directions.

Period of time	*% of annual average rate*
Max. monthly	14%
Max. daily	15% to 180%
Max. hourly	200% to 300%

For rough estimating of minimum flow :

$$\frac{\text{Average flow}}{\text{Min. flow}} = \frac{\text{Max. flow}}{\text{Average flow}}$$

$$\therefore \quad \text{Max. flow} = \frac{(\text{Average flow})^2}{\text{Max. flow}}$$

$$\text{Min. hourly flow} = \frac{100 \times 100}{200} = 50\% \text{ of average flow}$$

Some use of following formula :

$$M = 1 + \frac{14}{4 + P^{1/2}} \text{ (by Haimon)}$$

where P = Population in thousands

i.e. if population = 9,000

Then $P = 9$

$$\therefore \quad M = 1 + \frac{14}{4 + \sqrt{9}} 1 + \frac{14}{4 + 3} = 1 + 2 = 3$$

∴ Max. sewage flow = 3 times the average flow, if the population is 9000.

Rate of sewage flow to be used in the design of sewer:

According to the specification by American State Health Deptt.

Lateral and Submain—to carry 400% of average flow.

Mains or trunks, outfalls —to carry 250% of average flow.

The above percentage takes care of possible infiltration, to this industrial waste should be added.

Design Period

1. For collection system that is Sewerage—40 years

2. Pumping plant—10 years
3. Sewage treatment plant—15 to 20 years.

Quantity of Storm Sewage

Rational method : $Q = AIR$

where Q = rate of flow

A = area drained

I = run off co-efficient

R = average rainfall intensity during a period t

where t = duration of storm.

In Indian conditon $R = \dfrac{2.54a}{t+b}$

where

R = rainfall intensity in cm/hour

a, b = constant

t = duration of rainfall in min.

For t = 5 to 20, a = 30 and b = 10

For t = 20 to 100, a = 40 and b = 20

Time of Concentration

Time taken by water or sewage to reach the measuring point from the remotest point of the area.

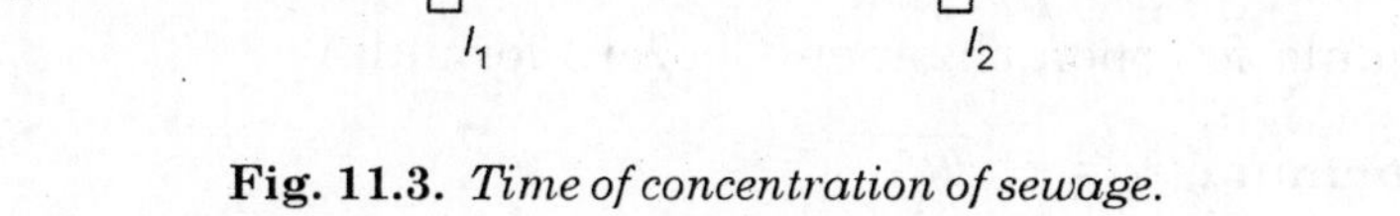

Fig. 11.3. *Time of concentration of sewage.*

P = Point at which flow is to be estimated

I_1, I_2 = inlets from blocks

Time of concentration T = time of inlet from block A + time of flow from I to P, i.e. from block B.

Run off Co-efficient

In this rate of run off to the rainfall and depends on the imperviousness of the surface.

Type of surface	Co-efficient
1. Water tight rock surface	0.70 to 0.93
2. Asphaltic pavement	0.85 to 0.90
3. Stone, bricks, wood block pavements with highly cemented joints	0.75 to 0.85
4. With uncemented joints	0.50 to 0.60
5. Maccadam road way	0.25 to 0.60

11.6 SEWER PIPES

Materials. For sewer pipes clay and concrete are most common material, though cast iron, thought iron and steel pipes are sometimes used under special conditions. In early days bricks sewers were also used, but nowadays brick is not usually used. Sewers may be constructed in factory or built in place.

Flow in Sewers. Sewage consists of about 99.9% of water and 0.1 % of solid. Specific gravity of sewage is about 1.00. So in hydraulic design of sewers, sewage is considered as water. The condition of flow in sewer:

1. Open channel flow,
2. Pressure flow.

11.7 OPEN CHANNEL FLOW

Hydraulic grade line coincides with the surface of the flowing water.

Pressure flow. Flow in sewer occurs under a pressure above or below the atmospheric pressure.

Most sewer are designed as open channel. In inverted syphones and discharge line from pumps flow occurs under pressure.

Flow formula for open channel (Chezei's formula)

Chezei's formula, $V = C\sqrt{RS}$

where V = Velocity of flow in m/sec

R = Hydraulic mean depth in m

S = Slope of the hydraulic grade line

C = Co-efficient depending on the nature of the surface, hydraulic depth, velocity, and slope of hydraulic grade line.

C is given by Kutter's formula,

$$C = \frac{\left(23 + \frac{0.00155}{5}\right) + \frac{1}{n}}{1 + \left(23 + \frac{0.00155}{5}\right)\frac{N}{\sqrt{R}}}$$

where N = rugosity co-efficient (co-efficient of roughness)

S. No.	*Types of surface*	*Value of n*
1.	Well laid stone-ware pipe, good concrete, rejected steel pipe, well laid brick pipe	0.013
2.	Stone-ware pipe and concrete pipe poorly jointed and unevenly settled	0.015
3.	Cast iron pipe of ordinary roughness	0.012
4.	Neat cement	0.010
5.	Rough brick wall, tuberculated iron pipe	0.017
6.	Smooth earthen channel	0.020

Darcy's Formula

$$H = f\frac{LQ^2}{3D^5}$$

where

H = Head loss in m

L = Length of pipe in m

Q = Rate of flow m²/sec

D = Diameter of pipe in m

f = Friction factor = 0.03 (usually).

Manning's Formula

$$V = \frac{1}{n}R^{2/3}S^{1/2}$$

where

V = Velocity in m/sec

R = Hydraulic mean depth in m

S = Slope of hydraulic grade line

n = Rugosity co-efficient and same as in Kutter's formula.

Hazens William's formula

$$V = 0.85\ CR^{0.63}\ S^{0.54}$$

where V = Velocity in m/sec

R = Hydraulic mean depth in m

S = Slope in hydraulic grade line

C = Co-efficient depend on pipe material.

Type of pipe	*Value of C*
Extremely smooth and straight	140
Cast iron pipe : New	130
5 years old	120
10 years old	107–113
20 years old	90–400
30 years old	75–90
Concrete or Cement line	120–140
Welded Steel	120
Riveted Steel	107–113
Brick	130

11.8 VELOCITY OF FLOW IN SEWERS

A velocity of flow which gives the sewage solids in suspension and does not allow deposition in pipe is called self-cleaning velocity.

In sewer velocity of flow should not be less than the self-cleaning velocity. Thus the minimum velocity is in sanitary sewers 60 cm/sec and in storm sewer 75 to 90 cm/sec.

Maximum velocity—to avoid abressional sewer material, the velocity should not exceed 200 cm/sec. For same velocity of flow larger pipe require flatter slope than smaller pipe but this is true only if the pipe is not less than half full.

11.8.1 Cross-section of Sewers

1. Circular section
2. Egg-shaped section
3. Rectangular section
4. Semi-elliptical section
5. Horse-shoe section

11.8.1.1 Circular section

This type of section is commonly used in sewer lines. The main advantages of this section are as follows.

1. Maximum cross-sectional area for a given wall material can be used and it is very economical.
2. Fairly good hydraulic qualities.
3. Easy to manufacture.
4. Fairly stable.

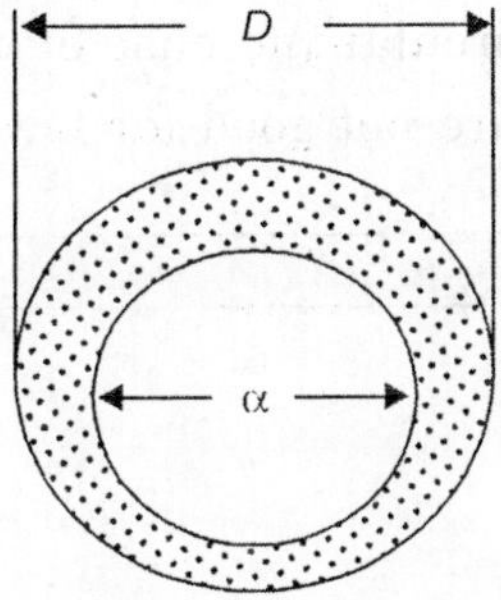

Fig. 11.4 *Circular section.*

The only disadvantage is that the excavation should be shaped to the barrel of pipe and at bending may have some trouble.

11.8.1.2 Egged-shaped section

Generally this type of sewer is used for carry combined flow (Fig. 11.5). The advantage of this type of sewer is that it gives higher velocity in low flow than a circular sewer of the same capacity. The disadvantages are (1) unstable in position, small end being drowned, (2) difficult to construct.

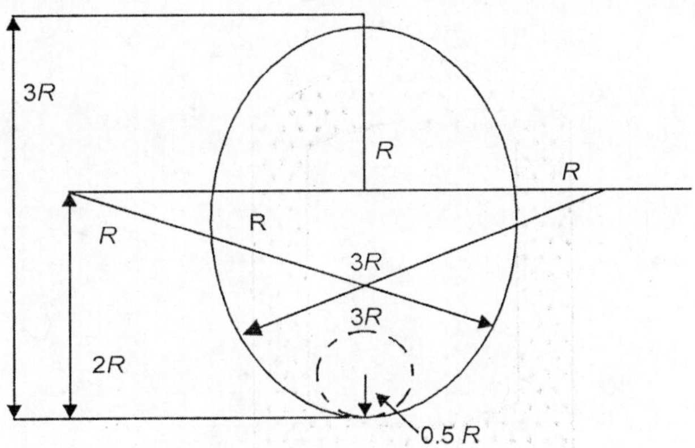

Fig. 11.5 *Egg-shaped section.*

11.8.1.3 Rectangular section

This type of sewer is used for storm sewer of larger size. Sometime this type of sewer is known as a box type sewer. The main advantages are:

1. Good hydraulic properties until it is filled.
2. Easy to construct
3. Economical

The disadvantages are:

(1) Hydraulic radius decrease sharply from nearly full to completely full condition because of circular increase of wetted perimeter by roof.

(2) Hydraulic properties are not good for low flow.

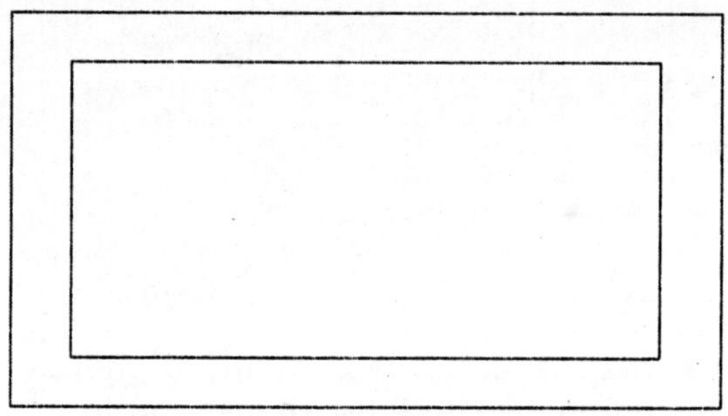

Fig. 11.6 *Rectangular section.*

11.8.1.4 Semi-elliptical section

This type of sewer is generally used for soft soil, because it is more stable (Fig. 11.7). This is suitable for carrying large amount of flow. Easy to

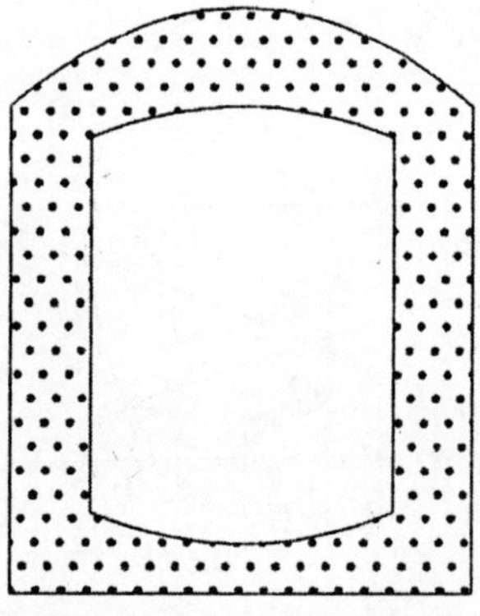

Fig. 11.7. *Semi-elliptical section.*

construct and economical too. The disadvantage of this type is the hydraulic properties are not good for low flow.

11.8.1.5 Horse-shoe section

This type of sewer is used as the out fall or at the main sewer line. This is not good for low flow at all. This is mostly used in tunnels. It is quite stable in position and also easy to construct (Fig. 11.8).

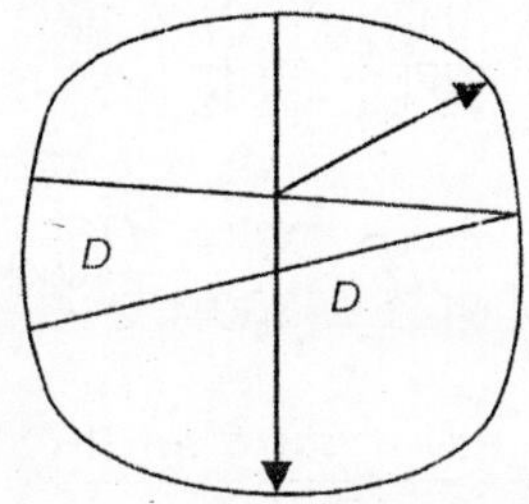

Fig. 11.8. *Horse-shoe section.*

11.9 PARTIAL FLOW IN SEWER

Very often sewer flow partially fills and it becomes necessary to determine velocity and depth of flow.

0 = Central angle

D = Diameter of sewer = Depth of flow when flowing full

d = Depth of flow.

For completely filled section

$$\text{Area} = \frac{\pi D^2}{4}$$

$$\text{Perimeter } P = \pi D$$

$$\text{Rate of flow } Q = AV$$

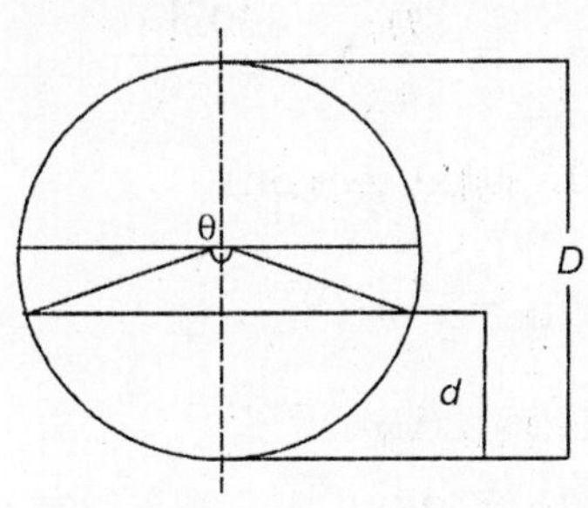

Fig. 11.9. *Partial flow in sewer.*

where V = Velocity when flowing full

$$\text{Hydraulic mean depth} = R = \frac{\pi D^2/4}{\pi D} = \frac{D}{4}$$

For partially filled section

$$d = \text{depth of flow} = \frac{D}{2} - \frac{D}{2}\cos\frac{\theta}{2}$$

$$= \frac{D}{2}\left(1 - \cos\frac{\theta}{2}\right)$$

$$\text{Area of flow} = \frac{\pi}{4}D^2 \times \frac{\theta}{360} - \frac{D}{2}\cos\frac{\theta}{2} \times \frac{D}{2}\sin\frac{\theta}{2}$$

$$= \frac{D^2}{4}\left(\frac{\pi\theta}{360} - \frac{\sin\theta}{2}\right)$$

Wetted perimeter (partially filled section):

$$p = \frac{P}{360}(\theta)\frac{P\theta}{360} = \frac{\pi D\theta}{360}$$

where P = Perimeter of full section

For practically filled section

Hydraulic mean depth

$$r = \frac{a}{p} = \frac{\frac{\pi}{4}D^2\left(\frac{\theta}{360} - \frac{\sin\theta}{2\pi}\right)}{\frac{\pi D\theta}{360}}$$

$$= \frac{D}{4}\left(1 - \frac{360\sin\theta}{2\pi\theta}\right)$$

Velocity of flow (partially filled section)

$$v = \frac{1}{n}r^{2/3}S^{1/2}$$

Rate of flow (partial filled section)

$$q = av$$

Ratios (θ in degree) for same slope and same value of n

$$\frac{d}{D} = \frac{1}{2}\left(1 - \cos\frac{\theta}{2}\right)$$

$$\frac{a}{A} = \frac{\frac{\pi}{4}D^2\left(\frac{\theta}{360} - \frac{\sin\theta}{2\pi}\right)}{\frac{\pi}{4}D^2} = \left(\frac{\theta}{360} - \frac{\sin\theta}{2\pi}\right)$$

$$\frac{r}{R} = \frac{\frac{D}{4}\left(1 - \frac{360\sin\theta}{2\pi\theta}\right)}{\frac{D}{4}} = 1 - \frac{360\sin\theta}{2\pi\theta}$$

$$\frac{v}{V} = \left(\frac{r}{R}\right)^{2/3} = \left(1 - \frac{360\sin\theta}{2\pi\theta}\right)^{2/3}$$

$$\frac{q}{Q} = \frac{av}{AV} = \left(\frac{\theta}{360} - \frac{\sin\theta}{2\pi}\right)\left(1 - \frac{\sin\theta}{2\pi}\right)^{2/3}$$

$$= \frac{\theta}{360} = \left(1 - \frac{360\sin\theta}{2\pi\theta}\right)^{5/3}$$

For various values of θ ratios can be determined in a graph.

				$\frac{N}{n} = 1$		
$\frac{d}{D}$	$\frac{a}{A}$	$\frac{p}{P}$	$\frac{r}{R}$	$\frac{v}{V}$	$\frac{q}{Q}$	$\frac{N}{n}$
1.0	1.0	1.0	1.0	1.0	1.0	1.0
0.90	0.949	0.857	1.192	1.124	1.066	0.93
0.8	0.858	0.705	1.217	1.14	0.988	0.89
0.7	0.748	0.631	1.185	1.12	0.838	0.85
0.6	0.626	0.564	1.110	1.072	0.671	0.82
0.5	0.501	0.500	1.00	1.00	0.500	0.80
0.4	0.373	0.44	0.857	0.902	0.337	0.79
0.3	0.252	0.32	0.684	0.776	0.196	0.78
0.2	0.143	0.21	0.482	0.615	0.088	0.75
0.1	0.052	0.05	0.254	0.401	0.21	0.81

N = Rugosity co-efficient when flow in full condition, the rugosity co-efficient is not constant, it varies with velocity and type of the material.

Design of Non-circular Section

For given condition a circular section is first designed and then a hydraulically equivalent non-circular section is determined which will discharge at the same rate on the same hydraulic gradient.

Let

a, v, r = area, velocity and hydraulic mean depth of a non-circular section.

A, V, R = area, velocity and hydraulic mean depth of a non-circular section.

Q = rate of flow = $AV = av$

From Manning's formula,

$$V = \frac{1}{n} R^{2/3} S^{1/2}$$

$$v = \frac{1}{n} r^{2/3} S^{1/2}$$

$$A \times \frac{1}{n} R^{2/3} S^{1/2} = a \times \frac{1}{n} r^{2/3} S^{1/2}$$

$$AR^{2/3} = ar^{2/3}$$

If d = diameter of circular section

$$r = \frac{\frac{\pi}{4} d^2}{\pi d} = \frac{d}{4}$$

$$\therefore \quad AR^{2/3} = \frac{\pi}{4} d^2 \left(\frac{d}{4}\right)^{2/3} = \frac{\pi (d)^{8/3}}{(4)^{5/3}} \quad \text{... (1)}$$

From this relation the equivalent non-circular section will be determined egged-shape section

$$A = 0.051\ D$$

$$R = 0.193\ D$$

where D = Vertical height

D = 1.25d.

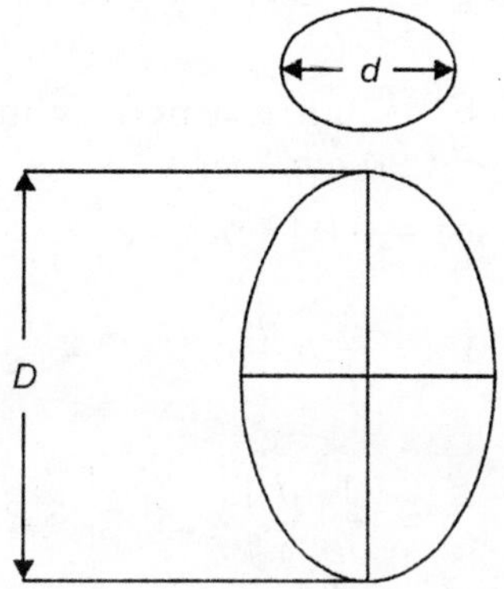

Fig. 11.10

11.10 DESIGN CONSIDERATION OF SEWERS

1. Average flow of sanitary sewage is taken to be the same as average rate of water supply.

2. Design period is about 25 years.

3. Design rate of sewage flow in sanitary sewers is 400% at the average flow for lateral and sub mains and 250% for mains and trunks.

4. Sewers in combined system should be capable of carrying at least two times the Dry Weather Flow (D.W.F.) in addition to storm water.

5. Sewers are normally designed to flow full when discharging the max. flow.

6. Sanitary sewers should be designed for self creating velocity of at least 60 cm/sec and combined sewers for a velocity of 75 to 90 cm/sec.

7. Minimum size of sewers to be used is 15 cm.

8. Manholes should be provided at intersections at changes of direction of gradient and at interval not exceeding 120 m of straight length.

9. A deep of about 12 mm is given in manhole where only direction changes.

10. At points where size changes crowns arė kept at same elevation.

11.10.1 DESIGN OF CIRCULAR SEWERS

Following example described for design of circular sewers.

Example (a) A circular sewer is to carry 48 lit/sec when flowing full. Slope of the ground and pipe is 0.0022 and $n = 0.013$. Determine the size of the sewer and velocity of flow.

(b) If the velocity of 90 cm/sec is to be attained what size and slope is required?

(c) If the slope is 0.0040 is maintained, what size has to be used to attained the velocity of 90 cm/sec?

Solution, (a) Q = 48 lit/sec = 0.048 m³/sec

Manning's formula $$V = \frac{1}{n} R^{2/3} S^{1/2}$$

$$= \frac{1}{n}\left(\frac{d}{4}\right)^{2/3} S^{1/2} \left[\because R = \frac{d}{4}\right]$$

$$Q = A.V. \ \frac{\pi}{4} d^2 \times \frac{1}{4}\left(\frac{d}{4}\right)^{2/3} S^{1/2}$$

$$= \frac{\pi}{4^{2/3}} \times \frac{1}{n} \times S^{1/2} d^{8/3}$$

$$= \frac{\pi}{10.03} \times \frac{1}{n} \times S^{1/2} d^{8/3}$$

Substituting the values

$$0.048 = (0.3125)\left(\frac{1}{0.013}\right)(0.0022)^{1/2} d^{8/3}$$

$$d^{8/3} = \frac{0.048 \times 0.013}{0.3125\ (0.0022)^{1/2}}$$

$$= \frac{0.000624}{0.3125 \times 0.047} = 4.25 \times 10^{-2}$$

$$d = (425 \times 10^{-2})^{3/8}$$

$$= 4.25 \times 10^{-3/8} = 0.306 \text{ m}$$

$$\therefore \quad d = 0.306 \text{ m}$$

$$V = \frac{Q}{A} = \frac{0.048}{\frac{\pi}{4}(0.306)^2} = \frac{0.048 \times 4}{\pi \times 0.093636}$$

$$= 0.655 \text{ m/sec}$$

$$= 65.5 \text{ cm/sec}$$

$$\left.\begin{aligned} d &= 30.6 \text{ cm} \\ V &= 65.5 \text{ cm/sec} \end{aligned}\right\} \textbf{Ans.}$$

(b)

$$Q = 0.048 \text{ m}^3/\text{sec}$$

$$V = 90 \text{ cm/sec}$$

$$S = ?$$

$$d = ?$$

$$Q = AV = \frac{\pi}{2} d^2 V$$

$$0.048 = \frac{\pi d^2}{4} \times 0.9$$

$$d^2 = \frac{0.048 \times 4}{\pi \times 0.9} = \frac{0.192}{\pi \times 0.9} = 6.8 \times 10^{-2}$$

$$\therefore \quad d = 2.61 \times 10^{-1} = 0.261 \text{ m} = 26.1 \text{ cm}$$

Again

$$V = \frac{1}{n} \times R^{2/3} \times S^{1/2}$$

$$0.9 = \frac{1}{0.013} \left(\frac{0.26}{4} \right)^{2/3} S^{1/2}$$

$$\therefore \quad S^{1/2} = \frac{0.09 \times 0.013 \times 4^{2/3}}{(0.261)^{2/3}} = 7.21 \times 10^{-2}$$

$$S = 52 \times 10^{-4} = 0.0052$$

$$\left.\begin{aligned} d &= 26.1 \text{ cm} \\ s &= 0.0052 \end{aligned}\right\} \textbf{Ans.}$$

(c)

$$Q = 0.048 \text{ m}^3/\text{sec}$$

$$V = 90 \text{ cm/sec}$$

$$S = 0.040$$

$$= 0.9 \text{ m/sec.}$$

Wetted area

$$a = \frac{Q}{V} = \frac{0.048}{0.9} = 0.0533 \text{ m}^2$$

$$V = \frac{1}{n} = R^{2/3} S^{1/2}$$

$$0.9 = \frac{1}{0.013} \times R^{2/3}(0.004)^{1/2}$$

$$\therefore \quad R^{2/3} = \frac{0.09 \times 0.013}{(0.004)^{1/2}} = \frac{0.0117}{0.0633} = 0.185$$

$$R = (0.185)^{3/2} = 0.0795$$

But $R = \dfrac{d}{4} \quad \therefore d = 4 \times R = 4 \times 0.0795 = 0.3180 \text{ m}$

$\therefore \quad d = 0.3180 \text{ m} = 31.80 \text{ cm}.$ **Ans.**

If this problem is treated by partially filled section then it will be as follows.
If D = diameter

Then $a = \dfrac{\pi D^2}{4}\left(\dfrac{\theta}{.360} - \dfrac{\sin\theta}{2\pi}\right)$

$$= \frac{\pi D^2}{4} \cdot \frac{\theta}{360}\left(1 - \frac{360\sin\theta}{2\pi\theta}\right)$$

$$R = \frac{D}{4}\left(1 - \frac{360\sin\theta}{2\pi\theta}\right)$$

Here

$$\frac{\pi D^2}{4} \cdot \frac{\pi}{360}\left(1 - \frac{360\sin\theta}{2\pi\theta}\right) = 0.0533 \quad \text{... (1)}$$

$$\frac{D}{4}\left(1 - \frac{360\sin\theta}{2\pi\theta}\right) = 0.0795 \quad \text{... (2)}$$

By squaring Eq. (2)

$$\frac{D^2}{16}\left(1 - \frac{360\sin\theta}{2\pi\theta}\right)^2 = 0.00635 \quad \text{... (3)}$$

Dividing Eq. (1) from Eq. (3) we get

$$\frac{1}{4}\frac{\left(1 - \dfrac{360\sin\theta}{2\pi\theta}\right)^2}{\dfrac{\pi\theta}{360}} = \frac{0.00635}{0.0533} = 0.119$$

$$\frac{360}{4\pi\theta}\left(1-\frac{360\sin\theta}{2\pi\theta}\right)=0.119$$

i.e. $$1-\frac{360\sin\theta}{2\pi\theta} = 0.119\times\frac{\pi\theta}{90}=0.00416\ \theta$$

$$1-0.00416\ \theta = \frac{360\sin\theta}{2\pi\theta}=\frac{57.4\sin\theta}{\theta}$$

By trial and error

Try $\theta = 90°$ L.H.S = 56.3

R.H.S = 57.4

0 = 93° L.H.S = 57

R.H.S = 57

Substituting the value of = 93° in Eq. (2)

$$\frac{D}{4}\left(1-\frac{360\sin 93°}{2\pi\times 93°}\right)=0.0795$$

i.e. $$\frac{D}{4}\left(1-\frac{360\times 0.9986}{186\times\pi}\right)=0.0795$$

i.e. $$\frac{D}{4}(1-0.615)=0.0795$$

$$\frac{D}{4}\times 0.385 = 0.0795$$

$$D = \frac{0.0795\times 4}{0.385}=0.825\text{ m}$$

$$= 82.5\text{ cm}$$

D = 82.5 cm. **Ans.**

11.10.2 Design of Sewer System

The following steps are followed in the design of sewer system:

1. Preliminary investigation.
2. Detail survey.
3. Layout of the system.
4. Actual design.
5. Preparation of final maps, plans, profiles and specification etc.

11.11 PRELIMINARY INVESTIGATION

A map of the city or a position of the city to be surveyed is to be obtained. If a map is not available, then a survey must be done. The map should show the built up areas street and topographical features such as contours etc. Preliminary investigation will include.

1. Present and future areas to be served.
2. Present and future population to be served.
3. Expected amount of sewage to be handled.
4. Character of sewage.
5. Proposed size for sewage treatment plant.

11.11.1 Detail Survey

This is done mainly as the underground survey. This includes the underground services, such as presence and location of existing sewers if any, water main, gauge line, telephone and telegraph conduct, foundation of structure, position and location of underground rock, depth of water table etc.

This information should be shown on plans and profiles for each sheet.

Surface survey includes —Number and type of buildings, elevation of their plinth level and basement floor if any.

Profiles of all streets, elevation of stream bed if any, culverts, with elevation of maximum and normal water lines etc. A contour map also is to be prepared on the general map of the city.

A system of permanent Bench Mark should be established at convenient intervals to be used during construction of sewer line.

This information will be entered into map and other drawings as far as possible and also kept in field note books for use during design.

11.11.2 Layout of the System

A tentative layout of sewer is made on the map by drawing lines. Considering the usual slope of the ground and other topographical and underground consideration. During layout the following points should be considered. (1) The slope of the ground is to be followed. (2) Underground rock should be avoided. (3) Pumping should be avoided or kept minimum if unavoidable. (4) In case of presence of ridge, separate system should be adopted. (5) Interference with water main and other underground installation should be avoided. (6) House connection should be short. (7) Cost of excavation should be kept minimum. (8) Sewer line should not be laid within at least 3 m horizontally and 2 m vertically from water mains.

After the tentative layouts are finished manholes are located and numbered. A sewer line is the lane of sewer between the successive manholes. Sewer lines are numbered. Then the area tributary to each line is indicated on map.

11.11.3 Design of Sewer Lines and Appurtenance

This includes determination of grades, sizes, location of sewer lines, design of pumping plant, design of other sewer appurtenance. The design should be made according to the standard practice. The design criteria should be different for different system. The design proceeds from the uppermost point of the system to downwards. Where a branch joins the line of a new design is started, at the upper end of the branch end is completed at the inner section.

11.11.4 Final Map, Plan and Profile

After the design a final plan is drawn for each street showing the position of sewers and appurtenance. A provision of each street is also showing position of sewers, of their size, grade, elevation and other appurtenance.

A general map is also drawn showing all streets and layout of sewer system.

Design criteria of sanitary sewer system:

1. The average sewage flow = The average rate of water consumption. If infiltration is expected in considerable amount an allowance may be given for that.
2. The period of design is at least 25 years.
3. The design rate of sewage flow is 400% of the average flow per person ultimately to be served for laterals and submains and 250% for main trunks etc.
4. The sewer should carry the design load when flowing full and under free flow condition.
5. Manholes should be provided at intersections at changes of direction, grade and sizes, of sewers and at intervals not exceeding 120 m on straight portion.
6. A drop of 12 mm $\left(\text{or } \frac{1}{2}''\right)$ is given in the manhole where only the direction changes. Where the size changes, crowns of the inlets and outlets are kept at same elevation.
7. Minimum sizes of the sewers to be used 15 cm (or 6″).
8. Self cleaning velocity of at least 75 cm/sec (or 2.5 ft/sec) should be assumed at minimum flow.

9. Minimum cover over the sewer is 1.8 m (or 2′).
10. Syphons should consist of at least of 2 pipes of minimum diameter of 15 cm (or 6″).
11. Branches carrying to a manhole shall have their crown at the same elevation and as that of the large sewer. If in doing so the invert of the branch has to be lowered by 60 cm or more a drop manhole may be used.

11.12 Sewer Construction

It consists of the following works :

1. Excavation and preparation of trenches.
2. Sheeting and bracing if needed.
3. Dewatering.
4. Pipe laying.
5. Jointing.
6. Backfilling.
7. Construction of appurtenance.

11.12.1 Sewer Construction Procedure

1. Sewer centre line is transferred to the ground from plans in which it was carefully located considering all other installation underground.

2. An offset line is established parallel to the centre line at a distance so that it will not be disturbed or covered with the excavated material.

3. Then the trench edge lines are set out.

4. After setting trench edge lines excavation started. Excavation may be by hand or machined.

5. If the trench walls are unstable sheeting and bracing are used. Sheeting and bracing consists of supporting by means of wooden bars in contact with trench wall and supported planks by cross wooden truss called strut wall. If the walls are stable sheeting or bracing are not needed.

6. If ground water table is above the bottom of trench, the water will flow into it and so it should be pumped out then.

7. When the excavation is nearly completed sight rails are placed across the trench at 8 m to 16 m intervals. The center line of the sewer is marked by putting a nail on the sight rails help of offset line drawn in the beginning. A tough rope passing through these nails on sight rail will give the centre line of sewers. Then this centre line of sewer is transferred to the bottom of trench by plumb bobs.

8. The next thing is done to move a point on each sight rail on the vertical line through the centre line nail at a predetermined height above the inverd of the sewer. It could be convenient if the top of the sight rail are kept at a fixed distance above the inverd of the sewer, so that the top of the sight rod touches the bottom of the trench to have the same grade as the sewer inverd. The vertical distance of the rope above the sewer inverd should be equal to the height of the boning rod. The required elevation of the top of a sight rail is calculated from designed rate.

9. Then the inverd level is checked with the help of boning rod and the sight rail are placed at short intervals along the length of the trench.

10. The bottom of the trench is shaped as required by the design. In case of excess cutting the refilling must be done with selected material, where concreting is required because of soft material, the trench bottom should be dipper by an amount equal to the thickness of concrete bed and pipe level. In case of rock bottom the excavation shown should be made wider and deeper by 10 cm (4 ´´) which will be refilled with selected materials or concrete before laying.

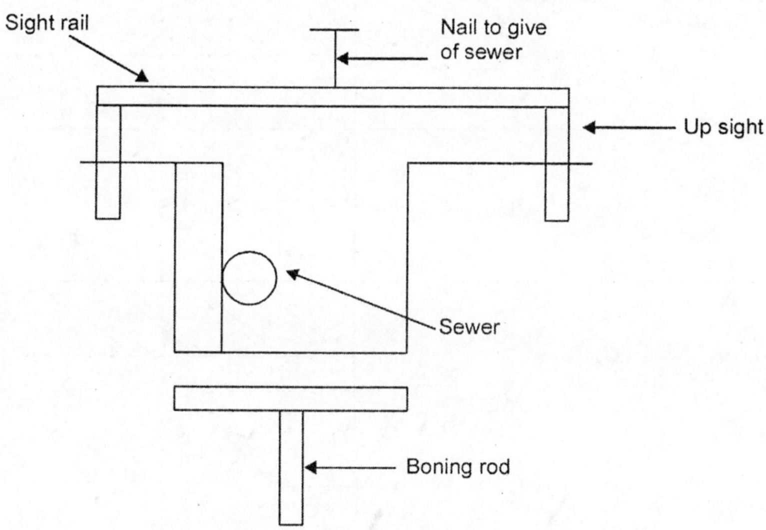

Fig. 11.11 *Sight rails and boning rod.*

11. *Laying.* After the preparation of the trench bottom pipes are laid. The pipes when smaller, may be lowered by hand or when larger and heavier, lowered by rope or chain from a derrick or crane. The belt of socket ends are on the up stream side. A spur of earth is taken out from trench bottom under the belt to make room for it. After laying of the pipe, inverd level is checked. All junctions such as Y, T etc., are placed as per design and plan.

12. *Jointing*. After laying of sewer pipe and checking of inverd levels jointing is done. In stone wave, salt glazed pipes, usually, the belt and spigot joint is used. (It is called socket and spigot joint).

A gasket is made of twisted hemp of such thickness that it will support the spigot end at proper height to make truly concentric. The lane of gasket should be sufficient to end circle the spigot with proper overlaying. It is soaked in cement grout. It is laid in bell for a lower trade of its circumference and covered with mortar. Then the spigot is placed in position over the gasket. Then a small amount of mortar is inserted in the annular space around the entire circumference. Then the remaining portion of the gasket is inserted round the upper 2/3rd of the circumference. Then the remaining portion of the annular space of the joint is filled into cement mortar or and bevelled off at the end at an angle of 45°.

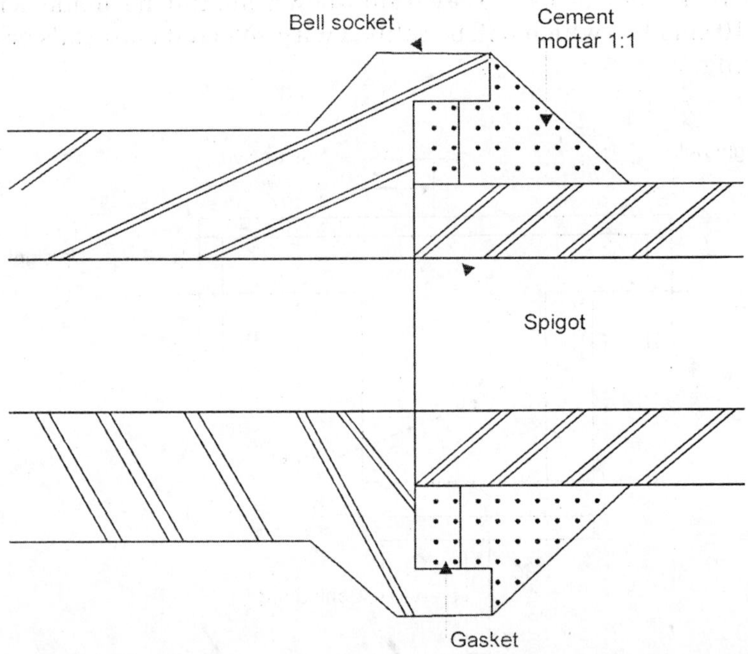

Fig. 11.12

Disadvantage of cement mortar joint is that any settlement will cause in the joint and permit infiltration.

Use of bitumen at above 400°F with jute gasket will make the joint flexible but overheating will make it inflexible and joint may be leaking.

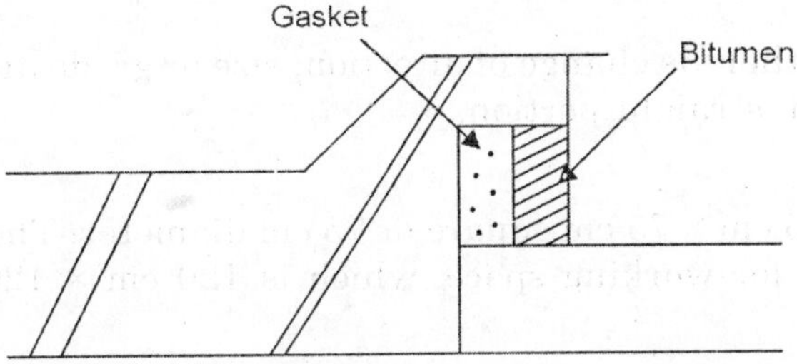

Fig. 11.13

13. *Backfilling.* Sheet material or selected material usually fine earth or sand is carefully dropped in trenches and uniformly tamped on both sides of the pipe in 15 cm (6″) layer up to 0.6 m (2′) above the pipe top. Taming should be evenly done on both sides to avoid high pressure on one side. All precautions have to be taken not to disturb the laid pipe as otherwise joint may break. No lump or large rock pieces or spongy material should be used in backfilling up to 0.6 m (2′) height from top of the pipe.

Permanent pavement should be constructed after 6 months and before one year of backfilling.

11.13 Sewer Appurtenances

Sewer appurtenances are devices which are essential for proper operation of the system.

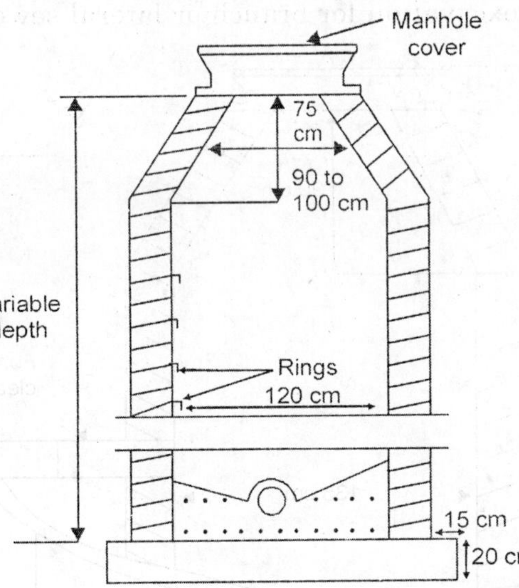

Fig. 11.14

Manhole

A masonry chamber on sewer lines providing access for inspection and cleaning. It may be called also inspection chamber.

Position

At points where there is change of direction, size or grade and at intervals of 90 m to 150 m on straight portion.

Size

Minimum size 75 cm × 75 cm square or 75 cm diameter. The bottom port on should be wider for working space, which is 120 cm × 120 cm or 120 cm diameter.

Opening

Manhole opening at top is 75 cm. It may be square or circular. It is kept covered by C.I. manhole covers.

11.13.1 Thickness of Brick Wall

Up to 3.6 m depth thickness of wall = 20 cm

Then thickness = $(5 + 4.1\ d)$ m

where d = depth in m.

Manhole Bottom

Bottom floor is usually made of concrete with open channel or channels semicircular in section. The floor is slope towards channel (1:10).

Drop Manhole

To connect two sewers at considerable level difference

(To save cast of excavation for branch or lateral sewers).

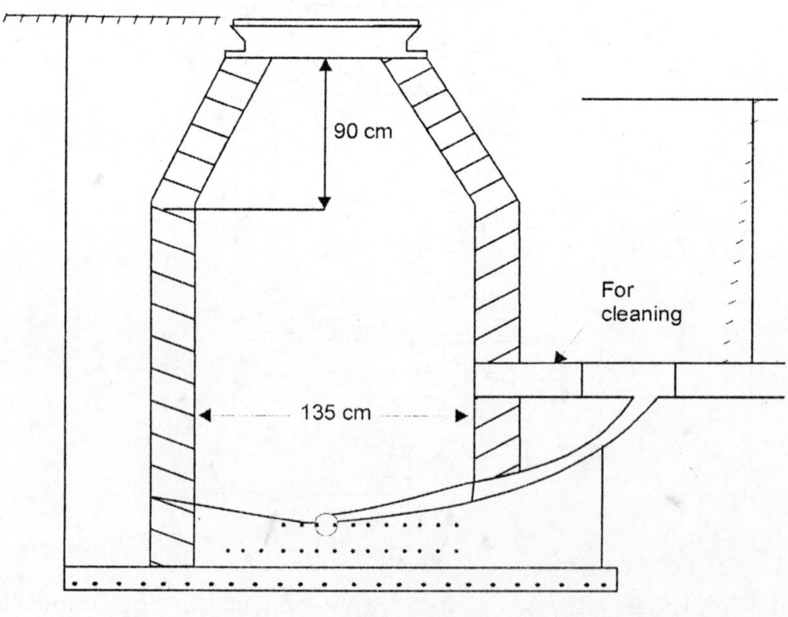

Fig. 11.15

Sand, Grease and Oil Trap

To intercept grease oil, sand etc. from sewage from kitchens, outlet hotels, garages etc. They tend to accumulated on sewers and clog it.

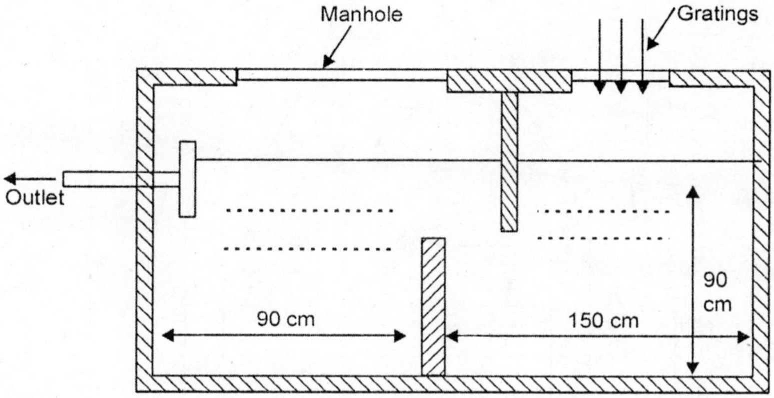

Fig. 11.16

Gulley Trap

It provides entry or rain water as well as drain water into sewer from the yard of a house. The water seal prevent escape of odour. It may be placed as road side inlet for rain water.

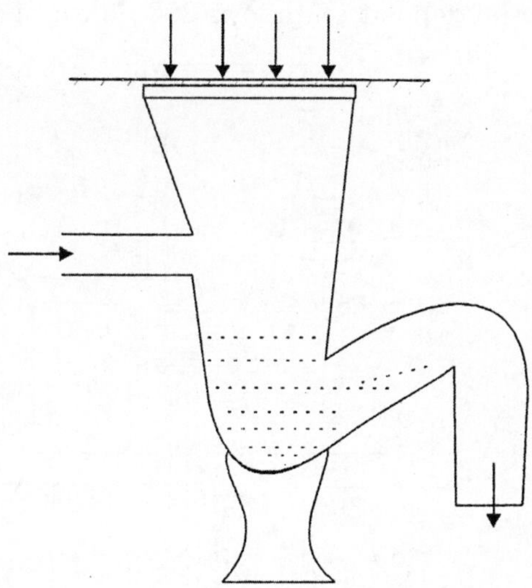

Fig. 11.17

Catch Pits or Catch Basins

Ills an inlet with a basin. The basin helps to separate the debris and floating matter. The water seal prevents to escape of gases from the sides.

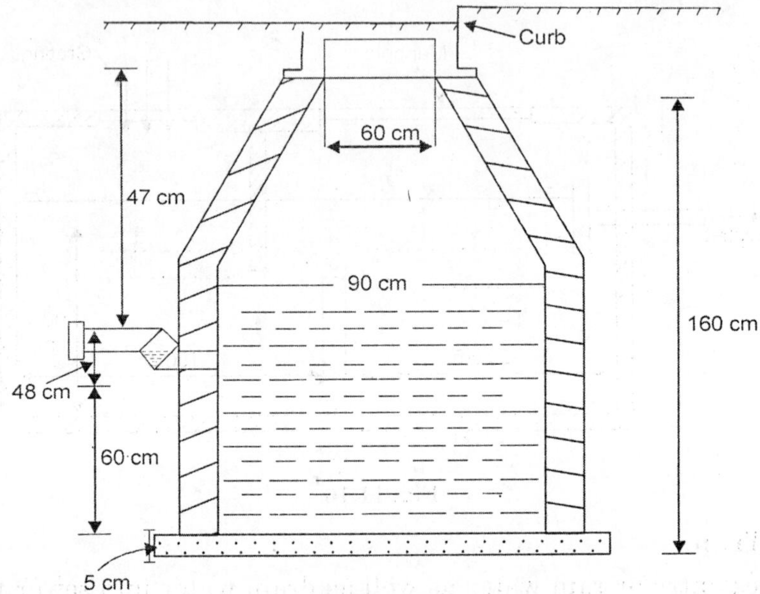

Fig. 11.18

Automatic Flash Tank

Where grades are low automatic flush tanks are placed at the upper ends of laterals to remove the deposition,[about 200 galan of water/flush]

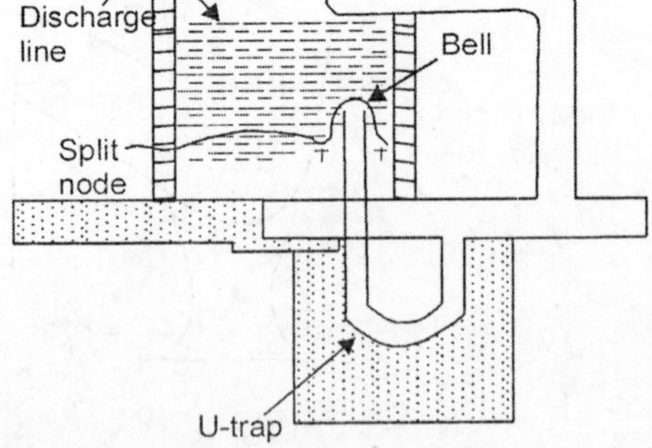

Fig. 11.19

11.14 REQUIRED VELOCITIES

In sanitary sewers, velocity should be such as to prevent any deposition of sewage solids in sewers. For this purpose a velocity of 60 cm/sec is at least required.

In storm sewers minimum velocity to prevent, deposition of heavier particles like grit, etc. The minimum velocity is 75 cm/sec to 90 cm/sec. Maximum velocity should not exceed 240 cm/sec to avoid abrasion of sewer material.

For same flow larger pipes require less slopes to give a specified velocity. So there may be tendency to use larger pipes in flat countries with less slope to maintain self-cleaning. But if the flow in the pipe is less than half full, velocity will be less. With bigger size of sewer velocity increases only when it flows full or half-full. So while using bigger size of pipe in flatter slopes, it should be checked that the depth of flow should not be less than half the diameter of the pipe.

If the grades too low for cleaning velocity are used, provision should be made for frequent flushing to remove deposition.

11.14.1 Superficial Loads on Buried Pipes

Sewer located in the yards of industrial plants may be subjected to superficial loads caused by stacks of coal, gravel etc. in addition to that caused by trench backfill. Such loads are called, "long", loads if they are longer than the trench width.

Sewer located under streets must have to withstand superficial loads caused by traffic, road-rollers etc. These are called "short", loads.

Marston also investigated the proportion of such loads that will be transmitted to the sewer, it depends on the depth to width ratio of trench and filling materials.

PROBLEMS

11.1. Define sewage, sewer and sewerage and classify each. Draw a sketch showing layout of sewers and dedicate thereon laterals, branches, mains and outfall sewers.

11.2. State the conditions under which separate and combined sewers are suitable. State the advantages and disadvantages of both the system.

11.3. How the quantities of sanitary sewage and storm sewage from a community estimated? Define "times of concentration".

11.4. How does the sanitary sewage flow fluctuate? What are the effects of such fluctuations on the design of sewers?

11.5. State the considerations for fixing minimum and maximum velocities of flow in a sewer? What are the usual values? What should be the gradient of a sewer?

11.6. State the design considerations for sanitary and storm sewers.

11.7. State the different materials and shapes of which sewers are made and state their merits and demerits.

11.8. Design an outfall sewer in the separate system for a town with a population of 100,000 persons with water supply at 180 litres. per head per day. The sewer is to be of brick work rendered smooth with cements mortar (n = 0.012) and the permissible slope is 1 in 1000. A self-cleaning velocity of 0.75 m/sec is to be developed. The DWF may be taken at 1/3rd of the maximum discharge.

What is the equivalent egg-shaped section of this circular sewer? Calculate the velocity of flow in the new section when flowing full and at DWF.

11.9. A combined sewer of circular section is to be laid to serve a particular area. Calculate the maximum combined discharge in liters per sec. Which will be accommodated by the sewer from the following detail:

(i) Area to be served = 100 hectares

(ii) Population = 90,000

(iii) Time of concentration = 20 minutes

(iv) Rate of water supply = 240 liters per capita

(v) Run off co-efficient = 0.50

(vi) Rate of rainfall = $\dfrac{76.20}{T+10}$ cm per hour

where T is the time of concentration.

11.10. If the maximum velocity of flow in sewers be 3 metres per sec. What will be the diameter of the sewer? Calculate using Manning's formula the required slope for the sewer to attain the above velocity.

11.11. (a) A circular sewer is to carry 48 liters per sec when flowing full. If the sewer is laid following the slope of the ground which is 0.0022. What size will be used and what will be the velocity? n = 0.013.

(b) If the velocity of 90 cm/sec is to be attained. What slope will be required if sewer size is kept the same as above?

(c) If ground slope is maintained what size has to be used to attain in the velocity of 90 cm/sec.

11.12. (a) A circular sewer is designed to carry a sewage flew of 60 liters per sec. When flowing full with a slope of 0.012. If the flow of sewage falls to 40 litres per sec. What will be the depth of sewage in the sewer and velocity of flow?

(b) What will be the rate and velocity of flow when the depth or sewer is 0.6 times the full depth?

Chapter

12

Analysis of Sewage

12.1 PURPOSE OF ANALYSIS

The purpose of analysis is as follows :

1. To determine those constituents which cause difficulties in treatment of disposal.
2. To select a type of treatment.
3. To control treatment operation.

Strength of sewage is referred to as an index of its nuisance producing properties.

12.2 DIVISION OF ANALYSIS

Complete analysis is divided into the following four categories:

1. Sanitary survey
2. Physical analysis
3. Chemical analysis
4. Biological analysis

12.2.1 Sanitary Survey

To get proper sanitary survey it is very important to realise few important point. It is useful to do the proper survey before going to any bigger project. It is also very useful to do the job step by step. The whole survey job should be divided into suitable zone. Care should be taken to the bye-laws of the localities. The following points should be carefully noted for any Public Sanitary Survey.

1. Financial coast
2. Present method of disposal
3. Population to be surveyed
4. Rainfall rate of the area
5. Quality of sewage to be handled
6. Source of sewage
7. Area topography
8. Treatment method to be applied
9. Sewage rate
10. Future development

12.2.3 Physical Analysis

The following tests are carried out—

1. Colour
2. Taste and odour
3. Temperature
4. Turbidity

1. Colour. The colour of fresh sewage is generally yellowish, grey or light brown. For septic or state sewage the colour will be black. Due to the presence of some industrial wastes the colour of the sewage may be different.

2. Taste and Odour. Fresh domestic sewage generally has no odour and it is not objectionable. Odour is present in the state sewage and this is due to the presence of hydrogen sulphide. This occurs due to the decomposition of sewage. Also it is found that some industrial wastes will create special type of odour to the sewage.

3. Temperature. In the sewage temperature has the effect on biological activities, viscosity of sewage and also on solubility of gases. The temperature of normal domestic sewage is little higher than that of water.

The presence of surface water or ground water will lower the temperature of sewage below normal. But due to the presence of hot industrial wastes the temperature of sewage will be above normal.

At higher temperature the biological activities will be greater and also the viscosity will be lower. Due to the lower viscosity of sewage, the treatment efficiency will increase.

4. Turbidity. Due to the presence of suspended particles normally the sewage is turbid. By analysing the water this can be determined.

12.2.4 Chemical Analysis

The chemical analysis of sewage is done as follows:

1. Chlorine
2. Nitrogen
3. Fats, greases and oils
4. Oxygen
5. pH value
6. Total solids.

1. Chlorine. Chlorination can be done in two ways:

(a) Chlorine demand test

(b) Chlorine residual test

(a) *Chlorine demand test.* Sometime it appears that though the sewage treatment is done, still some bacterias are present, to get rid of these bacterias chlorine is added. This addition of chlorine is called the chlorine demand.

(b) *Chlorine residual test.* After the reduction of bacteria from the sewage, if it is found that after 15 minutes the chlorine is deposited as residual chlorine, this is known as the chlorine residual test.

2. Nitrogen. Nitrogen presence in sewage in various forms, such as, free ammonia, albuminoid ammonia, nitrites or nitrates.

Presence of good amount of free ammonia means that the sewage is stale or old. If there are undecomposed nitrogenous are in sewage albuminoid ammonia will indicate that. In fresh sewage practically very little amount of nitrites and nitrates are present. Nitrites will show the intermolecule change of stale of the organic matter in sewage. Nitrates will stable the sewage. So nitrites will prominent in state sewage, whereas nitrates is prominent in well oxidized sewage.

3. Fats, Greases and Oils. The presence of these things is undesirable in sewage. They create lots of troubles. The main trouble creates by them, they will fill the voids of the filter material; and create an objectionable deposit in sewers and treatment units. To locate these materials, the following test is done some times.

Take a sample of water, and evaporate it, after the evaporation, it leaves behind the fats materials. Then add some ether, it will dissolve some material. Then ether will be evaporated, leaves behind the ether soluble matter which is the quantity of fats, greases and oils in sewage.

4. Oxygen. Presence of oxygen in a sewage in the following forms:

1. Dissolved oxygen
2. Oxygen consumed
3. Oxygen demand

1. *Dissolved oxygen.* This means that there is presence of oxygen in sewage in dissolved form. Practically sewage has no dissolved oxygen. This is an indication of a fresh sewage.

2. *Oxygen consumed.* This is sometimes found by a test on sewage, how much oxygen is consumed by carbonaceous matter in sewage.

The test—Standard amount of potassium permanganate with dilute sulphuric acid make a solution and to this solution adding some amount of sewage sample. The temperature is maintained 291°K (18°C) upto 15 minutes to 4 hours. After that the unused oxygen is calculated analytically from this the consumed oxygen is found.

3. *Oxygen demand.* In sewage oxygen is demanded by the inorganic and organic matters for their oxidation.

4. *Biological oxygen demand (B.O.D).* It is the amount of oxygen required for biological decomposition of organic solids to occur under aerobic condition at standardized time and temperature.

The quantity of O_2 required for complete stabilisation of water is an indication of amount of organic matter. The quantity of oxygen required depends on temperature and time. As efficient oxygen is available at 293°K (20°C) BOD is satisfied very rapidly descending 6 to 7 days and then slows down until the end of 20 days. Therefore it may accelerate and again slows down

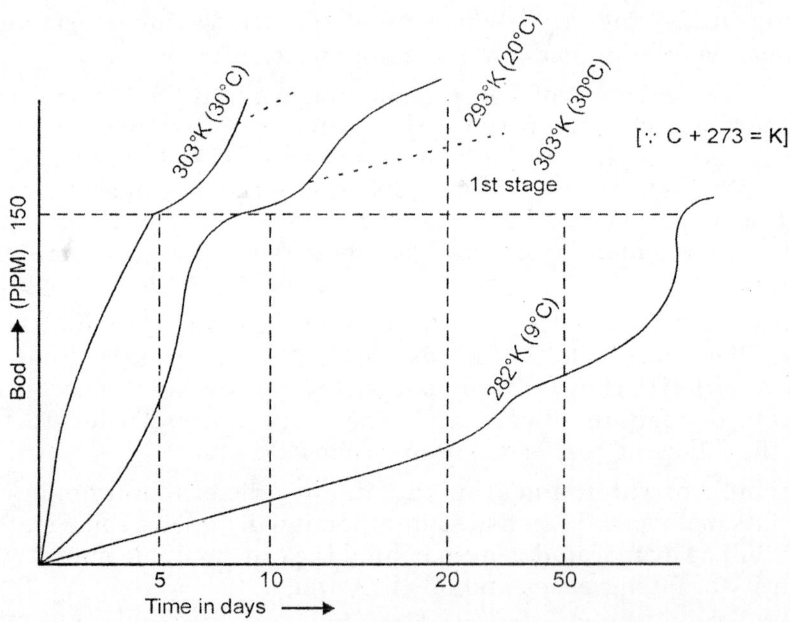

Fig. 12.1

to a very low rate for an indefinite period. The demand exerted is due to complete oxidation of carbonaceous matter and dissolved BOD. The latter part is due to Nitrogeneous matter. The BOD test consists of diluting the sewage of water containing known amount of D.O and noting loss of O_2 after a period of incubation. The usual incubating period is 5 days and at 20° C temperature. BOD test is the most important test to determine the strength of sewage.

12.2.5 Mathematical Formulation of BOD

The rate of Biological oxidation at any time is assumed to be proportional to the unoxidised organic matter present at that time.

If $L = BOD$ at any time, then

$$\frac{dL}{dt} = -K'L$$

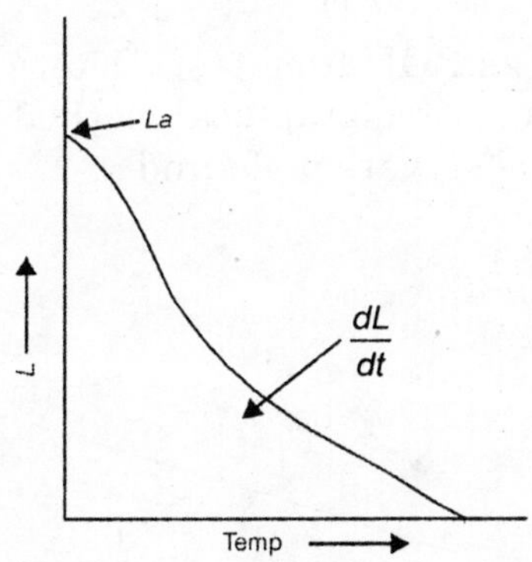

Fig.12.2

whence $$\frac{dL}{dt} = -kdt$$

Integrating log, $$L = -k't + C$$

where C = Const. of integration at $T = 0$, $L = La$ = the 1st stage of BOD.

Then $$\log La = C$$

Thus $$\log_e L = k't + \log_e La$$

i.e. $$\log_e L - \log_e La = -k't$$

$$\log_e = \frac{L}{La} = -k't$$

i.e. $$2.31 \log_{10} \frac{L}{La} = -k't$$

$$\log_{10}\frac{L}{La} = \frac{k}{2.31}t$$

$$\log_{10}\frac{L}{La} = -\,kt \qquad \left[k = \frac{k}{2.31}\right]$$

$$\frac{L}{La} = 10^{-kt}$$

$$L = (La)10^{-kt}$$

Amount of oxygen X required in time t is

$$X = La - L = La - (La)\ 10^{-kt}$$

$$= La\ [1 - 10^{-kt}]$$

X^x is generally reported as BOD from test. Determination of K and La – H.A. Thomas (1950) has suggested graphical method for determination of K and La from experimental observation of t and x.

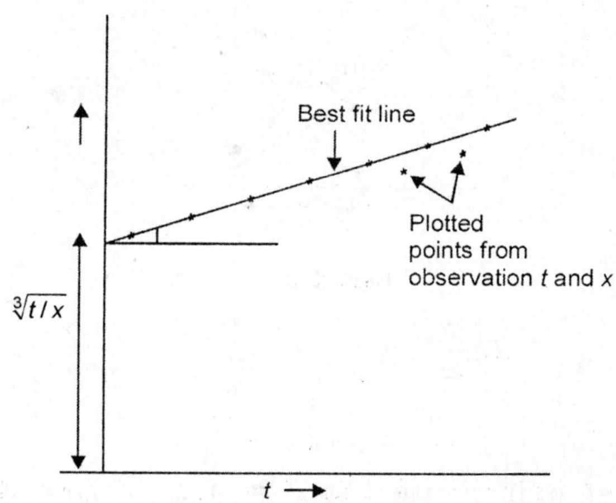

Fig. 12.3

A = interception on y axis by the best fit line.

$B = \tan \lambda$ = slope of the best fit line.

Then
$$K = (2.61)\frac{B}{A}$$

$$X = \frac{1}{2.3KA^3}$$

The deoxygenation constant K varies with temperature and may be given as

$$K_r = K_{20}[1.04^{(T-20)}]$$

when K_T = the value of K at $T°$

$$K_{20} = \text{the value of } K \text{ at } 20°$$

$$K_{20} = 1.000$$

The 1st stage BOD La also depends on temperature so that

$$(La)_T = (La)_{20}\,(0.02\,T + 0.6)$$

where $(La)_T$ = Value of La at K° ($T°$ C)

$(La)_{20}$ = Value of La at 293° K (20°C)

Example 12.1 *If 3 days 288°K (15°C) BOD of a sample is 100 mg/lit. What will be its 5 days 293° K (20°C) BOD ?*

Solution

$$K_{15} = K_{20}[1.047^{(15-20)}]$$

$$= (0.1)(1.047^{-5}) = \frac{1}{12.887} = 0.079433$$

$$X_{15} = 100 = (La)_{15}\,[1 - 10^{-(0.79466)(3)}]$$

$$= (La)_{15}\,[1 - 10^{-0.238299}]$$

$$= (La)_{15}\,[1 - 0.57771]$$

$$= (La)_{15}\,(0.42229)$$

$$(La)_{15} = \frac{100}{0.4229} = 236.47 \text{ mg./lit}$$

$$(La)_{20} = \frac{(La)_{15}}{[0.02(15)+0.6]} = \frac{(La)_{15}}{0.9} = \frac{236.47}{0.9}$$

$$= 262.74 \text{ mg/lit}$$

Then 5 day 20°C BOD is

$$X_{20} = L_{20}(1-10^{-K20\,t})$$

$$= 262.74\ [1-10^{-(0.1)(5)}]$$

$$= 262.74[1-10^{-0.5}]$$

$$= 262.74(0.68377) = \mathbf{179.66\ mg/lit}$$

Example 12.2 *If I day 30°C BOD of a sewage is 50 mg/lit. What will be its 5 day 293°K (20°C) BOD ? Assume deoxygenation constant K_{10} = 0.1.*

Solution

$$K_{30} = K_{20}\,[1.047^{(30-20)}]$$

$$= (0.1)[1.047^{10}] = 0.15849$$

$$X = La(1-10^{-kt})$$

where $50 = (La)_{30}[1-10^{-0.15849}]$

$$= (La)_{30} = [1-10^{-0.15849}]$$

$$(La)_{30}(0.30576)$$

$$(La)_{30} = \frac{50}{0.30576} = 163.54 \text{ mg/lit}$$

$$(La)_{20} = \frac{(La)_{30}}{[(0.02)(30)+0.6]} = \frac{163.54}{1.2}$$

$$= 136.2 \text{ mg/lit}$$

Then 5 day 293°K (20°C) BOD is

$$= (La)_{20}[1-10^{-(0.1)(5)}]$$

$$= (La)_{20}[1-0.31623]$$

$$= \mathbf{93.125 \text{ mg/lit.}}$$

Determine the percentage of BOD after 3 and 5 days at 20°C Assume = 0.1

$$X = (La)(1-10^{-kt})$$

% of BOD satisfy $= \dfrac{X}{La} \times 100$

$$= \frac{La(1-10^{-kt})}{La} \times 100 = (1-10^{-kt})$$

$$= (1- 10^{-0.1kt}) \times 100$$

After 3 days % of BOD satisfied

$$= [1-10^{-0.1(3)}] \times 100$$

$$= [1-10^{-0.3}] \times 100 = 49.88 \approx 50$$

After 5 days % at 293° K (20°C) BOD satisfied

$$= [1-10^{-0.1(5)}] \times 100$$

$$= [1- 0.31873] \times 100$$

$$= \mathbf{68.377 \approx 68.00}$$

$\therefore$ **For 3 days = 50%**

For 5 days = 60%

A logarithmic equation has been developed by H.E. Orfed in 1952 which fit the observation of BOD more closely. The equation is BOD (for time t and temp. T)

$$= S\ [0.85 \log \{t + 1.072^{(T-20)}\}(+ 0.41]$$

t = time of incubation in days at temp. T°K (T°C)with t between 0.5 and 20 and T between 10° and 30°C.

a = t axis intercept of the normal domestic sewage divided t, axis intercept of observed BOD curve.

S = BOD. When $t = \dfrac{\phi}{a}$ and it is independent of T.

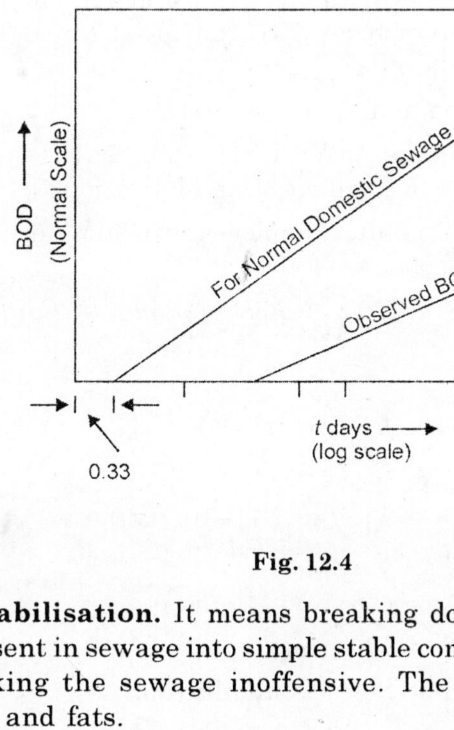

Fig. 12.4

Sewage Stabilisation. It means breaking down of completely organic compound present in sewage into simple stable compound by bacterial action and thus making the sewage inoffensive. The sewage contain protine, carbohydrates and fats.

Stabilisation may be brought about by an or aerobic process.

Anaerobic Process. In this process decomposition is caused by anaerobic bacteria which can live in absence of free oxygen. The products of anaerobic process are :

1. From decomposition of carbonaceous organic matter— CO_2 organic acid, and CH_4.
2. From decomposition of Nitrogenous organic matters or proteins — CO_2

organic protein. CH_4 NH_3 amino acid, amines indove, skatole of weak indove and skatole are unpleasantly, odour.

3. From decomposition of sulphurous matters—H_2S, and sulphur products.

Aerobic Process. Decomposition is caused by aerobic bacteria which live in presence of oxygen. For aerobic decomposition sufficient amount or free oxygen must be available. Under aerobic process initial decomposition is similar to that under anaerobic process, thereafter with adequate supply of free oxygen. Nitrogenous matters are oxidized 1st to Nitrites and then to Nitrates. Sulphureous matter to sulphate and carbonaceous matters to carbonates.

Some free N_2 are produced by both anaerobic and aerobic process. There are certain bacteria which Oxidized Nitrogen to Nitrates. This is called "Nitrogen fixation".

Comparison of aerobic and anaerobic process:

1. Anaerobic process will take longer time to stabilise the sewage. Aerobic process takes less time to complete the stabilization.
2. Anaerobic process will produce unpleasant odour. Aerobic process is free from odour.

Example 12.3 *3-day 288° K (15°C) BOD = 20 mg/lit.Find 7-day 298°K (25°C) BOD for K_{20}= 0.1.*

Solution

$$K_{15} = K_{20}(1.047^{T-20)}$$
$$K_{25} = K_{20}(1.047^{T-20)}$$
$$200 = X_{15}(La)_{15}\ (1-10^{-k_{15-t}})$$

Thus $(La)_{15}$ can be found.

$$X_{25} = (La)_{25}\ (1-10^{-k_{15-t}})$$
$$= \mathbf{500\ gm.}$$
$$S = \text{Sewage stabilisation}$$
$$= \frac{\text{Oxygen available}}{\text{Oxygen required}} \times 100$$
$$S = 100\ [1-0.794^{t_{20}}]$$
$$= 100\ [1-0.605^{t_{37}}]$$

where t_{20} t_{37} → time in days required for a standard value of sample to decolourise standard and standard value, of Methyle blue test will justify the stabilisation.

12.6 COMPOSITION OF SEWAGE

Average composition of domestic sewage in gm per capita/day is given below.

Sewage	*Minerals*	*Organic*	*Total matter*	*5 days 293°K (20°C) BOD*
Suspended solids	25	65	90	42
Settleable	15	39	54	19
Non settleable	10	26	36	23
Dissolve solids	80	80	160	12
Total solids	105	145	250	54

Decomposition of Sewage—Caused by bacteria. There are mainly three types of bacteria which causes the main decomposition

1. Anaerobic bacteria lies in absence of oxygen.
2. Aerobic becteria lies in presence of oxygen.
3. Facultative bacteria can lies in both the conditions.

Fresh Sewage. It contains a little amount of DO which is quickly exhausted. This type of sewage is called stable sewage. After the exhaustion of DO decomposition starts under anaerobic condition. The sewage at that time is called the "Septic Sewage".

5. pH Value. pH value test of a sewage indicate the nature of the sewage whether acidic or alkaline in nature. This test is very important in some treatment. Acid will kill the bacteria so the acidic sewage is not wanted. Generally the fresh sewage is alkaline in nature as the time passes on the sewage becomes acidic in nature. It prefer to have an efficient should have the pH value more than 7.0.

6. Total Solids. In sewage there are organic and inorganic solids. They are in different forms may be in suspension. Solution or in collodial manner. To decide the mature of treatment it is necessary to know the total solids in sewage. Through different fast, the following forms of solid may be determined.

1. Total solid
2. Suspended solids
3. Settleable solids.

Total solid. From a sample of sewage the liquid is evaporated, the residue left is dried and weight is taken, that will give the amount of total solid in sewage.

To know the amount of volatile and fined portion of the total solids it is generally heated or ignited. After that weight is taken. The loss of weight means the portion of the volatile solids. Generally the organic matters are volatile.

Suspended solids. The suspended solids are generally obtained by filtration through asbestos fibre. The suspended solids are in the fibre and the dissolved solids are in the filtered liquid. So we can say

Total solids = Suspended solids + Dissolved solids.

↓ ↓

From effluent after filtration   Remaining

Settleable solids. To get the quantity of settleable solids it is generally the imhoff cone is used which is conical at the bottom. The capacity of the vessel is one litre and from the bottom graduation upto 50 ml is marked. Sample of sewage is poured into the vessel and allow to stay at least two hours. Then watch the settlement and read it to get the exact proportion, then carefully drain off the top liquid and weigh the settled solids.

12.4 BACTERIOLOGICAL ANALYSIS

Bacteriological analysis is not that much of importance in sewage analysis. Because this analysis does not help to choose the treatment method. But the presence of biological life is essential for the efficient working of the treatment plant. Absence of bacteria means that the waste is from industry.

Bacteriological test results are used to determine the degree of pollution of natural sources of matter.

Population Equivalent. B.O.D test is the most important and reliable analysis of sewage. So its result is converted a sewage of particular strength into equivalent population. Generally the comparison is done with B.O.D of standard sewage. For normal domestic sewage, the average equivalent population per capita per day varies from 72 gm to 84 gm of 5-day B.O.D at 293°K (20°C). For combined sewers and sewage having good amount of industrial waste are 89 gm to 112 gm and 180 to 228 gm respectively.

This equivalent population can be based an oxygen demand and chlorine demand.

12.4.1 Uses of Population Equivalent

1. By this method, sewage from the various industries can be converted into equivalent population and each industry is then charged accordingly for the treatment of its sewage. This method is more useful than the ordinary method. The ordinary method of charging on the basis of volume of sewage.

2. This method serves as an index of the strength of industrial sewage and the treatment units are designed accordingly.

12.5 SEWAGE DISPOSAL

Sewage treated or untreated may be disposed of by (1) discharging it into a body of water — usually called dilutions (2) discharging on to a land surface— called irrigation or sewage farming and (3) discharging underground — called subsoil irrigation — which is often discouraged.

12.6 DILUTION

Disposal of sewage by dilution is mostly by used. With the discharged of sewage a body of natural water, will be polluted with the following results.

1. It will be unsafe for water supplies recreation purposes and fish culture because of the presence of disease organisms.
2. Dissolved oxygen in the receiving water will be depleted with fish life being impossible.
3. Acids, oils and toxic materials will make the water inedible.
4. Purification of organic matters will produce unpleasant odours and unsightly conditions.

After some time and at some distance from the point of pollution, normal balanced condition of the stream will be restored by the process called natural purification or self purification.

For fresh plain-sedimentation sludge, the head loss is about 1.5 to 4 times that of water.

For small rates of discharge, plunger and diaphragm pumps and compressed air ejectors are employed and for high rates of discharge centrifugal pumps not less then 10 cm in size are usually used.

Example 12.4 *Estimate for a 10% solids digested plain sedimentation sludge and a velocity of 60 cm/sec. The loss of head in 100 metre of 30 cm pipe.*

Solution

Assuming $$\frac{\eta}{\rho} = 5.375 \times 10^{-7} \text{ cm}^2\text{/sec}$$

and $$\frac{\tau_\gamma}{\rho} = 7.2 \text{ (cm/sec}^2\text{)}$$

$$\frac{h}{l} = \frac{28800}{g} \cdot \frac{\sigma}{d^2}$$

$$\frac{h}{l} = \frac{28800}{g} \cdot \frac{\sigma}{d^2} \left(930 \frac{\eta}{\rho} + \frac{1}{5400} \cdot \frac{\tau_y}{\rho} \cdot \frac{d}{\sigma} \right)$$

$$= \frac{28800}{981} \cdot \frac{60}{(30)^2} \left[930 \times 5.375 \times 10^{-7} + \frac{1}{5400} (7.2) \left(\frac{30}{60} \right) \right]$$

$$= 0.002283$$

where

h = loss of head in terms of sludge.
l = length of pipe
σ = velocity of liquid in cm/sec
d = diameter of pipe in cm
η = co-efficient of rigidity
τ_y = shearing stress at yield point
ρ = mass density
g = acceleration due to gravity = 981 cm²/sec

The terms in the parenthesis are analogous to the kinematic viscosity of a Newtonian liquid and possess the same dimension i.e. cm²/sec or stoke, and can be called co-efficient of kinematic rigidity σ_p. The Reynold's number is then

$$R = \frac{vd}{v\rho}$$

and laminar flow is expected below a value of R of about 2000 as it does also in the viscous flow.

For thick digested sludge (10% solids) the values of co-efficient are reported as

$$\frac{\eta}{\rho} = 5.375 \times 10^{-7} \text{ cm}^2\text{/sec (or stoke)}$$

$$\frac{\tau_y}{\rho} = 7.2 \times (\text{cm/sec})^2$$

For turbulent flow, loss of head of fairly homogeneous sludges (digested and activated sludges) is increased by not more than 1% for each percent of solids in sludge.

For purpose of comparison animal manures contain 1 to 4% nitrogen and about the same percentage of phosphoric acid and potash.

Sewage sludge may be used directly or used as a fertilizer base with other necessary elements added.

Use of sewage sludges involves hygienic hazards as pathogenic bacteria, viruses etc. can survive in sewage treatment process. Only heat dried sludge can be considered fully safe.

PROBLEMS

12.1. State the physical properties of sewage.

12.2. Define bacteria; state their function.

12.3. Why the analysis of sewage is necessary?

12.4. Discuss the physical tests of sewage.

12.5. What is relative stability?

12.6. Explain brief in the anaerobic and aerobic decomposition.

12.7. What will be the indication of gray sewage and dark sewage?

12.8. Define oxygen demand.

12.9. What is dissolved oxygen and how this could be determined?

12.10. Briefly explain the meaning of the term population equivalent and also mention its uses.

12.11. Briefly explain what is BOD?

12.12. Explain why BOD at 5 day 20° is giving the importance?

12.13. What is the indication of oxygen demand?

12.14. How the Settleable solids can be determined from a sample of a sewage.

12.15. (a) What is meant by "Strength of Sewage" ? How is the strength estimated ? What is meant by "Septic Sewage" ? How does septic sewage differ physically from fresh sewage?

(b) Give a list of tests which are usually performed to evaluate the physical and chemical characteristics of sewage. State the significance of these tests.

12.16. (a) What is meant by BOD ? How is it determined and state the importance of BOD in sewage treatment.

(b) Show that BOD satisfied in time t is given by $La\ (1-10^{-kt'})$, where La is the 1st stage of BOD.

(c) If 3-day 15°C ? BOD of a sample is 80 mg/lit, what will be its 5-day 20° C BOD Assume $K_{20} = 0.1$.

12.17. (a) What is meant by "Stabilisation of sewage"? Describe how carbonaceous, Nitrogenous and Sulphurous materials of sewage are stabilised in aerobic and anaerobic processes.

(b) If the colour of a sample in a relative stability test disappears in 4.5 days at a temperature of 37°C. What is its relative stability? In how many days would the colour disappear if the sample were stored at 20°C?

Chapter **13**

Sewage Treatment

13.1 PRIMARY TREATMENT

A. Removal of coarse suspended solids
   1. Grit chambers

B. Removal of fine suspended solids
   1. Sedimentation tanks
   2. Septic tanks
   3. Imhoff tanks
   4. Chemical Precipitation tanks.

13.2 SECONDARY TREATMENT

A. Oxidation by filtration
   1. Intermittent sand filters
   2. Contact filters
   3. Trickling filters, standard
   4. Trickling filters, high rate.

B. Oxidation by Aeration
   1. Activated sludge process
   2. Modified activated sludge process.

C. Oxidation by atmospheric oxygen and photosynthesis
   1. Oxidation ponds

D. Disinfection

1. Chlorination.

13.2.1 Sludge Treatment Method

1. Digestion
2. Conditioning
   (a) Elutriation (b) Chemical conditioning
3. Vacuum filtration
4. Drying.

13.2.2 Sewage Treatment Methods

Sewage is usually treated partially or completely before disposal in order to avoid dangerous degradation of receiving waters and associated health hazards. The choice of the treatment method requires consideration of certain factors such as (1) method of final disposal (2) character of sewage (3) degree of treatment necessary (4) initial cost and cost of operation etc.

The treatment methods are broadly classified into primary and secondary treatment methods.

Primary treatment methods are those which remove a part of the suspended and floating solids.

Secondary treatment methods are then which are aimed at the satisfaction of the BOD. The secondary treatments are usually preceded by one or more of primary treatments.

The sewage which has received only primary treatment is called partially treated whereas one receiving secondary treatment is called completely treated.

13.3 METHODS OF SEWAGE TREATMENT

I. Primary treatments:

A. Removal of floating solids

1. Racks
2. Screens
3. Skimming tanks.

13.4 BAR SCREENS OR RACKS (Fig. 13.1)

Functions. (1) To separate floating objects from sewage.

(2) To prevent damage of mechanical equipments like pumps.

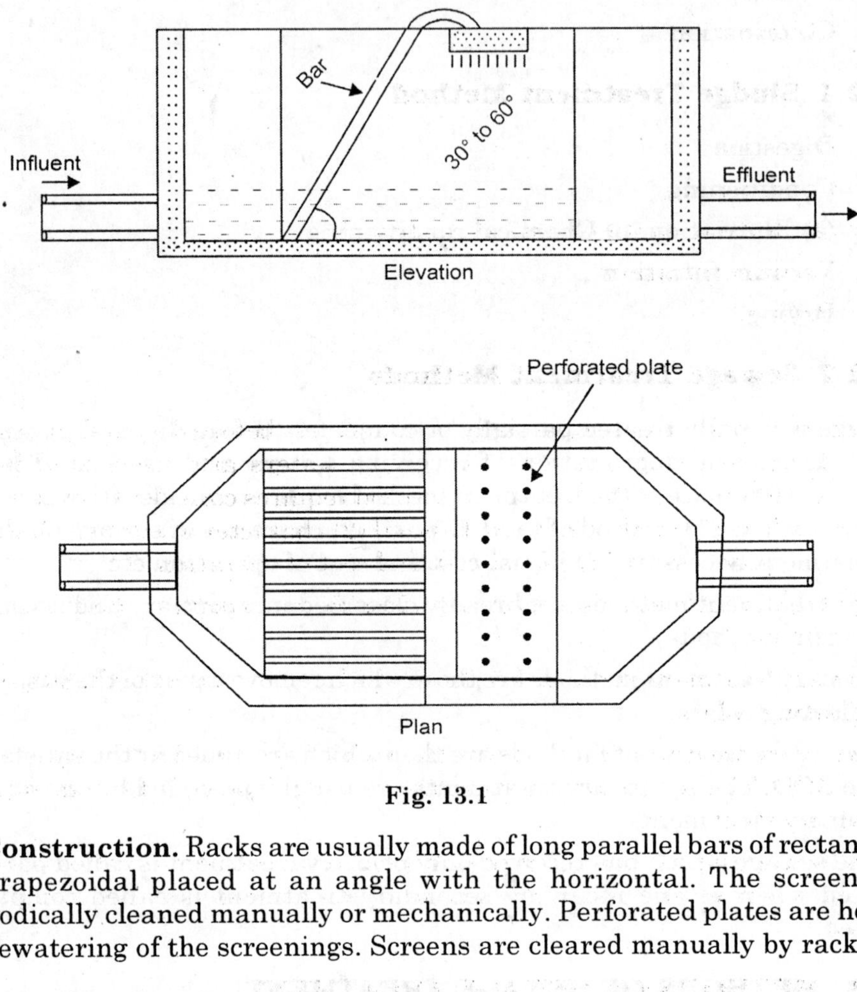

Fig. 13.1

Construction. Racks are usually made of long parallel bars of rectangular or trapezoidal placed at an angle with the horizontal. The screens are periodically cleaned manually or mechanically. Perforated plates are helpful in dewatering of the screenings. Screens are cleared manually by racks.

Design considerations

1. Size of openings—0.6 cm to 5 cm.
2. Angle of inclination—30° to 60° with the horizontal.
3. Approach velocity of flow in the screen chamber—not less than 60 cm/sec for grit bearing sewage and not less than 30 cm/sec for other sewages.
4. Velocity of flow through screen openings (projected area of screen—not more than 30 cm/sec at the average flow.
5. Velocity normal to screen surface—not more than 30 cm/sec.
6. Maximum head loss through the clogged racks—75 cm.
7. Width of the screen chamber is same as the width of screen. Length of

screen chamber should be sufficient to prevent formation of eddies around the screen. The width of the screen hamber will be increased gradually to the required value.

8. Quantity of screenings = about 2.25 to 35 litres/1000 m^3 of sewage.
9. Total weight of screenings is about 6 kg/litre.

13.5 HEAD LOSS THROUGH SCREENS

According to O. Kirschmer (1926), the head loss through racks when the rack is clean is

$$h = \beta\left(\frac{w}{b}\right)^{4/3} h_v \sin\theta$$

where h = Head loss in cm

w = Max. width of bars faceing flow in cm

b = Min. width of clear opening in cm

h_v = Velocity bead in cm in approach channel on the basis of face velocity

$$= \frac{v^2}{2g}$$

θ = angle of rack with horizontal

β = bar shape factor

β = 2.42 for sharp-edged rectangular bars

= 1.79 for circular bar

= 1.83 for rectangular bars with semi-circular upstream face

= 1.67 for rectangular bars with semi-circular upstream and downstream faces

Example 13.1 *Design a rack chamber for an average flow of 450 litres/sec. Number of openings is one more than that of bars. The depth of flow is 75 cm. Assume clear opening between bars is 2.5 cm and section of the bar as 6.0 mm × 25 mm.*

Solution

Assume—slope of bar with horizontal = 30°

velocity through projected area of screen = 30 cm/sec.

Flow = 450 litres/sec = 450,000 cm^3/sec.

$$\text{Submerged projected area of opening} = \frac{450,000}{30} = 15,000 \text{ cm}^2$$

Depth = 75 cm

$$\text{Clear width} = \frac{15,0000}{75} = 200 \text{ cm}$$

Width of opening between bars = 2.5 cm

$$\text{No. of openings} = \frac{200}{2.5} = 80$$

No. of bars = 79

Assume thickness of bar = 6 mm

Total with of bar chamber = 200 + (79) (0.6)

= 200 + 47.4 = 247.4 cm

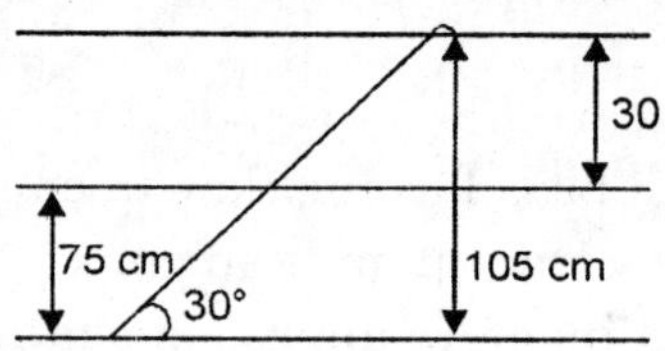

Fig. 13.2

Assume free board = 30 cm

Total depth = 75 + 30 = 105 cm

$$\text{Minimum length of bar} = \frac{105}{\sin 30°} = \frac{105}{1/2} = 210 \text{ cm.}$$

13.6 RACK CHAMBER DESIGN

Conditions involved in rack chamber design are (1) Rate of flow Q. (2) Dimensions and spacing of the bars (3) The velocity of flow normal to the screen V_s (4) Velocity of flow through the screen chamber V_c (5) Width of screen spacing W_s (6) Width of screen chamber W_c (7) The angle of screen with the horizontal α (8) Submerged length of the screen l_s (9) Depth of flow in screen chamber.

All of these factors are interrelated and assumptions of the magnitude of a few may fix the magnitudes of the others.

13.7 DESIGN CRITERIA

1. Clear openings — 2.5 cm to 5 cm
2. Inclination of bars — 30° to 60° with the horizontal.
3. Velocity normal to the screen openings —30 cm/sec at average rate of flow and should not exceed 75 cm/sec at maximum flow.

4. Horizontal velocity through the screen chamber—not less than 60 cm/sec for grit bearing sewage and not less than 30 cm/sec for other sewages.
5. The submerged area of the surface of a screen including bars and openings should be about 200% of the cross-sectional area of the approach sewer for separate sewers and 300% combined sewer.
6. The screen channel invert shall be 7.5 to 15 cm below the invert of the incoming sewer.
7. The channel preceding and flowing the screen shall be filleted to prevent standing and sedimentation of solids.

Example 13.2 *Design a bar screen chamber through which maximum, average and minimum rates of flow are respectively 450, 225 and 90 litres per sec. There is one more space than there are bars. The outlet is controlled by a proportional flow weir, such that depth of flow is directly proportional to the rate of flow. The inclination of the bars are 30° with the horizontal and the dimension of bars normal to the flow is 6 mm. The clear opening between bars is 2 cm. The depth of flow at maximum rate of flow is 75 cm.*

Solution

Assume velocity normal to the screen surface = 30 cm/sec.

Submerged area of clear openings at maximum flow

$$= \frac{450{,}000}{30} = 15{,}000 \text{ cm}^2$$

$$\text{Depth} = 75 \text{ cm (given).}$$

$$\text{Length of screen} = \frac{75}{\sin 30}$$

$$\text{Width of clear openings} = \frac{15{,}000 \sin 30°}{75} = 100 \text{ cm}$$

$$\text{Size of opening} = 2 \text{ cm}$$

$$\text{No. of openings} = \frac{100}{2} = 50$$

$$\text{No. of bars} = 50 - 1 = 49$$

Total width of chamber = 100 cm + (49) (0.6) = 100 + 29.4 = 129.4 cm

Then the velocity of flow through the chamber at maximum flow

$$= \frac{450 \times 10^3}{129.4 \times 75} = \mathbf{46.4 \text{ cm/sec}}$$

Since the flow is controlled by a proportional flow weir and depth varies with the rate of flow the velocity of flow will remain same at all rates of flow.

13.8 SKIMMING TANKS

Purpose. Removal of oil and grease which are lighter than water. These are long trough shaped tanks. Air is blown from diffusers placed at tank floor to keep heavy solids away from settling. The longitudinal baffles flanking the central channel create side compartments which serve as stilling chambers where oil and grease accumulate at the surface. The effluent conduit rises from the bottom of the tank to an effluent channel.

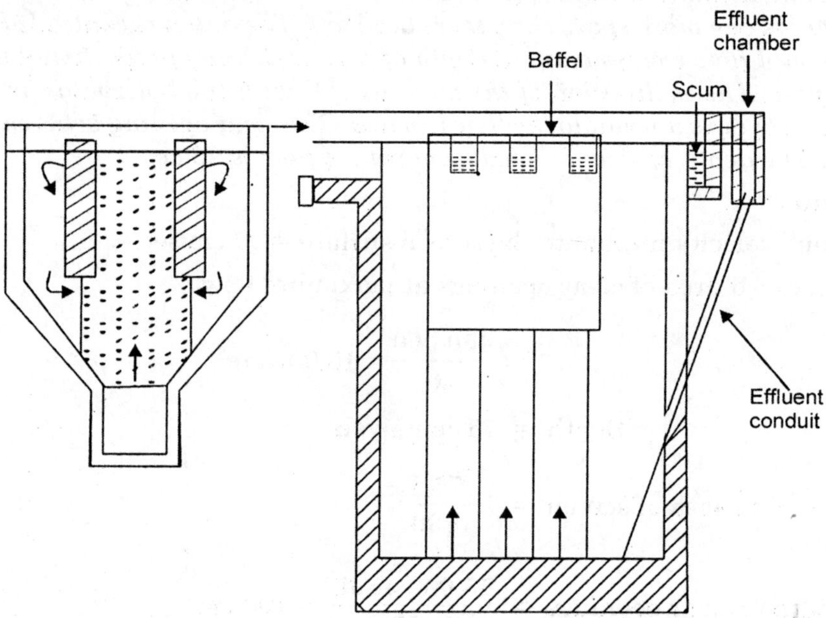

Fig. 13.3

Deign considerations

1. Detention period = 3 min.
2. Surface area A is given by

$$A = \frac{11.5\,Q}{v}$$

where A = Surface area in cm

Q = Rate of flow in m³/day

v = Rising velocity of grease in cm/sec

The value of v is in the neighbourhood of 0.47 cm/sec.

3. Air requirement = 0.22 litres/litre of sewage
4. L : W = 2 : 1 or $1\frac{1}{2}:1$
5. Volume of skimming = 0.75 cc to 45 cc per m^3 sewage
6. Addition of chlorine about 13 mg /lit of sewage increases removal of grease 2 to 4 times.

Example 13.3 *Design a skimming tank, the following data are given. Row of sewage = 250 litres/sec, detention period = 3 min and smallest rising velocity = 0.45 cm/sec.*

Solution

Flow of sewage

$$= 250 \text{ litres/sec}$$

$$= \frac{250}{1000} \times 60 \times 60 \times 24 \text{ m}^3\text{/day}$$

$$= 21{,}600 \text{ m}^3\text{/day}$$

Detention period = 3 min.

Vol. of skimming tank = 250 × 3 × 60 litres

= 45,000 litres.

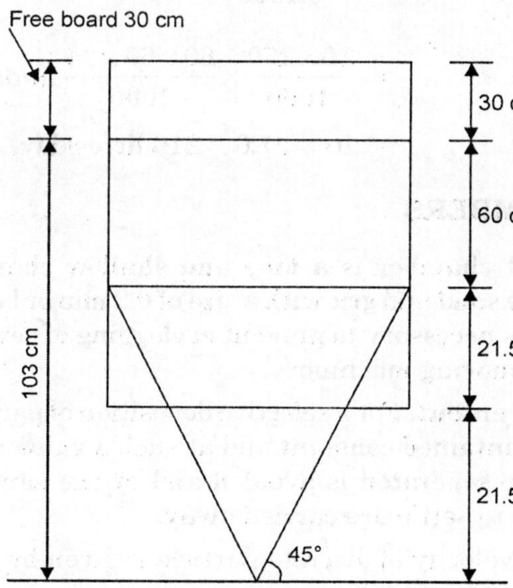

Fig. 13.4

Smallest rising velocity = 0.45 cm/sec

$$\text{Surface area} = \frac{10 \times 250}{1000} \times \frac{60 \times 60 \times 24}{1000} \text{ m}^2$$

$$= 5.52 \times 10^5 \text{ cm}^2$$

$$\text{Straight depth} = \frac{45000 \times 1000}{5.52 \times 10^5} \text{ cm} = 81.5 \text{ cm}$$

$$\text{Assume } L : W = 2 : 1$$

$$\text{Surface area} = 2\,W^2 = 5.52 \times 10^5 \text{ cm}^2$$

$$= 5.52 \times 10^4 \text{ m}^2$$

$$W^2 = 27.6 \times 10^4$$

$$W = 5.25 \times 10^2 = 525 \text{ cm}$$

$$= 5.25 \text{ cm}$$

$$L = 1050 \text{ cm} = 10.50 \text{ m}$$

$$\text{Air required} = 0.22 \times 250 \text{ litres/sec}$$

$$= 55.0 \text{ litres/sec.}$$

Assuming rate of skimming as 10 cc per m^3 of sewage

Volume of skimmings

$$= 10 \times \frac{250}{1000} \text{ cm/sec}$$

$$= \frac{10 \times 250}{1000} \times \frac{60 \times 60 \times 24}{1000} \text{ lit/day}$$

$$= 10 \times 21.6 = 216 \text{ litres/day.}$$

13.9 GRIT CHAMBERS

Description. A grit chamber is a long and shallow channel designed to separate from sewage sand and grit with a size of 0.2 mm or larger in diameter. The removal of grit is necessary to prevent of clogging of sewers and damage of pumps and other moving machinery.

Essential Requirement. For a selective deposition of particles the velocity of flow should be maintained constant and at such a value that the scouring of the particles to be separated is avoided and at the same time the finer particles not wanted to settle are carried away.

Theory. Settling velocity of discrete particle is given by,

$$v_s = \sqrt{4/3g/c_D(S_s - 1)d}$$

where v_s = settling velocity of particle of diameter d.

g = acc. due to gravity

C_d = drag co-efficient

S_s = sp. gr. of the particle

$$C_D = \frac{24}{R} + \frac{3}{\sqrt{R}} + 0.34$$

where R = Reynold's number = $\frac{vd}{v}$

where V = Kinematic viscosity = $\frac{\mu}{\rho}$

where μ = absolute viscosity

ρ = mass density

Note. Sp. gr. of sewage solids = 1.00 to 1.2.

Settiling velocities in cm/sec at 50°C or 283°K (10°C)

Diameter mm	1.0	0.5	0.2	0.1	0.05	0.01	0.005
Quartz sand	12.7	7.18	2.28	0.676	0.169	0.0085	0.0169
Coal	4.23	2.12	0.718	0.212	0.0423	.00212	.00004
Sewage	0.042	085–	.00004–	.00004–	–.0212	.000035	.00021
Solids	3.38	1.69	.0057	.085			

13.10 SCOURING OF DEPOSITED PARTICLE

The scouring velocity V that will start scouring of a deposited particle of diameter d is

$$V_0 = \sqrt{\frac{8k}{f} g (S_s - 1)\, d}$$

where K=0.04 unigranular sand

= 0.06 nonuniform sticky materials

f = friction factor = 0.03 (normally)

g = acc. due to gravity

S_s = Sp gr. of sewage solids

d = diameter

Using k = 0.04 and f = 0.3

$$V_0 = 3.27\sqrt{g(S_s - 1)\, d} = 3.3\sqrt{g(S_s - 1)d}$$

13.11 SURFACE AREA LOADING

$$\frac{Q}{A} = v_s$$

where Q = m³/sec

A = cm²

v_s = cm/sec

t = time taken to reach the bottom = $\frac{h}{v_s}$

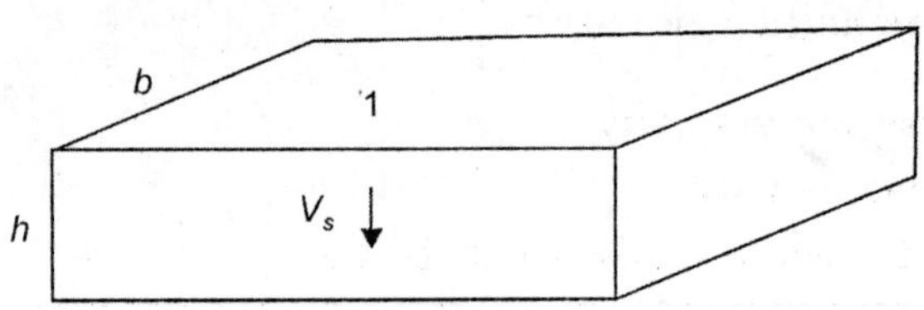

Fig. 13.5

$$\text{Vol. of tank} = Q.t = Q\frac{h}{v_s} = Ah$$

Considering inadequate performance of grit chamber half the above value is adopted i.e.,

$$\frac{Q}{A} = \frac{1}{2}v_s$$

The settling velocity of 0.2 mm particles at 10°C (i.e., 50°F) 2.1 cm/sec.

Surface area loading = 2.1 cm ³/sec/cm³

= 1800 m³/day/m²

Assuming $K = 0.06$ and $f = 0.03$ and $S_s = 2.65$

Scouring velocity is

$$V_0 = \sqrt{\frac{(8)(0.06)}{0.03}(981)(2.65-1)(0.02)} \text{ cm/sec.}$$

$$= 22.6 \text{ cm/sec}$$

Then the channel velocity should not exceed this velocity.

Design consideration of grit chamber

1. Usually sand and grit particles with a size of 0.2 mm or larger are to be deposited in grit chambers

2. Horizontal channel velocity should not exceed scouring velocity for the particle to be deposited and the channel velocity should be kept constant.

The scouring velocity V_0 that will start scouring or resuspension of deposited particles with size d is

$$V_0 = \sqrt{\frac{8k}{f} g(S_s - 1)d}$$

Example 13.4 *Determine dimensions of a grit chamber consisting of two channels of which the overflow is controlled by a proportional flow weir.*

Max. rate of flow = 800 litres/sec

Average rate of flow = 400 litres/sec

Mm. rate of flow = 200 litres/sec

Mm. size of particles to be removed = 0.2 mm

Sp. gr. of the particles = 2.65. Quantity of grit expected = 30 litres/1000 m³ of sewage.

Cleaning interval = 7 days.

Solution

1. *Surface area loading*

Settling velocity of the particle at 50° F from table v_s = 2.28 cm/sec.

$$\text{Surface area loading } \frac{Q}{A} = v_s = 2.28 \text{ cm/sec}$$

$$= 2.28 \text{ cm}^3/\text{sec/cm}^2$$

$$= 1970 \text{ m}^3/\text{day/m}^2$$

According to Imhoff and Fair because of Inadequacies of performance the surface area loading in grit chambers should be half of the above. Hence

$$\text{allowable surface area loading} = \frac{1970}{2} = 985 \text{ m}^3/\text{day}.$$

2. *Scouring velocity and channel velocity*

$$V_0 = 3.3\sqrt{g(S_s - 1)d}$$

$$= 3.3\sqrt{(981)(2.65 - 1)(0.02)}$$

$$[\because d = 0.2 \text{ mm}]$$

$$= 18.5 \text{ cm/sec.}$$

Let us use channel velocity = 18 cm/sec.

3. *Surface area required*

Consider one channel

Max. rate of flow Q = 400 litres/sec

$$= \frac{400}{1000} \times 60 \times 60 \times 24$$

$$= 34560 \text{ m}^2\text{/day}$$

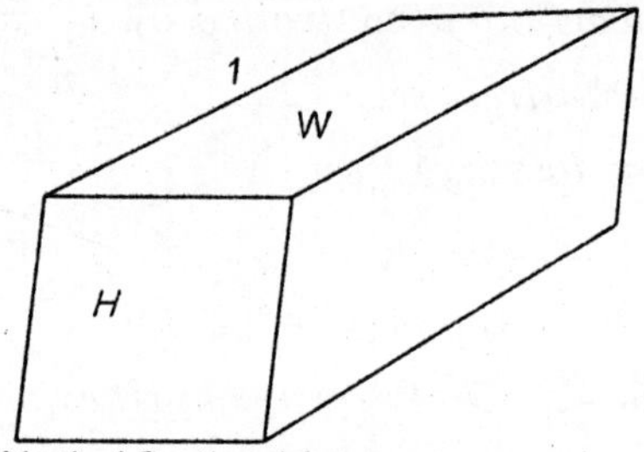

Fig. 13.6

$$\text{From step (1), } \frac{Q}{A} = 985 \text{ m}^3\text{/day/m}^2$$

$$\text{Suraface area } A = \frac{Q}{985} = \frac{34560}{985} = 35 \text{ m}^2$$

4. *Vertical sectional area*

h = height

w = width

l = length

$$\text{Vertical sectional area} = (wh) = \frac{Q}{V_c} = \frac{400 \times 1000}{18.0} = 22.2 \times 10^3 \text{ cm}^2$$

5. *Length, depth and width*

$$\text{Min. allowable, } \frac{l}{h} = \frac{V_c}{v_s} = \frac{18}{2.28} = 7.9$$

From step (3), surface area = $A = wl = 35 \times 10^4$ ...(1)

From step (4), vertical sectional area = $wh = 22.2 \times 10^3$ ... (2)

Dividing, $\frac{(1)}{(2)} = \frac{l}{h} = \frac{35 \times 10^4}{22.2 \times 10^3}$ = **15.8**

Assume depth of flow at maximum flow

$$h = 1\text{m}$$

Then length l (15.8) (h) = 15.8 m = 1580 cm.

Since vertical sectional area (wh) = 22.2 × 10^3 cm^2

$$w = 222 \text{ cm} = 2.22 \text{ m}$$

Check

$$\text{Volume of chamber} = 35 \times 10^4 \times 100 \text{ cm}^3$$
$$= 35 \times 10^6 \text{ cm}^3$$

$$\text{Detention period} = \frac{\text{Capacity}}{\text{Rate of flow}} = \frac{35 \times 10^6}{400 \times 10^3}$$
$$= 8.75 \times 10 \text{ sec}$$
$$= 87.5 \text{ sec}$$

Time taken by 0.2 mm particle to fall through $h = \frac{100}{2.28}$ sec.= 44 sec.

which is less than 87.5 sec. Hence all particles of size 0.2 mm and above must settle during the detention period.

$$\text{Velocity of flow} = \frac{Q}{wh} = \frac{400 \times 1000}{22.2 \times 10^3} = \textbf{18 cm/sec}$$

6. *Quantity of grit*

Quantity of flow per day = 400/litres/sec = 34560 m^3/day

$$\text{Volume of grit per day} = \frac{30}{1000} \times 34560 \text{ litres}$$
$$= 30 \times 34.56 = 1036.80 \text{ litres}$$

$$\text{Volume of grit in 7 days} = 1036.8 \times 7 \text{ litres}$$
$$= 7257.6 \text{ litres}$$
$$= 7.2576 \text{ m}^3$$

Hence depth required for grit storage

$$= \frac{\text{Volume of grit}}{\text{Surface area}}$$

$$= \frac{7.2576}{35} \text{ m} = \frac{725.76}{35} \text{ cm}$$

$$= 20.7 \text{ cm} = 21.0 \text{ cm}$$

Since the over flow is controlled by proportional flow weir, the depth of flow will vary with the rate of flow and the velocity of flow will remain same at all rates or flow.

Example 13.5 *A grit chamber 18 m long and 1.2 m wide treats a maximum sewage flow of 200 litres per sec. The grit chamber is controlled by a proportional flow weir. The depth of flow is 1.2 m at the maximum discharge. Determine the minimum size of the sand particles of the sp. gr. 2.65 that may be separated from sewage in the grit chamber.*

Solution

Surface area of the grit chamber = $18 \times 1.2 \text{ m}^2 = 21.6 \text{ m}^2$

$$= 21.6 \times 10^4 \text{ cm}$$

Surface area loadmg = $Q/A = \dfrac{200 \times 1000}{21.6 \times 10^4}$ cm³/sec

$$= 0.953 \text{ cm}^3/\text{sec/cm}^2$$

Expected settling velocity of smallest particle $u_s = 2(Q/A)$

$$= 0.953 \times 2 \text{ cm/sec}$$

$$= 1.906 \text{ cm/sec.}$$

Vertical cross section at maximum descharge = width × depth

$$= (1.2)\,(1.2)$$

$$= 1.44 \text{ m}^2$$

$$= 1.44 \times 10^4 \text{ cm}^2$$

$$\text{Velocity of flow} = \frac{Q}{\text{Cross sectional area}} = \frac{200 \times 1000}{1.44 \times 10^4} \text{ m/sec}$$

$$= 13.9 \text{ cm/sec}$$

Particle size

(1) Particle size corresponding to the settling velocity v_s equal to 1.906 cm/sec = obtained by trial and error form

$$v_s = \sqrt{\frac{4}{3}\frac{g}{C_D}(S_s - 1)d}$$

where $C_D = \dfrac{24}{R} + \dfrac{3}{\sqrt{R}} + 0.34$

and $R = \frac{vd}{\upsilon}$

At 283° K (10°C) $\upsilon = 1.3101$ centistoke

$= 1.3101 \times 10^{-2}$ cm²/sec

Assume $d = 0.15$ mm

$$R = \frac{1.906 \times 0.015}{1.3101 \times 10^{-2}} = 2.18$$

$$C_D = \frac{24}{2.18} + \frac{3}{\sqrt{2.18}} + 0.34 = 13.36$$

$$v_s = \sqrt{\frac{4}{3} \times \frac{981}{13.36}(2.65-1)(0.015)}$$

$$= \sqrt{2.42} = 1.56 \text{ cm/sec}$$

which is less than 1.906 cm/sec

Assume $d = 0.17$ mm

$R = 2.47$

$C_D = 11.98$

$v_s = 1.75$ cm/sec

which is less than given value.

Assume $d = 0.19$ mm $= 0.019$ cm

$R = 2.76$

$C_v = 10.85$

$v_s = 1.94$ cm/sec

Say $d = 0.185$ mm

Hence the surface area loading corresponds to settling of particles of size 0.185 mm and above

(2) Particle size corresponding to the scouring velocity,

$$V_0 = 3.3\sqrt{g(Ss-1)d}$$

whence $g(S_s - 1)d = \left(\frac{V_0}{3.3}\right)^2$

i.e. $d = \frac{V_0^2}{(3.3)^2 g(S_s - 1)} = \frac{V_0^2}{10{,}650(S_s - 1)}$

Velocity of flow $= 13.9$ cm/sec (calculated)

$$d = \frac{(13.9)^2}{(10650)(1.65)} = \frac{193}{17,600} = 0.11 \text{ mm}$$

which mean that the velocity of flow in the chamber will not suspend the particles of size greater than 0.11 mm.

Check for detention period

$$\text{Detention period} = \frac{\text{Capacity}}{\text{Rate of flow}} = \frac{1800 \times 120 \times 120}{200 \times 10^3} \text{ sec}$$

$$= 129.5 \text{ sec.}$$

Time taken by particles of size 0.185 mm through the depth of 1.2 m

$$= \frac{120}{1.906} \text{ sec} = 63 \text{ sec}$$

Hence the detention period is sufficient for settling of the particles of size 0.185 mm.

Thus min. size that may be separated = **0.185 mm**

13.12 INLET AND OUTLET DEVICES FOR GRIT CHAMBER

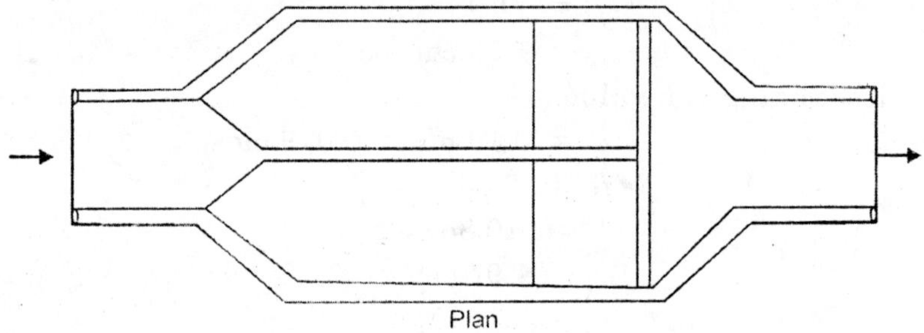

Plan
Inlet and outlet devices

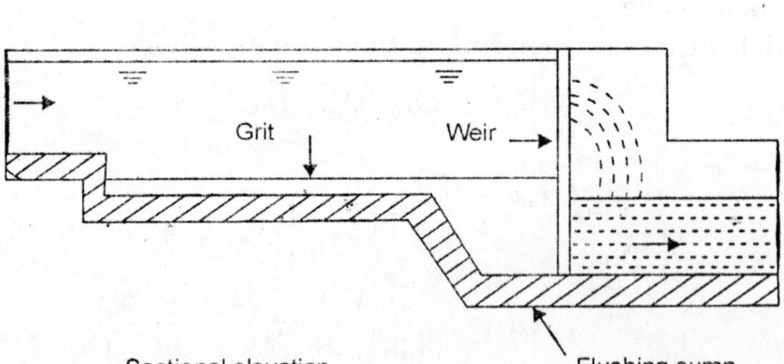

Fig. 13.7

The inlet pipe should be placed on the centre line of the grit chamber for the best results.

The outlet device should such that a constant velocity of flow at all rates of flow must be maintained. This is an essential requirement in grit chamber design. Such control of velocity can be achieved by (i) Proportional flow weir (ii) Control flume and (iv) Adjustable throat.

13.13 PROPORTIONAL FLOW WEIR

It is a combination of a weir and an orifice. The bottom of the orifice which comprises the weir is straight and horizontal, whereas sides of the orifice are so curved that the flow through the orifice is directly proportional to the head over the weir. This device necessitates the loss of head of the full depth of water over the crest of the weir.

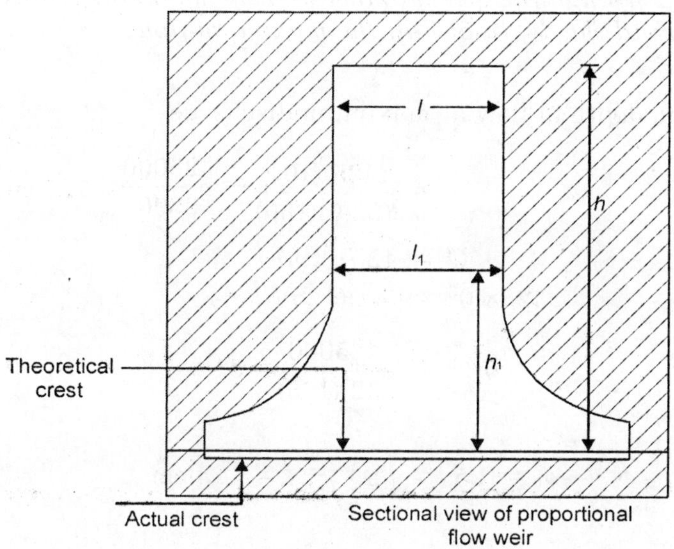

Fig. 13.8

Flow through such an orifice is given by

$$Q = 42\ lh^{3/2}$$

$$Q = 42\ lh^{3/2} = 42\ (lh^{1/2})\ (h)$$

where Q = rate of flow through the orifice in cm³ sec

l = width of orifice at the surface of sewage

h = height of water above the crest in cm.

For Q to be proportional to depth h, $lh^{1/2}$ must be a constant.

Design procedure

(i) With given velocity of flow through the grit chamber and the dimensions of the grit chamber, find max. depth of flow, h.

(ii) Compute the value of constant
$K = lh^{1/2}$ from $Q = 42\ (lh^{1/2})\ h$

(iii) Determine width of orifice l_1, l_2, l_3 etc. at heights h_1, h_2, h_3, etc. from $K = l_1\ h_1^{1/2} = l_2\ h_2^{1/2}$.

(iv) For $h = 2.5$ cm or less the side curves are terminated vertically to the weir crest, without any allowance for the area cut off, as it is negligible. If the side curves are terminated at h equal to more than 2.5 cm the actual weir crest should be lowered below the theoretical level to compensate for the area cutoff.

Example 13.6 *Determine the shape of the proportional flow weir for a grit chamber 1356 cm long and 150 cm wide. The designed velocity of flow is 22.6 cm/sec through the grit chamber and the maximum rate of flow is 425 litres/sec. Find the depth at half the maximum flow.*

Solution

Maximum depth of flow at maximum discharge

$$= \frac{(425)(10^3)}{(22.6)(150)} = \frac{425000}{3390} = 125 \text{ cm}$$

$$Q = 42(lh^{1/2})(h)$$

i.e. $$425\times10^3 = (42)(125)(lh^{1/2})$$

i.e. $$lh^{1/2} = \frac{425000}{(42)(125)} = 81$$

at $$h = 125, \quad = l = \frac{81}{\sqrt{125}} = 7.25 \text{ cm}$$

$$h = 100, \quad = l = \frac{81}{\sqrt{100}} = 8.1 \text{ cm}$$

$$h = 81, \quad = l = \frac{81}{\sqrt{81}} = 9 \text{ cm}$$

$h = 70$ cm $\quad l = 9.7$ cm

$h = 60$ cm $\quad l = 10.4$ cm

$h = 50$ cm $\quad l = 11.5$ cm

$h = 25$ cm $\quad l = 16.2$ cm

$h = 2.5$ cm $\quad l = 51.3$ cm

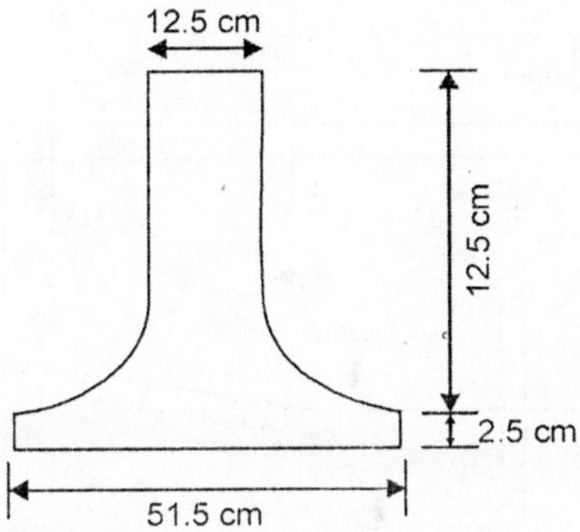

Fig. 3.9

Let the length of the weir be 51.5 cm and the sides are straight verticals for $h = 2.5$ cm

$$Q = 42(lh^{1/2})\ h = (42)\ (81)h = (3402)h$$

when
$$Q = \frac{425,000}{2} = 212,500 \text{ cm}^3\text{/sec}$$

$$h = \frac{212,5000}{3402} = 62.5 \text{ cm.}$$

13.14 PLAIN SEDIMENTATION OF SEWAGE

Plain sedimentation of sewage solids are effected in settling tanks where the sewage solids settle under gravity.

13.15 TYPES OF SETTLING TANKS

Settling tanks differ in size and shape, in arrangements of inlets and outlets and in the method of sludge removal.

They may be rectangular, square and circular in plan. The direction of flow through the tank may be horizontal or vertical and radial. Sludge can be removed manually, hydrostatically or mechanically.

Rectangular, horizontal flow tanks, hand cleaned with sludge hopper, sludge being with drawn hydrostatically.

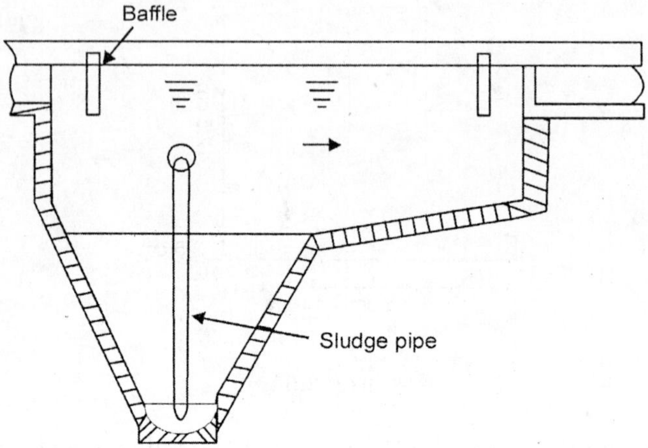

(a)

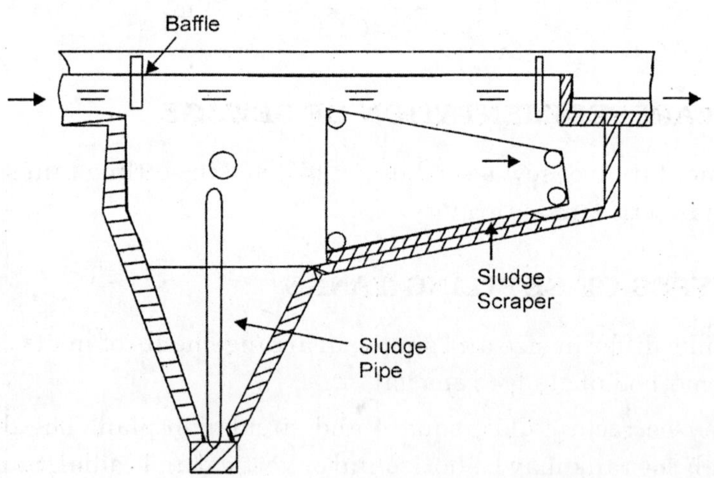

(b)

Horizontal flow, rectangular tanks with sludge scraper for continuous sludge removal. Sludge is withdrawn hydrostatically.

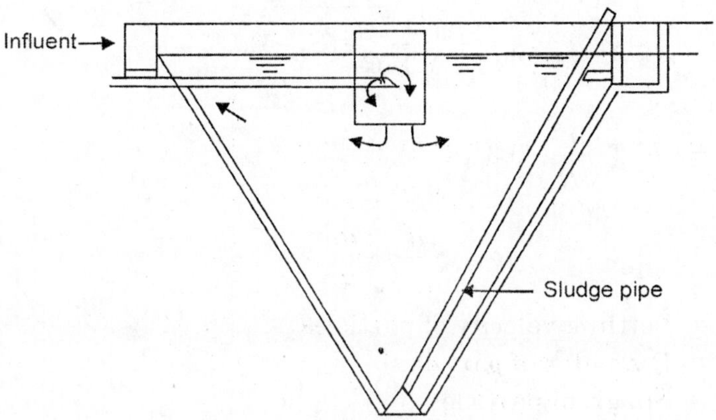

(c)

Vertical flow circular tank. Sludge is withdrawn hydrostatically.

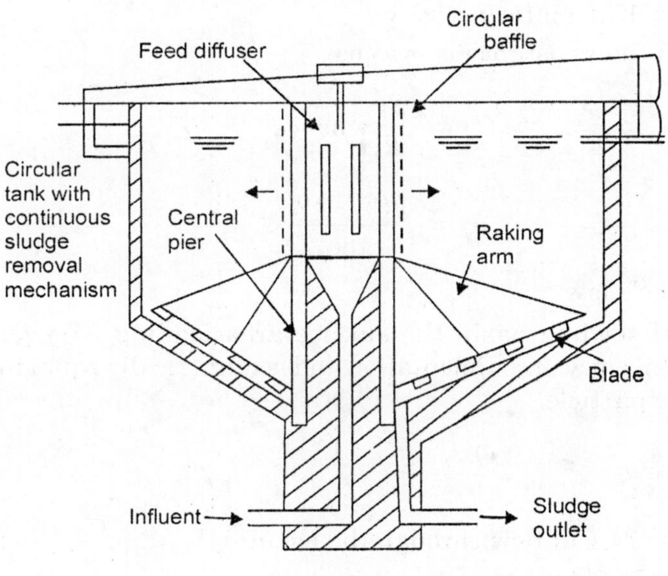

(d)

Fig. 3.10

13.16 PRINCIPLES OF SEDIMENTATION

Settling of suspended particles in an ideal basin is governed by gravitational force of particle and the drag force of the liquid resisting settlement. Rate of settling of granular discrete particles are given by

$$v_s = \sqrt{\frac{4}{3}\cdot\frac{g}{C_D}(S_s - 1)d}$$

where $C_D = \frac{24}{R} + \frac{3}{\sqrt{R}} + 0.34$

$$R = \frac{vd}{v} = \frac{vdp}{u}$$

where v = Settling velocity of particles
d = Diameter of particles
S_s = Sp. gr. of particle
g = Acc. due to gravity
R = Reynold's number
ρ = Mass density of liquid
μ = Absolute viscosity
v = Kinematic viscosity

For $R < 0.5$ the resistance is viscous

and $C_n = \frac{24}{R}$ whence

$$v_s = \frac{g(S_s - 1)d}{18v}$$

which is Stoke's law.

In an ideal settling basin the surface area loading is governing factor influencing efficiency of sedimentation and is numerically equal to the settling velocity of the particle.

i.e. $\frac{Q}{A} = v_s$

where Q = rate of flow through the basin
A = surface area.

Due to inadequacies of the tank performance due to turbulence, convection currents etc, 2/3 value of above is taken i.e., $\frac{Q}{A} = \frac{2}{3}v_s$

$$\text{Detention period} = \frac{\text{Tank volume}}{\text{Rate of flow}} = \frac{V}{Q}.$$

According to Imhoff and Fair.

Surface area loading ($m^3/day/m^2$)

$$= \frac{(24.1)\ (\text{tank depth in m})}{\text{Detention period (hr)}}$$

13.16.1 Design Considerations for Primary Settling Tanks

1. Detention period —1 to 3 hrs
2. Depth—3 m excluding sludge storage space. Mechanically cleaned basins as shallow as possible but not less than 2.1 m.
3. Surface area loading—24 tc 36 $m^3/day/m^2$
4. Velocity of flow—about 30 cm/min.
5. $\frac{\text{Length}}{\text{Width}}$ should be as great as possible and not less than 5 : 1 or 4 : 1
6. Maximum width of flowing through channel—6 m to 7.5 m
7. Overflow rate on out let weir—should not exceed 186 m^3 length/day/ metre/length/preferably between 62 and 124 m^3/ day/ metre.
8. Slope of the bottom of the basin towards sludge hopper—1 to 2%
9. Slope or sides of sludge hoppers:
   Min. 1.7 (V) : 1 (H) for Pyramidal hopper
   and Min. 1.5 (V): 1 (H) for conical hoppers.
   The floor of the hopper should not be wider than 0.6 m.
10. The minimum distance between outlet and in let is 3 m.

13.17 INLET AND OUTLET DEVICES

13.17.1 Inlet Arrangements

1. Inlet pipes evenly spaced across the end of basin discharging horizontally against hanging baffles the pipes being submerged 30 to 90 cm below the surface.

2. Perforated baffle walls.

3. Over flow weir—most common.

In radial flow hopper bottom settling tanks, the inlet should be in the centre at least 1.8 m below the water surface and 0.5 m above the bottom of the hopper.

Outlets are designed as overflow weir or troughs, sometimes called launders. In horizontal flow rectangular tanks the outlets are placed opposite the inlet and in circular radial flow tanks, outlet is located on the periphery with inlet at the centre of the tank.

The minimum distance between outlet and inlet of a basin is 3 m.

13.17.2 Baffle

Baffles are placed 0.6 to 0.9 m in front of the inlet and submerged 45 cm to 60 cm.

Baffles are also provided in front outlet weir to held back floating solids, grease and oil.

If the length required of the outlet trough is greater than the troughs may be arranged as shown in Figs. 13.11 and 13.12.

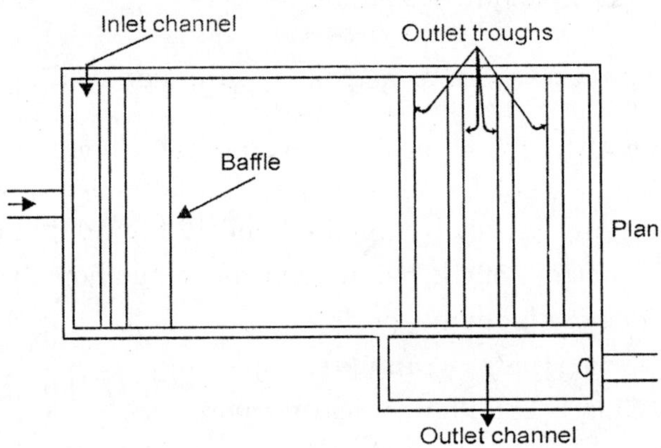

Fig. 13.11

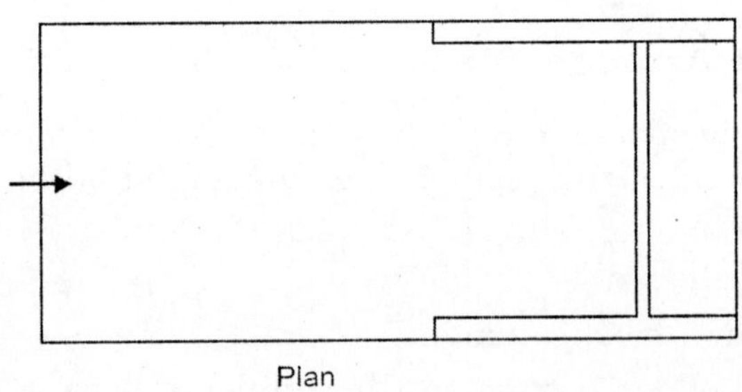

Fig. 13.12

13.17.3 Removal of Seum in Sedimentation Basin

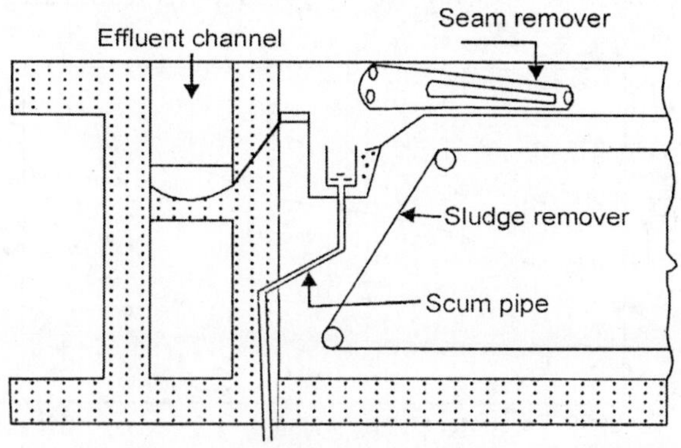

Fig. 13.13

13.18 SETTLING EFFICIENCY

Settling efficiency in plain sedimentation tank depends on the nature and concentration of suspended solids and period of detention.

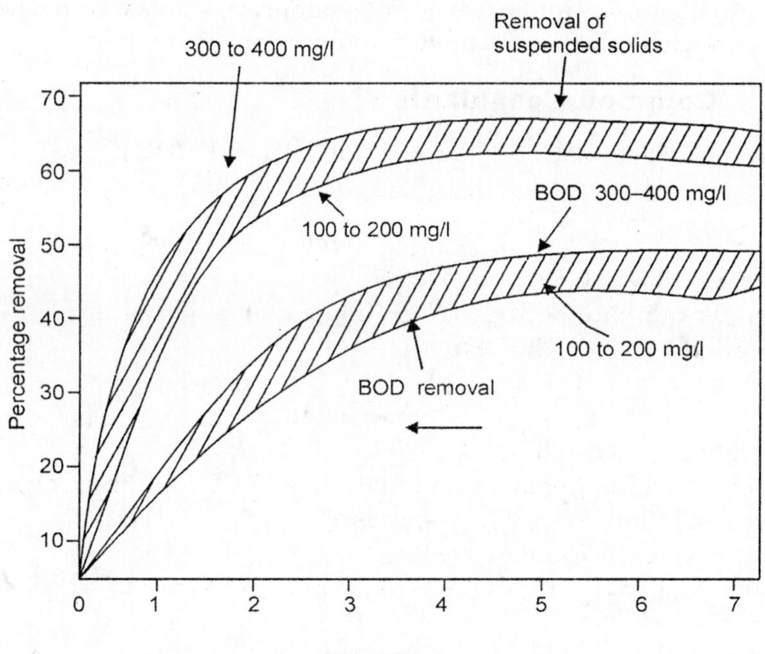

Fig. 13.14

The removal of suspended solids in 2 hours and varies from 45% for weak sewage having less than 100 mg/1 of suspended solids to 65% for domestic sewage having more than 300 m/1 solids.

The corresponding BOD removals are 25 to 35%.

The removal of BOD is dependent on surface area loading also.

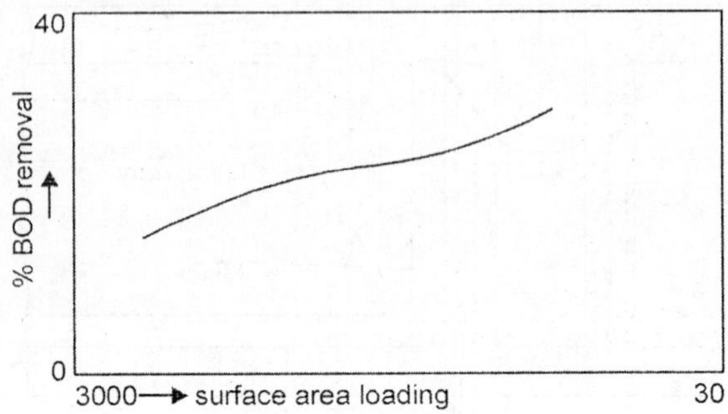

Fig. 13.15

13.19 CHEMICAL PRECIPITATION OF SEWAGE

Purpose. To remove finely divided colloidal matters from sewage.

Process. Some chemical when added to water react with one another and with sewage substances and produce flatulent precipitates which when settle combine with each other, grow in size and are removed quickly.

13.19.1 Common Coagulants

1. Ferric Chloride $FeCl_3 \rightarrow FeCl_3 + Ca(OH)_2 \rightarrow Fe(OH)_3$ ($Ca(OH)_2$: Lime; $Fe(OH)_3$: Floc)

   pH range 3.5 to 6.5 and above 85.
2. Ferrous sulphate $FeSO_4 . 7H_2O$ (Copperas) $\rightarrow FeSO_4 . 7H_2O + Ca(OH)_2 + 0_2 \rightarrow Fe(OH)_3$ ($Fe(OH)_3$: Floc)

   pH range — above 8.5.
3. Chlorinated Copperas – Combination of $FeCl_3$ and $Fe_2 (SO_4)_3$.
4. Ferric sulphate $Fe_2 (SO_4)_3$ —ferrisul.

   Ph range 3.5 to 7 and above 9.
5. Alum. $A1_2(SO_4)_3 . 18H_2O \rightarrow Al_2 (SO_{4)3} . 18H_2O + Ca(OH)_2 \rightarrow Al(OH)_3$ ($Al(OH)_3$: Floc)

Lime and sulphuric acid for pH adjustment of these iron compounds are used most economically. Lime is usually used as auxiliary chemicals along with main coagulants.

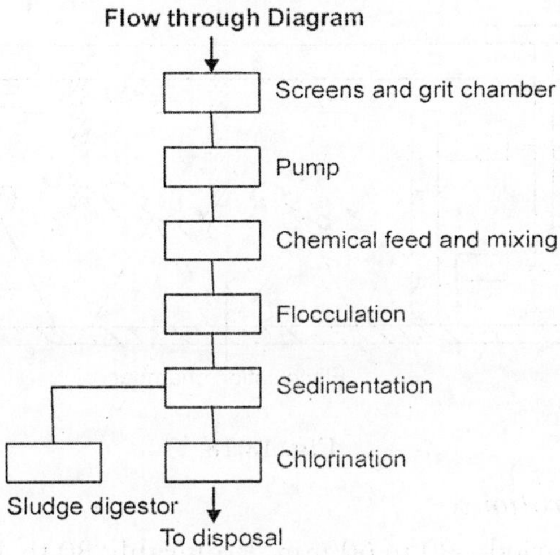

Fig. 13.16

Dose of Coagulant. It can be best determined experimentally by jar test.

Feeding Device. Mixing is done in flush mixers in which the sewage is violently agitated by paddles driven by electric motors.

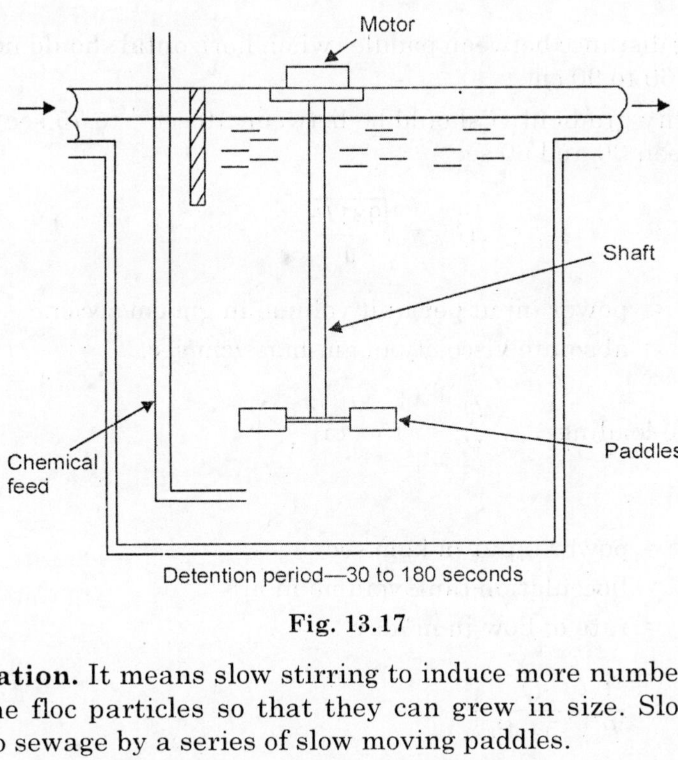

Fig. 13.17

Flocculation. It means slow stirring to induce more number of contacts between the floc particles so that they can grew in size. Slow motion is imparted to sewage by a series of slow moving paddles.

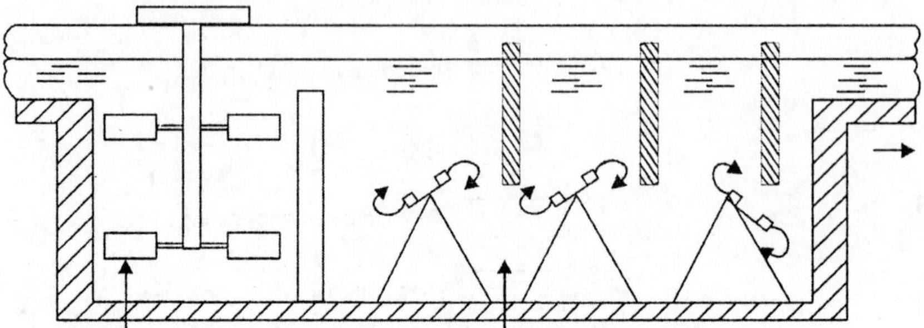

Fig. 13.18

Design Considerations

1. Detention period—20 to 60 mm. Preferably 30 to 40 min.
2. Velocity of flow through the flocculation basin = 0.9 to 1.5 m/min.
3. Peripheral speed of the paddles—45 to 75 cm/sec.
4. Velocity of liquid near the paddles is $\frac{1}{4}$th the speed of paddle.
5. Clear distance between paddles when horizontal should not be greater than 60 to 90 cm.
6. Velocity gradient G should be between 10 sec^{-1} to 75 sec^{-1} preferably between 30 and 60 sec^{-1}

$$G = \sqrt{\frac{981P}{\mu}}$$

where P = power input per unit volume in gm cm/sec/cm^3

and μ = absolute viscosity in gm-mass/cm/sec.

7. Volume loading $\frac{Q}{V} = 154.45\left(\frac{P'}{V}\right)^{1/2}$

where

P' = power input in kgm/sec.

V = flocculation tank volume in m^3

Q = rate of flow in m^3/day

or $$V = \frac{1}{P'}\left(\frac{Q}{154.45}\right)^2$$

In terms of paddle areas

$$V = \frac{Q^2}{1245 A\gamma^3 N^3}$$

where V = flocculation basin volume in m^3
Q = rate of flow in m^3/day
A = area of paddles in m^2
γ = distance between shaft and paddle centre
N = revolutions of shaft in r.p.m.

where sewage is flocculated by defused air then

$$V = \frac{Q^2}{7111\,\beta^3 q}$$

where V = volume in m^3
Q = rate of flow in m^3/day
q = air supply in litres/min
β = diffuser depth factor.

Diffuser Factors

Depth in m	1.83	2.13	2.44	2.74	3.05	3.35	3.66	3.96	4.27	4.57
β	0.27	0.29	0.30	0.32	0.33	0.35	0.36	0.37	0.39	0.40

Setting Basins

These are designed as primary settling tanks loading 45.5 m in 3 m deep tanks.

Efficiency

With a chemical dose of 30 mg/lit of ferric chloride removal of suspended solids —about 80% and removal of BOD about 65%. With double the amount of the chemical the % removal of suspended solids and BOD are 90% and 80% respectively.

PROBLEMS

13.1. (a) What is meant by "stabilisation of sewage"? Describe how carbonaceous, nitrogenous and sulphurous materials of sewage are stabilised in aerobic and anaerobic processes.

(b) If the colour of a sample in a relative stability test disappears in 4.5 days at a temperature of 37°C, what is its relative stability? In how many days would the colour disappear if the sample were stored at 20°C?

13.2. (a) Define 'relative stability' and 'population equivalent', in connection with sewage treatment.

(b) A town with a population of 10,000 is designing a sewage treatment plant to handle both the domestic and industrial wastes. There are three industrial plants, a dairy, a packing plants, and a cannery.

A sanitary survey revealed the followings :

Dairy plant—2.5×10^6 litres per day waste with BOD of 1200 mg/ lit.

Packing plant—3.0×10^6 litres per day waste with BOD of 500 mg/lit.

Cannery plant—5.0×10^6 litres per day waste with BOD of 250 mg/lit

Domestic sewage is 350 litres per capita per day with BOD of 80 mg per capita per day expansion factor of 20% is to be provided. Find (a) design population equivalent for the plant, (b) BOD loading and (c) Hydraulic loading.

13.3. (a) What do you mean by 'primary treatments' and 'secondary treatments' of sewage? Name the various methods of treatment under each of the above categories.

(b) Determine dimensions of a Rack Chamber through which maximum average and minimum rates of flow are respectively, 900, 450, and 200 litres per sec. There is one more space thin there are bars. The outlet is controlled by a proportional flow weir such that depth of flow is directly proportional to the rate of flow. The inclination of the bars is 30° with the horizontal and the dimension of the bars normal to the flow is 6 mm clear opening between the bars is 2 cm. The depth of flow at maximum rate of flow is 75 cm.

13.4. (a) What is the function of a grit chamber ? State the principles for designing a grit chamber.

Determine dimensions of a grit chamber consisting of two channels to separate particles of size 0.2 mm and above of sp. gr. 2.65. The settling velocity of 0.2 mm particles may be taken as 2.28 cm/sec. The average rate of flow is 400 litres/sec. maximum and minimum rates being 2 times and 0.5 times the average flow. The outlet is controlled by a proportional flow weir with a discharge co-efficient of 150 in c.g.s. units. The quantity of grit is expected to be 30 litres per 1000 m of sewage and cleaning interval is 7 days.

(c) Determine the share of the proportional flow weir.

13.5. (a) Determine the length, width, and depth of sedimentation basins for a flow of 6000 m^3 per day if the detention period is 1.5 hours, surface area loading is 36 m^2 per day per m^2 and length to breadth ratio is 4:1.

(b) What will be the length of the outlet weir assuming an over flow rate of 120 m per day per m?

(c) Calculate the depth of sludge if 60% of the raw solids of 200 mg/lit settle in the basins. Assume moisture content and sp. gr. of sludge as 95% and 1.02 respectively. The cleaning interval is 7 days.

13.6. Why is coagulation used? Describe the process. What are the common coagulants? State the reactions that occur when coagulants are added to sewage. How much solids and BOD may be removed by coagulation?

13.7. The suspended solids in a raw sewage are 270 mg/lit of which 90% is removed by sedimentation following chemical precipitation by addition of 54 kg of commercial alum containing 17 percent aluminium sulphate per million litres.

(i) How many litres of sludge containing 95% moisture and having a sp. gr. of 1.05 will be produced per million litres of sewage treated, if all, insoluble precipitates resulting from the coagulation are included in the sludge?

(ii) How many litres of sludge are formed by the added chemical alone. Assume that 100 gm of aluminium sulphate will produce 45.60 gm of aluminium hydroxide.

13.8. Define "aerobic" and "an aerobic" decomposition. Explain each briefly. Show the comparison of these two. Which process is useful for domestic purpose?

13.9. State the classification of sewer. Describe the design methods of sewer.

13.10. On what basis the sewage is classified? Write the classifications of sewage.

13.11. How many types of sewers are used? State their merits and demerits.

Chapter 14

Sewage Filtration

14.1 OBJECT

Oxidation of putrescible matter remaining offer the primary treatments. Oxidation is done by the aerobic bacteria and BOD is thus satisfied with the production of stable liquid.

14.2 METHODS OF FILTRATION

1. Intermittent sand filter
2. Trickling filters:
    (a) Low rate (standard)
    (b) High rate.

14.2.1 Intermittent Sand Filter

Description. An intermittent sand filter is a specially prepared bed of sand or other fine grained materials on which settled sewage is applied intermittently. The suspended solids are held by sand grains and acted upon by aerobic bacteria and other organisms, where by the putrescible matters are stabilised. The effluent from the filter is removed by a system of under drains. Intermittent application of sewage is effected by using dosing tanks and it permits penetration of air into the bed and maintains aerobic condition.

14.2.2 Construction of the Bed

It consists of a layer of sand of a specified size placed in a shallow cutting the bottom of which is made slopping towards the under drains The under drain consist of open-jointed tile-pipes placed in trenches below the sand bed.

14.2.3 Sand Bed

Thickness of sand bed—45 to 75 cm

Effective size of sand—0.20 to 0.5 mm

Uniformity co-efficient of sand—2 to 5

Area of bed—1000 to 4000 m^2

Shape — rectangular

Rate of filtration—4 to 8 litres/min/m^2

Underdrainage System. It consists of tile pipes laid in open joint. The underdrain is placed in a trench. It is surrounded by a gravel layer of 45 to 60 cm thickness of which 5 to 7.5 cm should be above the under drain. The size of the gravel is 2.5 cm to 5 cm. Above this layer there is a layer of 7.5 cm thickness of gravel of 6 mm to 25 mm and then a layer of 7.5 cm thickness of gravel having size less than 6 mm as shown in Fig. 14.1.

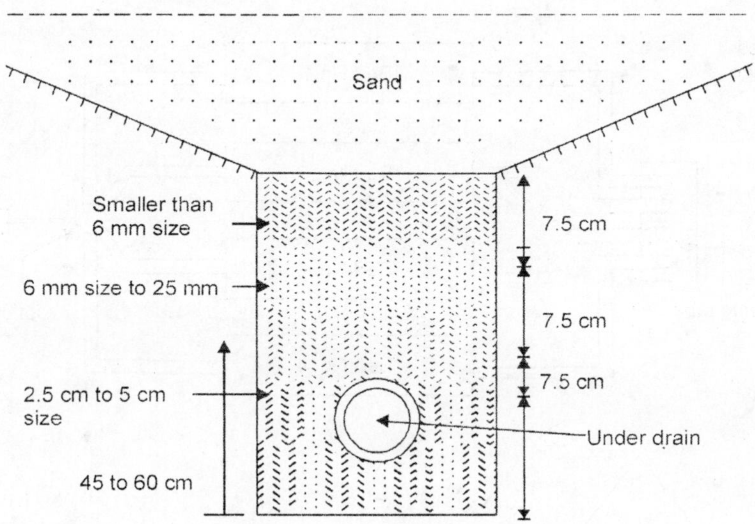

Fig. 14.1

1. Spacing of underdrains ≯ 4.5 m.

2. Slope of the underdrain—sufficient to develop a velocity of 90 to 120 cm/sec when flowing full.

3. Minimum size of under drain—15 cm in dia.

Distribution System. The influent is distributed on to a bed through a system of pipes terminating in outlets at the bed surface. The outlet discharge on a flat concrete slab or 45 to 60 cm diameter to prevent errosion of sand surface and spacing of such outlets is not more than 9 to 18 m. The piping system is designed for velocities of 30 cm/sec.

Intermittent dosage is effected by means of dosing syphons or by operating valves.

No. of doses —1 dose daily common, Max. upto 3 doses daily.

Depth per dose—2.5 to 7.5 cm.

Rate of dosing—61 litres/1000^2 m surface.

Time of disappearance of each dose on the bed—20 to 30 min.

Capacity of dosing tank—equal to one dose on a bed.

Operation. If septic condition prevails, the bed should be put off of operation for 1 to 4 weeks. If clogged the top 5 to 7.5 cm thick sand may be removed and replaced by new fresh sand.

Loading. American Society of Civil Engineers recommended the following hydraulic rates on the basis of the effective size of sand.

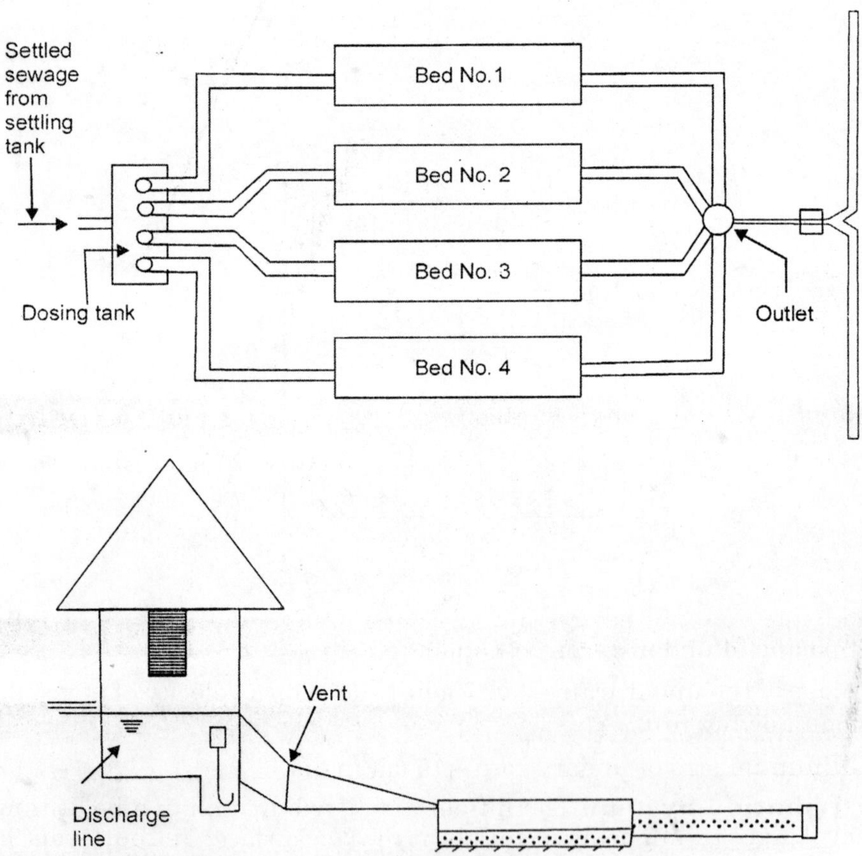

Fig. 14.2

Effective size	Hydraulic loading in litres/day/m^2		
	0.20 mm	0.3 mm	0.4 mm
1. Raw sewage	51.5	61.0	67.4
2. Septic sewage	75.0	93.5	112
3. Fresh settled sewage	112	131	150

BOD Loading. From studies in Florida by Grantham, Emerson and Henzy: 16.5 gm/day/m^2 for sand of effective size 0.25 mm and 18.5 gm/day/m^2 for sand of effective size 0.3 to 0.45 mm.

14.3 TRICKLING FILTERS

Description. A trickling filter is essentially a bed of crushed stone or gravel enclosed usually by a wall. Sewage is sprayed or sprinkled over the bed. It trickles down the filtering media; and is collected in an under drainage system which carries the effluent to final settling tanks also called humus tanks.

Theory. A sewage trickles down the surfaces of the filtering media in thin layers, there will soon develop, at the interface between the sewage and the contact media, growths of bacteria and other micro-organisms. They soon create slimy, gelatinous or zoological films on the surface of the medium. These films are inhabited by bacteria and other organisms. The finely divided substances colloids and even dissolved solids of the sewage that come in contact with these films are held and acted on by the micro-organisms present in the film. Through their life processes they oxidize the sewage solids and stabilisation is brought about. This process goes on continuously as long as sufficient food (organic substances) and oxygen are available for the bacteria and other micro-organisms. This process is sometimes called biological oxidation. The end products resulting from biological activities accumulate on the films which thus become thick and heavy and are dislodged by the flushing action of the trickling sewage. These sloughed off films appear in the effluent as humus like substances and need be removed in a settling tank. The purification effected by trickling filters is not due to filtration but due to contact.

The time required for the formation of the zoological film in a new bed is about 2 weeks.

The organisms which are present in the films are bacteria, protozoa and coliforms.

A thin transparent film is indicative of favourable condition and for efficient working of a filter requires for continuous unloading of worn-out film by sloughing off.

Temperature effects efficiency of trickling filters and BOD removal is more in summer months than in winter.

Pretreatment. Pre-sedimentation of the applied sewage is essential for good performance of filters. This is done in primary settling tanks with a detention period of 2 to 2.5 hours, surface area loading of 26.5 to 36.5 m^2/day/m^2 and depth 2.4 m to 3 m.

14.4 TYPES OF FILTERS

1. Low rate filters also called standard or conventional filters.
2. High rate filters.

Low rate filters are usually without recirculation of the filter effluent, though, sometimes recirculation is provided. *High rate filters* are characterised by almost continuous recirculation. The rate of filtration in high rate filters are much higher than that in low rate filters.

Recirculation. Recirculation is seldom used in standard trickling filter but is essentially used in high rate filters.

Advantages

1. Maintain continuous working of the filter regardless of fluctuations in influent sewage.
2. Freshens the sewage and reduces odour.
3. Seeds the filter with organisms and enzymes.
4. Increases flushing action which removes worn-out films.
5. Minimises filter fly breeding.
6. Equalizes and reduces loading to the filter.
7. Keeps self propelled distributors turning.

14.5 EFFECT OF RECIRCULATION ON EFFICIENCY

Let F = filter load in kg of BOD per day

L = additional load in recirculation flow in kg of BOD per day.

E = BOD in kg per day in final effluent after recirculation take off.

Let the removal of BOD is $\frac{2}{3}$ of applied loading and $\frac{1}{3}$ remains in the effluent. Then BOD in settled filter effluent = $\frac{F+L}{3}$

$$E = \frac{F+L}{3} - L = \frac{F}{3} - \frac{2L}{3}$$

which shows the effluent contains BOD loss by $\frac{2}{3}$ than that which would be there in case of no recirculation.

Amount of Recirculation. The recirculation flow may be upto 7 times the inflow.

14.5.1 Essential Features of Filter Construction

These are as below:

(1) Nature and depth of filtering material.

(2) Under-drainage and ventilation system.

(3) Method of distribution of sewage.

14.5.1.1 Filtering Material

Material. Crushed stone, gravel, blast-furnace slag, broken bricks, clinkers etc.

Requirements. Should be hard, weather resistant, strong enough to support its minimum weight.

Size. For low rate filter—6.25 to 8.25 cm uniform size. For high rate filter –100% passing 10 cm square mesh, 30 to 100% passing 8.75 cm and not more than 10% passing 5.0 cm mesh and not more than 2% passing 3.75 cm mesh.

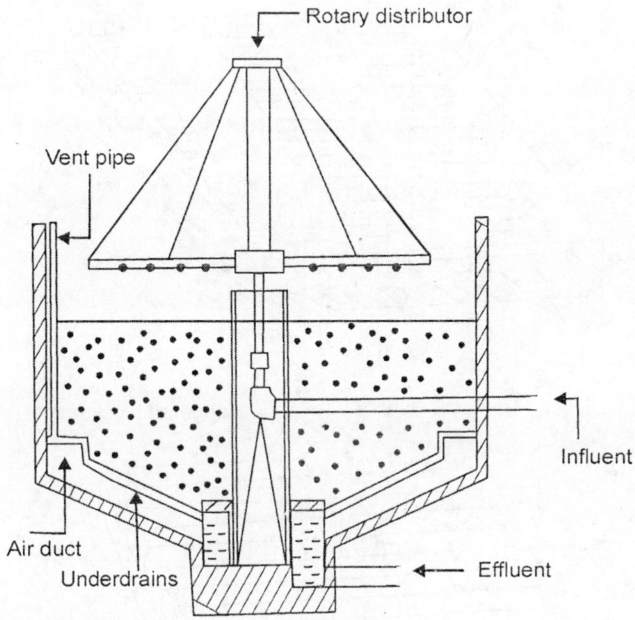

Fig. 14.3

14.5.1.2 Depth of Bed

Low rate — Varies between 1.5 m to 2.1 m
High rate — Varies from 0.9 m to 1.2 m
For Bio-filter, not less than 1.8 m

For Accelo-filter, 1.8 to 2.4 m for Aero filters.

14.6 UNDER DRAINAGE SYSTEM

14.6.1 Functions

1. Collection of effluent.
2. To permit passage of air upward and downward the bed — ventilation.

The system consists of laterals discharging into a main drain.

Lateral under drains generally consist of hall-round channels cast into concrete 1 floor. The channels are bridged by a false floor which supports the filtering materials above. The false floor may be constructed of (1) Perforated concrete slabs, (2) narrow concrete or brick blocks which are staggered in position above adjacent channels.

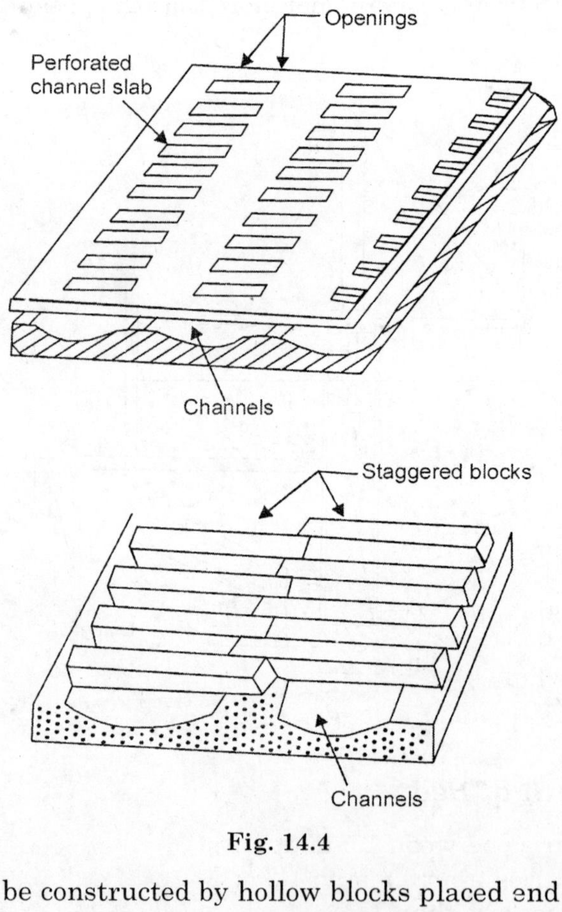

Fig. 14.4

Laterals may be constructed by hollow blocks placed end to end so that a long channel is formed, the entry of water being through the slots on the top of the block.

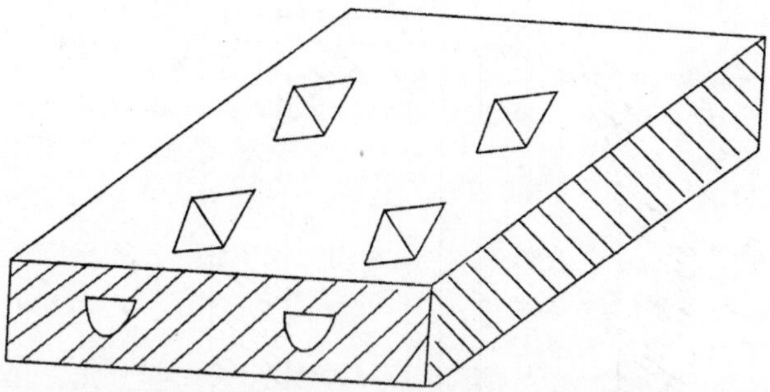

Fig. 14.5

(i) The slope of invert of laterals towards main drain is 1 to 2%.

(ii) The laterals should be designed to flow half-full and minimum velocity is 60 cm/sec.

(iii) Total inlet area into under drainage should be at least 5% of filter area.

Main-drain. In circular filter main drains are made parallel to diameter and placed slightly off the centre to avoid central column. It may run along the periphery also. It should be designed to flow partly full at a velocity of 60 to 90 cm/sec.

The outlets of laterals and main drain should not be submerged in order to allow ventilation. To facilitate ventilation vent pipes may be provided at the upper end of laterals.

14.6.2 Method of Distribution

(1) Fixed Nozzles — old system — In low rate filter.

(2) Rotary Distributors — modem system — in both low and high rate filters.

Fixed Nozzles. Sewage is sprayed through fixed nozzles fitted to a pipe grid. The get issuing from the nozzle opening impinges against a deflector and spreads out to cover an area lying between concentric rings. In order to equalise distribution the head on the nozzle is varied from a maximum to a minimum value. This is done by feeding the nozzle by a dosing tank under falling head. To cover the entire filter surface the maximum spray limits are over lapped.

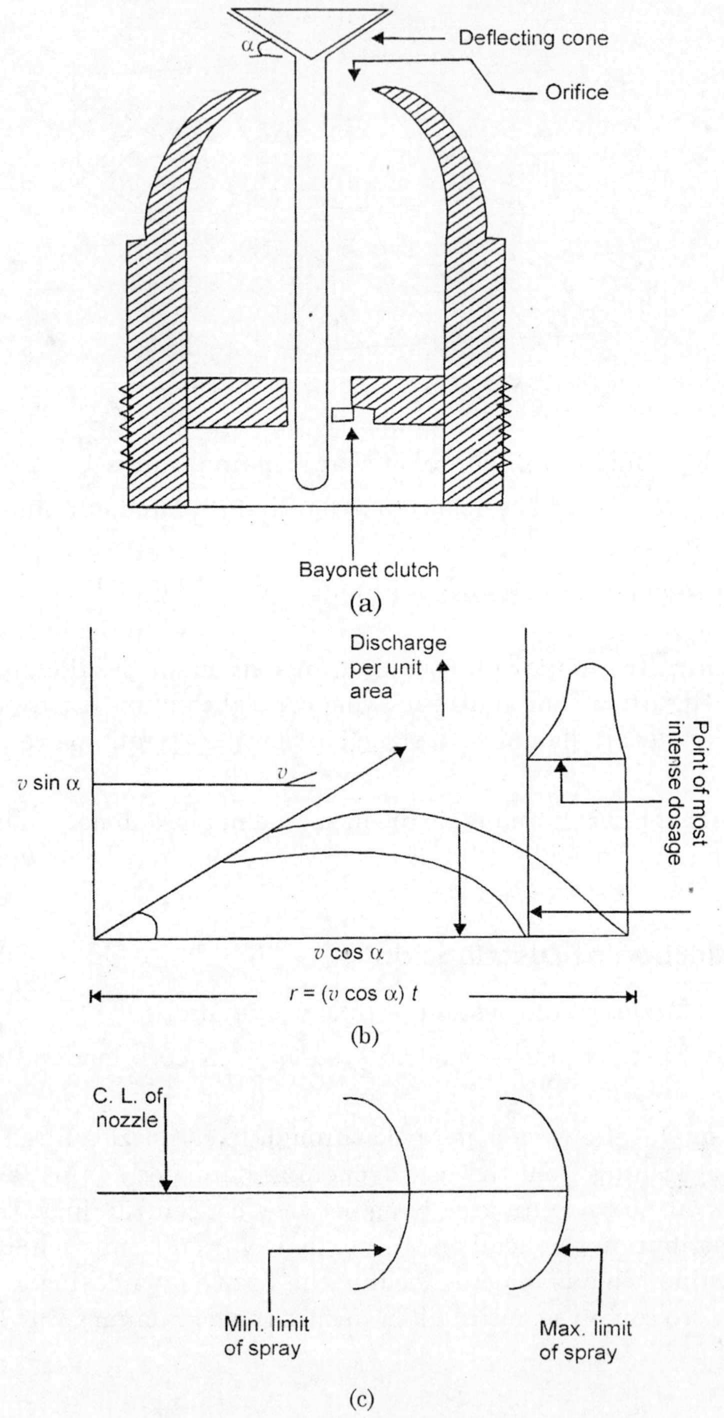
α
Deflecting cone
Orifice
Bayonet clutch
(a)
Discharge per unit area
Point of most intense dosage
v sin α
v
v cos α
r = (v cos α) t
(b)
C. L. of nozzle
Min. limit of spray
Max. limit of spray
(c)

Fig. 14.6

The nozzles are normally placed at a pieces of equilateral triangles by being staggered in position on the adjacent parallel supply pipes.

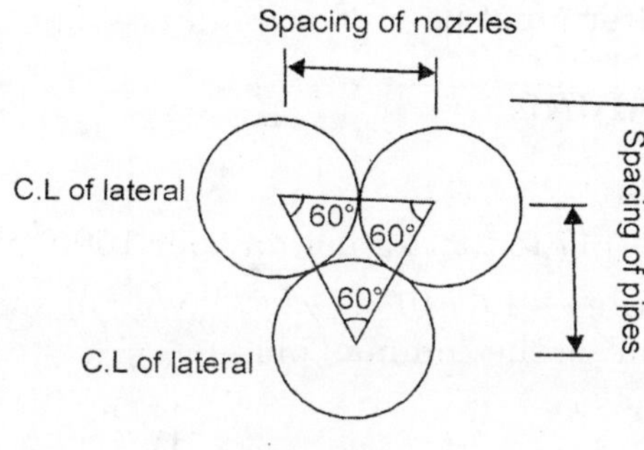

Fig. 14.7

14.6.3 Common Characteristics of Fixed Nozzles and Dosing Tanks

1. Diameter of nozzle outlet—18 mm to 25 mm
2. Diameter of nozzle base—36.0 mm to 50 mm
3. Spacing of nozzles—3.3 to 4.5 m
4. Spacing of pipes—(0.866) (nozzle spacing)
5. Total head of nozzles (difference in elevation of tank flow line and nozzle spacing)—1.8 to 3 m.
6. Loss of head in distribution (commonly distributed equally between frictional resistance of siphon and drop in head associated with getting the sewage in the system into motion i.e., dosing tank loss –1/4 of the total head on nozzle.
7. Overall head requirement of tank and filter—3.6 to 6.6 m.

Rotary Distributions. It consists of two or four horizontal pipe arms which rotate in a horizontal plane about a central hollow shaft. The sewage flows through the shaft and arms and is sprayed onto the bed through nozzles all fitted on the same side of the pipes. The rotation of the arms are caused by the back push of the emerging jets. They may also be rotated by an electric motor. Minimum clearance between arms and bed surface is 15 cm. The rates of revolution vary from 2 r.p.m for small distributors to less than 0.5 r.p.m for large ones. Head required for self propelled distributors — 30 to 75 cm.

The rotary distributors are normally fed by dosing tanks which add 0.3 to 1.5 m to the head requirement.

Advantages

1. More even distribution of sewage.

2. Less head loss than fixed nozzle.
3. Elemination of pipe distribution.

Disadvantage. 1 Filter must be made circular.

14.7 FILTER LOADING

(A) Low Rate Filter:

(i) *BOD loading* — 147 to 220 kg per day per 1000 m^3 of material.

(ii) *Hydraulic loading* — 1.87 to 3.74 $m^3/day/m^2$ of surface area. But 1.87 to 2.34 $m^3/day/m^2$ is the common one.

(B) High Rate Filter

(i) *BOD loading*—736 to 1840 kg/day/1000 m^3 of material

(ii) *Hydraulic loading*—9.36 to 28.1 $m^3/day/m^2$. But 10 to 25 m^3 day/m^2 is the common one.

Efficiency. According to National Research Council (America).

(a) *For single stage filtration*

$$E_1 = \frac{100}{1+0.01401\sqrt{\dfrac{w_1}{V_1 F}}}$$

where E_1 = % removal of 5-day BOD in second stage fillter and final sedimentation tank

w_1 = BOD loading to the filter in kg/day.

F = Number of effective passes of sewage through the filter.

$$= \frac{1+R}{(1+0.1R)^2}$$

where R = Recirculation ratio

$$= \frac{\text{Recirculated flow}}{\text{Average flow of raw sewage}}$$

V_1 = Volume of filter in 100 m^3

(b) *Two stage treatment*

In two-stage treatment efficiency of the first-stage is given by the formula for the single stage treatment and the efficiency of the second stage is given by

$$E_2 = \frac{100}{1+\left[\dfrac{0.01401}{1-E_1}\right]\sqrt{\dfrac{w_2}{V_2 F}}}$$

where

E_2 = % removal of BOD in second stage filter and final settling tank

w_2 = BOD load to the second filter in kg/day.

V = volume of filter in 1000 m^3

F = number of effective passes of sewage through the filter.

$$= \frac{1+R}{(1+0.1R)^2}$$

Example 14.1 *Determine the volume of a low-rate (standard) trickling filter to treat 15000 m^3 of settled sewage per day. 5-day BOD in the raw sewage is 200 mg/l and BOD removal in primary settling tanks is 35%. It is desired that the BOD in the final effluent should not exceed 20 mg/l.*

Solution

$$\text{BOD load to filter} = (1-0.35)\left(\frac{200}{10^6}\right)(15000 \times 10^3) \text{ kg/day}$$

$$= 1950 \text{ kg/day}$$

BOD permissible in the final effluent

$$= \left(\frac{200}{10^6}\right)(15000 \times 10^3) \text{ kg/day} = 300 \text{ kg/day}$$

BOD to be removed by the filter

$$= 1950 - 300$$

$$= 1650 \text{ kg/day.}$$

$$\text{Then efficiency desired} = \frac{1650}{1950} \times 100 = 84.6\%$$

Now
$$E = \frac{100}{1+0.01401\sqrt{\dfrac{w_1}{V_1 F}}}$$

For this problem, $E = 84.6$

$$w = 1950 \text{ kg}$$

$$E = \frac{1+R}{(1+0.1R)^2} = 1 \qquad [\because R = 0]$$

Then
$$84.6 = \frac{100}{1+0.01401\sqrt{\dfrac{1950}{V_1}}}$$

$$= \frac{100\sqrt{V_1}}{\sqrt{V_1} + (0.1401)(44.2)}$$

$$= \frac{100\sqrt{V_1}}{\sqrt{V_1} + 0.62}$$

i.e. $(100 - 84.6) = (84.6)(0.62)$

whence $V_1 = 11.6$ i.e., Volume $= 11.6 \times 10^3$ m^3

Check for BOD loading

$$\text{BOD loading} = \frac{1950}{11.6} = \mathbf{168\ kg/1000\ m^3}$$

$$\text{Assume depth} = 1.5 \text{ m}$$

$$\text{Then surface area} = \frac{11600}{1.5} = 7750 \text{ m}^2$$

$$\text{Liquid loading} = \frac{15000}{7750} = \mathbf{1.94\ m^3/day/m^2}$$

Example 14.2 *Design a two-stage high rate filters to treat 15000 m³ of sexoage daily. BOD of raw sexoage is 200 mg of which 35% is removed in primary settling tank. It is desired that the BOD in the final effluent should not exceed 20 mg/l.*

Solution

$$\text{BOD applied to filters} = (1 - 0.35)\frac{200}{10^6} \times 1000 \times 10^3$$

$$= 1950 \text{ kg}$$

Let there be two filters of equal size in series

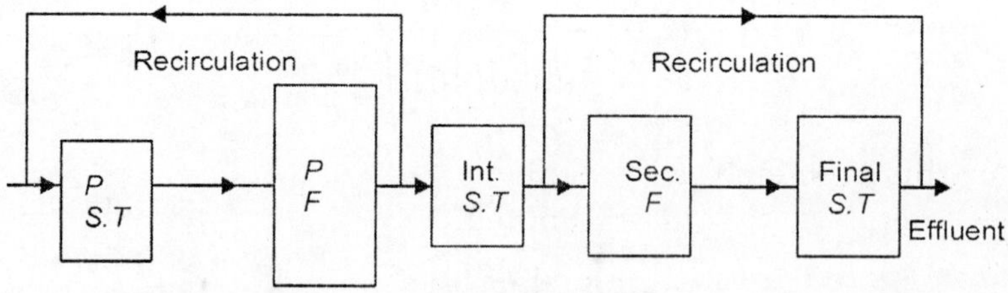

Fig. 14.8

where P = Primary
$S.T$ = Settling tank
F = Filter
Int. = Intermediate
Sec. = Secondary

Assume a BOD loading = 2000 kg/1000 m³/day to the first filter.

$$\text{Volume of first filter} = \frac{1950}{2} = 975 \text{ m}^3$$

$$\text{Let recirculation ratio } R = 1$$

$$F = \frac{1+R}{(1+0.1R)^2} = \frac{2}{(1.1)^2} = \frac{2}{1.21} = 1.65$$

$$E_1 = \frac{100}{1+(0.0140)\sqrt{\dfrac{1950}{(0.975)(1.65)}}} = \frac{100}{1.417}$$

$$= 67\%$$

Thus BOD removed = (1950) (0.67) = 1310 kg

BOD remaining = 1950 – 1310 = 640 kg

Then BOD loading to 2nd filter = 640 kg

$$E_2 = \frac{100}{1+\dfrac{0.01401}{1-E_1}\sqrt{\dfrac{w_2}{V_2 F}}}$$

Here

$$E_1 = 0.067$$

$$w_2 = 640 \text{ kg}$$

$$V_2 = 0.975 \qquad [\because \text{ on } 1000 \text{ m}^3]$$

$$F = 1.65$$

$$E_2 = \frac{100}{1+\dfrac{0.01401}{1-0.67}\sqrt{\dfrac{640}{(0.975)(1.65)}}}$$

$$= \frac{100}{1.85} = 54\%$$

$$\text{BOD in final effluent} = (640)\ (1-0.54)$$

$$= 294 \text{ kg/day}$$

Concentration of BOD in effluent is

$$= \frac{294 \times 10^3 \times 10^3}{15 \times 10^6} \text{ mg}/l$$

$$= 19.6 \text{ mg/I}$$

which is within the desired limil i.e., 20 mg/l

Assume depth of filter material = 0.9 m

$$\text{Area per filter} = \frac{975}{0.9} \text{ m}^2 = 1080 \text{ m}^2$$

Liquid load = inflow + recirculated flow

$= (15000 + 15000)^3 = 30000 \text{ m}^3/\text{day}$

$$\text{Rate of hydraulic loading} = \frac{30000}{1080} \text{ m}^2/\text{day/m}^2$$

$$= \mathbf{27.18 \text{ m}^2 /\text{day/m}^2}$$

$$\text{Diameter of filter} = \frac{\sqrt{4A}}{\pi} = \sqrt{\frac{4 \times 1080}{\pi}} = 37.2 \text{ m}$$

Biofilter, Accelo filter, and Aero filter are all patented high rate trickling filter.

Methods of Operation

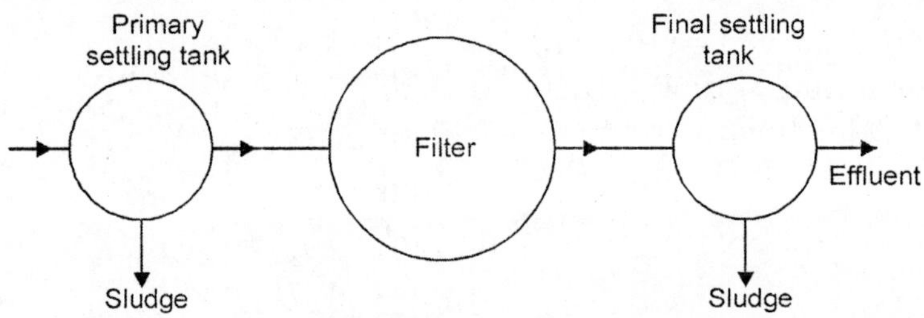

(a) For low-rate or standard trickling filter

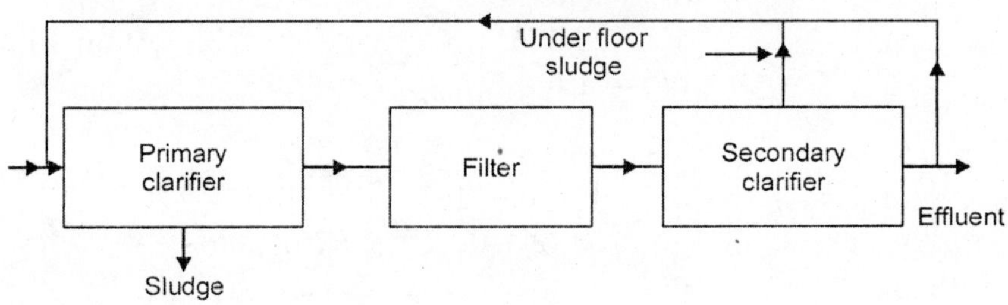

(b) Single stage treatment

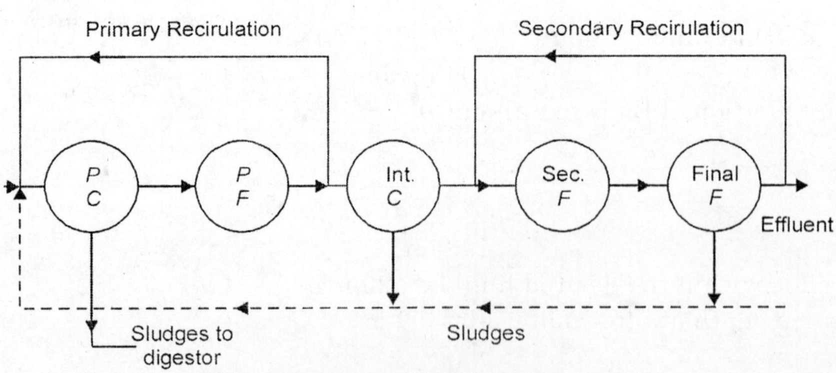

(c) Two stage treatment

where

P = Primary F = Filter
Sec = Secondary C = Clarifier
Int = Intermediate

N.B. Figures 14.9 (b) and 14.9 (c) are Biofilter flow diagrams.

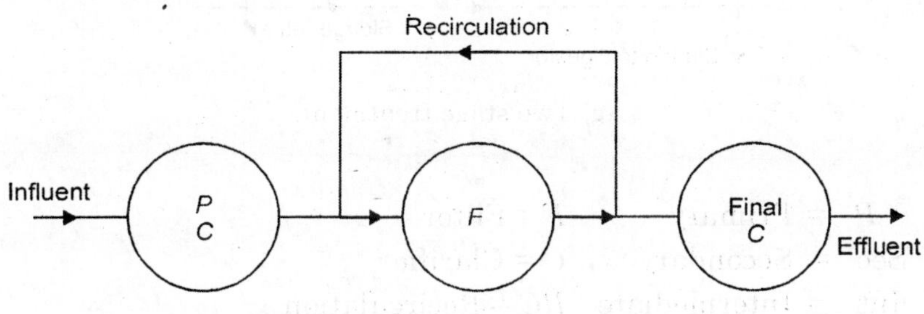

(d) Single stage treatment

where

P = Primary
C = Clarifier
F = Filter

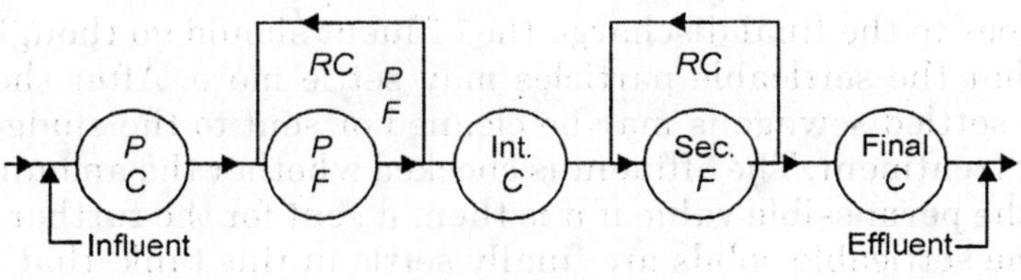

(e) Two stage treatment

where

P = Primary F = Filter
Sec = Secondary C = Clarifier
Int = Intermediate RC = Recirculation

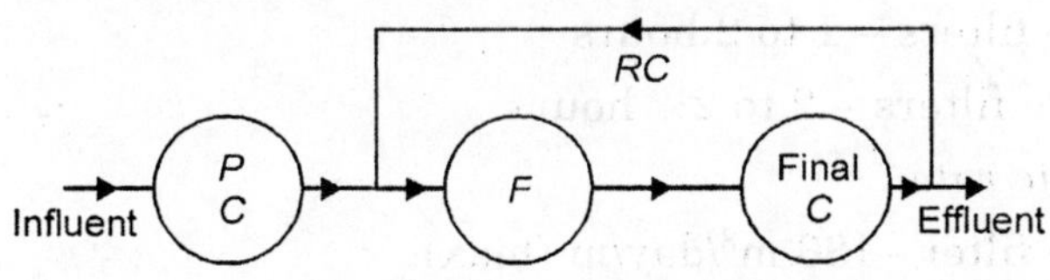

(f) Single stage treatment

where

P = Primary
F = Filter
C = Clarifier

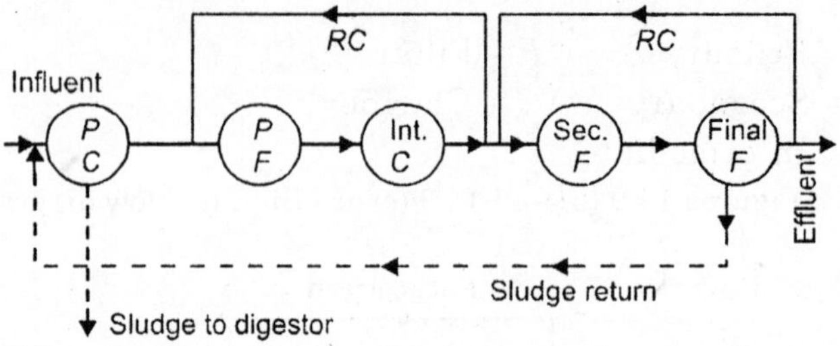

(g) Two stage treatment.

where

P = Primary		F = Filter
Sec = Secondary		C = Clarifier
Int = Intermediate		RC = Recirculation

Fig. 14.9

N.B. Figures 14.9 (f) and (g) are Aero filter flow diagrams.

14.8 FINAL SETTLING TANKS (HUMUS TANKS)

The effluent from trickling filter contains good amount of settleable solids. Before it goes to the final discharge the effluent should go though a settling tank, so that the settleable particles may settle more. After the detention period the settled sewage is may be cleaned or sent to the sludge disgracer for further treatment. The effluent is checked whether the amount of BOD is less than the permissible value if it is then, it sent for the further treatment. Because the settleable solids are finally settle in this tank, that is why it is called find settling tank.

Surface area loading

For low rate filter – 40 $m^3/day/m^2$

For high rate filter – 32 $m^3/day/m^2$

Detention period

For low rate filters – 1 to 2 hours

For high rate filters – 2 to 2½ hours

Weir overflow rate

For low rate filter –186 $m^3/day/m$ (max)

For high rate filter – can be taken same as low rate filters.

14.9 ACTIVATED SLUDGE PROCESS

Process. This is one of the most important and efficient sewage treatment methods. It involves aeration of settled sewage mixed with small volume of active sludge in an aeration tank. The activated sludge is sludge settled out of a sewage which has been aerated and agitated. It is composed of synthetic

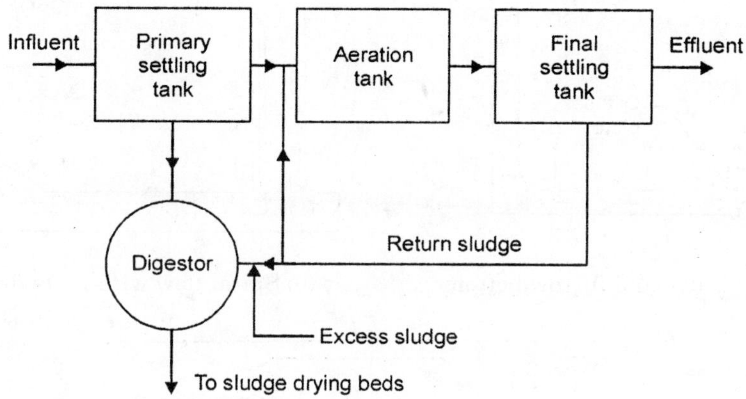

Fig. 14.10

gelatinous matrix containing a large number of aerobic bacteria, protozoa and metazoa and the organic substance separated from the sewage. The mixture of sewage and activated sludge is called mix liquor. The mixed liquor after being aerated in the aeration tank is settled in find sedimentation tanks. The sludge formed there is partly returned for use in aeration tanks and the remaining is wasted. The sludge serves as a vehicle for the bacteria which oxidize sewage solids.

Aeration Units

Depending on method of aeration, there are two types as follows.

1. Diffused air units — most common
2. Mechanical aeration units — rare

Function

1. Transfer of oxygen to sewage
2. Circulation of floc through sewage for contact between floe and sewage
3. Keep away the floc from settling.

14.10 DIFFUSED AIR TANKS

Oxygen is transferred to sewage from the compressed air which is bubbled through the sewage and also from atmosphere at the sewage surface.

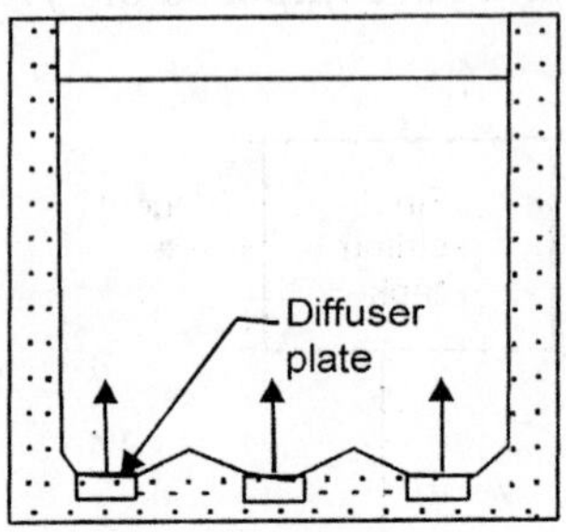

(a) Ridge and furrow bottom

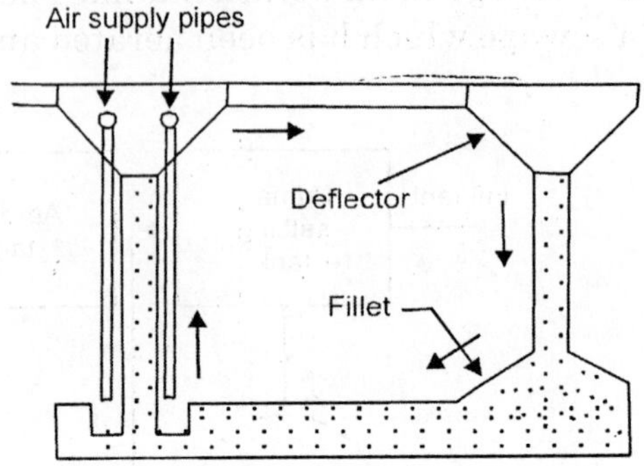

(b) Spiral flow with plate diffuser

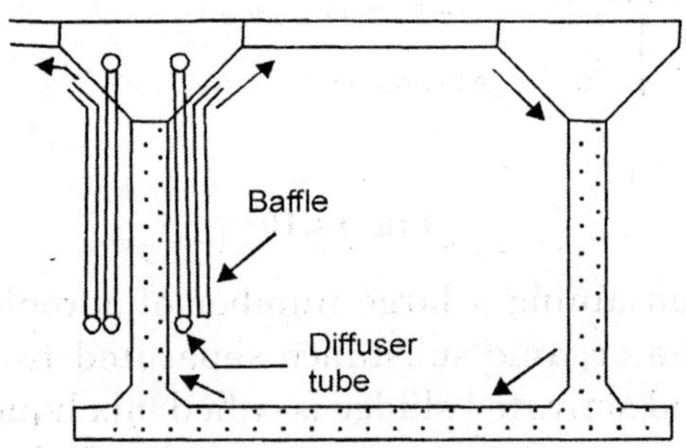

(c) Spiral flow with tubular diffusers and baffle.

Fig. 14.11

(1) Depth — 3.0 to 4.5 m; Width — 3.0 to 9.0 m; Length — 30 to 120 m } Common dimensions

(2) Length, L is related to depth D and width B as $L = \dfrac{Qt}{24BD}$

where Q = rate of flow in m^3/day

t = detention period in hours

B = width in m

D = depth in m

L = length in m

(3) $L : B > 5 : 1$.

(4) Detention period = 4 to 8 hours average 6 hours on the basis of influent sewage.

(5) Deflectors are inclined at 40° to 45° with the horizontal and the height of fillets is 30 to 60 cm. Deflectors reduce the free sewage surface by 5 to 20%. Diffuser tubes are placed half—the depth of tank below the water surface.

(6) Longitudinal furrows may be as close as 1.5 m on centres and sides steep 2 to 3 vertical to 1 horizontal.

(7) No. of tanks — two or more aeration tanks if their volumes exceeds 140 m^3 and two or more final settling tanks when their volume exceeds 70 m^3.

(8) Effective velocity of circulation of the sewage within the aeration tank must not be below 15 cm/sec to prevent floc from settling. For this the mean channel velocity is about 45 cm/sec.

(9) Spiral flow in the aeration tank is well maintained if width is not more than three times the depth. Common values of $\frac{B}{D}$ is 1.5:1.

Air Requirement

Oxygen transfer efficiency of bubble aeration = 5 to 15%.

One litre of air contains about 273 mg of oxygen. Then oxygen absorbed from each litre of air = 14 to 41 mg of oxygen.

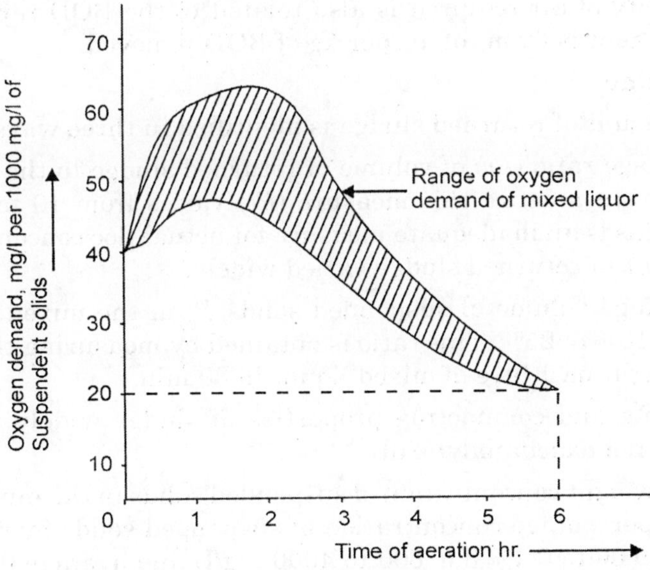

Fig. 14.12

Rate of oxygen demand of mixed liquor diminishes as the time of aeration increases. A maximum demand of 40 to 64 mg/1 per hour per 1000 mg/1 of suspended solids is exerted in the beginning and it drops to about 16 mg/1 per hour per 1000 mg/1 of suspended solids in course of 6 hours.

Let the mixed liquor contain 0.25% of solids i.e., 2,500 mg/l

Assuming the demand curve as parabolic the oxygen required

$$= (5)\ (2.5)\left[16 + \frac{1}{3}\{(40-16) \text{ to } (64-16)\}\right]$$

i.e., oxygen demand for 5 hours aeration = 300 to 400 mg/l

Hence air required for 5 hours aeration

$$= \frac{300}{(273)(0.05 \text{ to } 0.15)} \text{ to } \frac{400}{(273)(0.05 \text{ to } 0.15)} \text{ litres}$$

$$= (7.3 \text{ to } 22) \text{ to } (9.7 \text{ to } 29.3) \text{ litres}$$

$$= 7.3 \text{ to } 29.3 \text{ litres.}$$

It is assumed that half of this amount is derived from atmosphere at the surface of the sewage.

Thus the amount of diffused air required

$$= \frac{7.3}{2} \text{ to } \frac{29.3}{2} \text{ i.e., 3.7 to 14.7 litres}$$

1. Thus air required = 3.7 to 14.7 litres of air per litre of mixed-liquor.

2. The quantity of air required is also related to the BOD removed and normal requirement is 62 m^3 of air per kg of BOD removed.

Returned Sludge

The relative amount of returned sludge is identified in three ways.

1. As percentage ratio (*Pr*) of volume of returned sludge to the volume of influent sewage. In practice this measure (*Pr*) varies from 10 to 30% and average 20%. This is an inadequate measure for actual floc concentration as the solids content in returned sludge varied widely.

2. As percentage volume of suspended solids P_v in the mixed liquor. P_v varies between 10% to 25%. This ratio is obtained by measuring the volume of sludge settling from 1 litre of mixed liquor in 30 min.

As the settling and compacting properties of sludge widely vary, this measure is also not exact analytically.

3. As the dry weight concentration of suspended solids in the mixed liquor. In practice the percentage concentration of suspended solids by weight, P_w varies between 0.06 to 0.4% (i.e. 600 to 4000 mg/I) and average 0.25% (i.e., 2500 mg/l).

The condition of returned sludge is reflected by (1) Sludge age and (2) Sludge volume index.

(1) Sludge Age. Sludge age represent average time in days that a particle of suspended solids remain under aeration. Sludge age is calculated as a ratio of weight of suspended solids in the mixed liquor to the daily wt. of suspended solids in the influent i.e.

$$\text{Sludge age (days)} = \frac{\text{Wt. of suspended solids in mixed liquor}}{\text{Wt. of suspended solids in daily flow}}$$

$$\text{or } = \frac{\text{Suspended solids in mixed liquor (mg/l)} \times \text{aeration period (hr.)}}{(24) \times \text{(suspended solids in infulent mg/l)}}$$

Normal derivable sludge age = about 3 days.

(2) Sludge Volume Index. It is defined as volume in ml occupied by one gram of sludge (daily weight) after 30 min. of settling. This is also called Mohlmann Index.

Sludge Volume Index

$$= \frac{\text{Vol. of sludge settled in 30 min. present in mixed liquor}}{\text{Suspended solids, percent in mixed liquor}}$$

Good sludge has a sludge volume index of 50 to 100.

An index of 200 indicates bulking of sludge and poor settling characteristics of sludge.

The amount of return sludge required to maintain a desired percentage of solids, P_w in the mixed liquor can be expressed in terms of sludge volume index as

$$\frac{R}{I} = \frac{P_w I_v}{100 - P_w I_v}$$

where R = Volume of returned sludge

I = Volume of influent

P_w = Percentage concentration of suspended solids by weight in mixed liquor

I_w = Sludge volume index.

This expression was suggested by D.E. Bloodgood and is based on the assumption that solids in the waste sludge an influent flow and solid lost by oxidation are very small and negligible in comparison with solids in the return sludge i.e. solids in the mixed liquor is practically same as the solids in the return sludge.

If P = Percentage solid concentration in returned sludge = $\frac{100}{I_V}$

i.e., P_s = Solids in gm per 100 gm of sludge

I_v = Sludge volume index = gm of sludge per gm of solid.

I_v gm of sludge contains 1 gm of solid

100 gm of sludge contains $\frac{1}{I_v} \times 100 = \frac{100}{I_v}$ gm of solid

Thus of sludge contains $P_s = \frac{100}{I_v}$

Then from above assumption

$$P_w.(I + R) = (R)P_s$$

i.e.

$$P_s = P_w \cdot \frac{I+R}{R} = P_w\,(1 + I/R)$$

$$\frac{100}{I_v} = P_w\left(1+\frac{I}{R}\right)$$

$$\frac{100}{P_w I_v} = 1+\frac{I}{R}$$

$$\frac{100 - P_w I_v}{P_w I_v} = \frac{I}{R}$$

or

$$\frac{R}{I} = \frac{P_w I_v}{100 - P_w I_v}$$

Treatment Efficiency

According to the Committee on Sanitary Engineering of National Council

$$E = \frac{100}{\left[1 + 0.03\left(\frac{y_0}{Wt}\right)^{0.42}\right]}$$

were E = Percentage removal of BOD by aeration unit and secondary settling tank.

y_0 = 5-days 20°C (or 293° K) load to aeration unit in kg per day.

W = Weight of suspended solids in the mixed liquor in 1000 kg

t = Period of aeration in hour.

The equation can be written after rearranging as

$$Wt = \frac{y_0}{4200}\left(\frac{E}{100-E}\right)^{2.38}$$

$$= \frac{y_0}{4200}\left(\frac{y}{y_0-y}\right)^{2.38}$$

where y = BOD removed daily in kg

If V = Volume of the aeration tank in m^3

Then $$W \times 1000 = \frac{(V)(1000)P_w}{100}$$

i.e. $$W = \frac{VP_w}{100}$$

Again period of aeration t in hours is

$$t = \frac{24V}{I}$$

where I = In flow in m^3 per day.

Then $$Wt = \frac{VP_w}{100}\cdot\frac{24V}{I} = \frac{24}{100}\cdot\frac{P_w}{I}\cdot V^2$$

Whence $$V = \sqrt{\frac{100(Wt)I}{24P_w}} = 2.04\sqrt{\frac{(Wt)I}{P_w}}$$

$$t = \frac{24V}{I} = 48.94\sqrt{\frac{(Wt)}{IP_w}}$$

$$t \approx 49\sqrt{\frac{(Wt)}{IP_w}}$$

Example 14.3 *Determine the size of the aeration rank and the volume of air to be supplied per second in an activated sludge plant from the following data:*

(*i*) *Flow = 40 mid (million litre/day)*

(ii) *BOD of raw sewage = 200 mg/l*

(iii) *BOD removal in primary settling tank = 30%*

(iv) *Final effluent in BOD desired = 20 mg/l*

(v) *Row sewage suspended solid = 300 mg/l*

Solution

$$\text{BOD load to aeration tank} = \frac{(1-0.30)(200)\times 4\times 10^6}{10^6}\ \text{kg/day}$$

$$= (0.7 \times 800)\ \text{kg/day}$$

$$= 560\ \text{kg/day}$$

$$\text{BOD in the final effluent} = \frac{20\times 4\times 10^6}{10^6}\ \text{kg/day}$$

$$= 80\ \text{kg/day}$$

$$\text{BOD removal} = 560 - 80 = 480\ \text{kg/day}$$

$$\therefore \quad \text{Efficiency} = \frac{480}{560}\times 100 = 85.7\%$$

If

W = Wt. of suspended solids in mixed liquor in 1000 kg.

t = Period of aeration in hours

y_0 = BOD load to aeration lank in kg/day

Then,

$$Wt = \frac{y_0}{4200}\left(\frac{E}{100-E}\right)^{2.38}$$

$$= \frac{5600}{4200}\left(\frac{85.7}{100-85.7}\right)^{2.38}$$

$$= \frac{8}{60}\left(\frac{85.7}{14.3}\right)^{2.38} = \frac{2}{15}(6.00)^{2.38} = 9.48$$

If V = Volume of the aeration in m³

$$V = 2.04\sqrt{\frac{(Wt)(2)}{P_w}}$$

where I = Volume of influent in m³/day

P_w = % of suspended solids by wt. in mixed liquor

I = 4000 m³

Assuming $P_w = 0.25$

$$\therefore \quad V = 2.04\sqrt{\frac{(9.48)(4000)}{0.25}}$$

$$= 2.04\sqrt{16 \times 9480} = 780 \text{ m}^3$$

$$t = \frac{V}{I} \times 24 \text{ hr.} = \frac{780}{4000} \times 24 = 4.68 \text{ hours.}$$

Air requirement @ 62 m^3 per kg of BOD removal = 62 × 320 m^3/day

$$= \frac{62 \times 320 \times 1000}{24 \times 60 \times 60} \text{ litres/sec}$$

$$= \frac{62 \times 32 \times 100}{24 \times 36} \text{ litres/sec}$$

$$= \textbf{230 litres/sec}$$

Example 14.4 *Find (a) the allowable loading intensity of a diffused air activated sludge plant that is effect 30% removal of BOD from 38,000 m^3 of sewage per day. The sewage contains 130 mg/l of BOD.*

(b) Necessary time of aeration and volume of aeration tank if the solids concentration is to be maintained at 0.25% in the aeration unit.

(c) Necessary solid concentration if the detention period available in the tank is 6 hours.

(d) The approximate percentage of sludge to be returned to the influent under (b) and (c) if the sludge volume index is 80.

(e) The required approximate solid content of returned sludge under (b) and (c).

Solution

$$\text{BOD applied per day} = (130)(38000 \times 10^3) \times 10^{-6} \text{ kg}$$

$$= 4940 \text{ kg}$$

$$\text{(a) } Wt = \frac{y_0}{4200}\left(\frac{E}{100-E}\right)^{2.38} = \frac{4900}{4200}\left(\frac{80}{100-80}\right)^{2.38}$$

$$= 31.8$$

where W is in 1000 kg and t in hour, then the load intensity is 31800 kg hr.

$$\text{(b)} \quad t = 49\sqrt{\frac{(Wt)}{IP_w}}$$

$$P_w = 0.25, I = 38{,}000 \text{ m}^3\text{/day.}$$

$$t = 49\sqrt{\frac{31.8}{(3800)(0.25)}}$$

$$= 49\sqrt{\frac{31.8}{9500}} = 2.84 \text{ hours.}$$

$$V = 2.04\sqrt{\frac{(Wt)I}{P_w}}$$

$$= 2.04\sqrt{\frac{(31.8)(38000)}{0.25}} = 4488 \text{ m}^3$$

$$\left.\begin{array}{r}\text{Depth} = 3 \text{ m}\\ \text{Width} = 5 \text{ m}\\ \text{Lenght} = 20 \text{ m}\end{array}\right\} 15 \text{ No.}$$

(c) If time of aeration 6 = hours

$$t = 49\sqrt{\frac{(Wt)}{IP_w}}$$

i.e.

$$6 = 49\sqrt{\frac{31.8}{P_w(38000)}}$$

$$\left(\frac{6}{49}\right)^2 = \frac{31.8}{38000P_w}\left(\frac{49}{6}\right)^2 = 0.0558\%$$

(d) If $I_v = 80$, the percentage returned sludge would be as follows.

For $P_w = 0.25$ %

$$\frac{R}{I} = \frac{P_w I_v}{100 - P_w I_v} = \frac{(0.25)(80)}{[100-(0.25)(80)]} = \frac{20}{80}$$

$$= 0.25 = 25\%$$

For $P_w = 0.0558\%$

$$\frac{R}{I} = \frac{(0.558)(80)}{100-(0.558)(80)}$$

$$= \frac{4.47}{95.53} = 0.0468 = 4.68\%$$

(e) Approximate solids content in returned sludge will be

$$P_s = P_w\left(1+\frac{1}{R}\right)$$

For $P_w = 0.25,\ P_s = 0.25\left(1+\frac{1}{0.25}\right) = 1.25\%$

For $P_w = 0.0468,$

For $P_w = 0.0468,\ P_s\left(1+\frac{1}{0.0468}\right) = 1.25\%$

Other method of Aeration tank design BOD loading to diffused air units:

Not more than 50 kg of BOD per day per 100 kg of suspended the aeration units.

Return sludge = 20% to 25% of influent sewage by volume.

Aeration period = 4 to 8 hours, commonly 6 hours.

Concentration of suspended solid in mixed liquor = 1200 to 3000 mg/l, commonly 2500 mg/1 i.e. 0.25%.

Empirical method. Period of aeration and returned sludge are prescribed in this method.

Example 14.5 *Volume of sewage to be treated = 400 m³/day BOD after primary sedimentation =160 mg/l.*

Amount of returned sludge = 25%

Aeration period = 6 hours.

Solution

$$\text{Volume of aeration tank} = \frac{400 \times 1.25}{24} \times 6$$

[1.25 including returned sludge]

$$= 125 \text{ m}^3$$

$$\text{Total BOD per day} = \frac{400 \times 10^3 \times 160}{10^3 \times 10^3} = 64 \text{ kg.}$$

$$\text{Total volume provided} = \frac{125}{64} = 2 \text{ m}^3 \text{ per kg of BOD.}$$

Rational method. BOD load and concentration of suspended solids are given or assumed.

Example 14.6 *Volume of sludge to be treated = 400 m³/day*

BOD after primary sedimentation = 160 mg/l.

Solution

Assume BOD load = 40 kg per 100 kg of solid in mixed liquor and concentration of solid in mixed liquor = 2500 mg/l

$$\text{Total BOD per day} = \frac{400 \times 10^3 \times 160}{10^3 \times 10^3} = 64 \text{ kg.}$$

$$\text{Total suspended solid required} = \frac{64}{40} \times 100 \text{ kg} = 160 \text{ kg}$$

If V = Volume of tank in m³

$$\text{Then total solid} = \frac{V \times 10^3 \times 2500}{10^3 \times 10^3} \text{ kg} = 2.5 \text{ V kg}$$

$$\therefore \quad 2.5\,V = 160$$

$$V = \frac{160}{2.5} = 64 \text{ m}^3$$

14.11 AIR SUPPLY FOR DIFFUSED AIR UNITS

It includes the following component parts (1) diffusers, (2) air piping, (3) air compressors and (4) air-filters.

Air diffusers

Two kinds of diffusers are as follows

(1) Porous diffusers → (i) plate type (ii) tubular type

(2) Jet diffusers.

Plate diffuser — 30 cm × 30 cm and size × 2.5 cm thick.

Tubular diffuser — 60 cm long, 45 to 75 mm in internal dia, and 10 to 16 mm wall thickness.

Porous diffuser — made from grains of crushed quartz, aluminium oxide or carbon fixed or bonded together having porosity 30 to 36%.

Permeability of porous diffuser — defined as the quantity of air at 70°F and 10 to 25% relative humidity that is discharged per mm to the atmosphere by unit area of the diffuser surface with pressure loss through pressure of 5 cm of water when the diffuser is tested dry in a room kept at 70° F and relative humidity between 30 to 50%. Permeability of 11 to 22 $m^3/min/m^2$ are commonly specified submerged diffusers when new delivery 0.3 to 2.4 $m^3/min/m^2$ of air with pressure losses 5 to 12.5 cm of water. Common rate of operation is 0.9 $m^3/min/m^2$. Diffusers are cleaned when the head loss is twice the initial.

Mounting of Diffuser. The diffuser plates are mounted in metal or pre-cast concrete blocks or set over depressions cast in the floor. Upto 10 plates are held in a single frame. They are supplied with air from centre or from one side. Air flow is controlled by a valve serving a single or two adjacent frames.

Area of diffuser plates	—	5 to 25% (commonly 10 to 20% of the maximum tank area).
Tubular diffuser	—	Held in pair attached to a central pipe. The bottom surface (1/3 of area) of circular diffuser is considered ineffective.
Jet diffuser	—	These direct a stream of air downward against a metal bowl immediately below jet nozzle.

Maintenance of diffuser. To prevent clogging air should not contain more than 17.7 mg of dust per 1000 m^3 of air. To prevent clogging by sewage solid diffuser should always be kept under air pressure in excess of the head of sewage above them. Shoots are removed by burning.

Iron-oxide removed by hydrochloric acid (30%).

Sewage solid and growths are removed by nitric acid (50%).

Grease removed by gasoline and carbon tetrachloride.

Air Piping

Pressure drop in pipe due to resistance is $\Delta P = 0.05635 \dfrac{flTQ^2}{pD^5}$

where ΔP = Pressure drop in kg/cm^2

f = friction factor $= 0.0276 \dfrac{D^{0.027}}{Q^{0.148}}$ for $D < 25$ cm

T = absolute temperatue of compound air $= 519.6\left(\dfrac{P}{1.035}\right)^{0.283}$

p = absolute blower pressure in kg/cm^2 (i.e. 1.035 + gauge pressure)

l = length of pipe in metre

D = Diameter of pip in cm

Q = rate of free air flow in m^3/min.

The length of pipe that will produce a pressure drop to that of an elbow is

$$l = \frac{0.9D}{1+\frac{9.15}{D}}$$

where l is in M and D is in cm.

Similarly for a globe valve

$$l = \frac{1.35D}{1+\frac{9.15}{D}}$$

Air Compression. Rotary and centrifugal blowers are commonly employed for air compression. Resiprocating compressors are undesirable because of pulsating discharge. Compression is adiabatic and efficiency in rotary and centrifugal blowers ranges from 75 to 85%.

The required air pressure = Distribution losses + Head of sewage in tank.

Distribution losses is about 25% of sewage head. Thus the air pressure at

= (1.25) (depth of sewage) in terns of head of water.

Air pressure at blower (kg/cm^2 gauge)

= (0.125) (depth of sewage in m)

Blower horse/power is

$$P = \frac{800}{E} Q\left[\left(\frac{P}{1.035}\right)^{0.283} - 1\right] HP$$

where P = horse power required of blower

E = efficiency of blower in precent

Q = rate of free air flow m^3/min

P = absolute pressure at blower in kg/cm^2 (i.e. 1.035 plus the blower gage pressure).

14.12 SECONDARY SETTLING TANK

The following items are required to fulfil the requirement of a proper secondary settling tank.

(1) Detention period = $1\frac{1}{2}$ to 2 hours

(2) Surface area loading = 32 to 40 m^3 /day/m^2

(3) Weir loading—not over 186 m^3/m of weir

(4) $\frac{\text{Length}}{\text{Depth}}$ = 7:1 for rectangular tank

$$\frac{\text{Diameter}}{\text{Depth}} = 10 : 1 \text{ for circular tank}$$

(5) Minimum depth = 3 m to 3.6 m.

This is more or less the same as the primary settling tank. The inlet and outlet arrangements, shapes all are in primary settling tank. In the secondary settling tank, the scum removal and scum baffles are not necessary.

Overflow rate for secondary settling tank is generally taken as 1100 to 1500 litres per hour per m^3.

PROBLEMS

14.1. What is the function of a primary filter? Discuss briefly the secondary filter.

14.2. Discuss briefly the method of filtration.

14.3. How the filter beds are arranged? Show with sketches.

14.4. What type of materials are generally used in the filter.

14.5. Explain briefly the bacteriological function in the filter. How can you be able to judge the filter is not functioning properly?

14.6. (a) State the theory of trickling filtration. Give flow diagrams of conventional and high rate trickling filter.

(b) State the advantage of recirculation and its effect on efficiency of a trickling filter.

(c) Draw a sketch of a trickling filter unit and name the parts.

(d) How are the underdrains constructed?

14.7. (a) Describe how sewage is applied to a trickling filter.

(b) Determine the diameter of a standard rate trickling filter with depth 1.5 m to treat the sewage of a city with population of 100,000. The rate of water usage is 300 litres per capita per day and 5-day BOD production is 80 gm per capita per day. 35% of BOD is removed in the primary settling tank. It is desired that the BOD in the final effluent should not exceed 20 mg/l.

14.8. (a) Give the flow diagram of a two-unit two-stage high rate filter.

(b) Determine the volumes of a two-unit two-stage high rate filters to treat 15,000 m^3 of sewage daily. BOD of raw sewage is 200 mg/l of which 35% is removed in primary settling tanks, it is desired that the BOD in the final effluent should not exceed 20 mg/l and BOD loading should not exceed 1800 kg per 1000 m^3 of sewage per day.

14.9. Discuss the activated sludge process briefly with flow diagram.

14.10. (a) Describe with a flow diagram the working of an activated sludge plant.

(b) An aeration tank using conventional activated sludge is to treat 4×10^6 litres per day of a settled sewage containing BOD of 150 mg/l. The final effluent is to have a BOD of 10 mg/l.

(i) Find the volume of the aeration tank if the BOD loading is 535 gm per 1000 litres of volume.

(ii) Find the air required if 2.25 gm of air is needed for each mg/*l* of BOD removed.

(c) If the return sludge is 25 percent of the influent, what will be the aeration period?

14.11. (a) A settled sewage of 4×10^6 litres per day with BOD 150 mg/*l* is treated in an aeration tank where the suspended solids concentration is maintained at 2500 mg/l. If the BOD loading is 35 kg per 100 kg of suspended solids in the mixed liquor, find the size of the aeration tank and the aeration period of the returned sludge is 25% of the influent.

(b) A conventional activated sludge plant treats 16×10^6 litres of sewage in an aeration tank of capacity 3.2×10^6 litres. The suspended solids contents of the settled sewage and of mixed liquor are 100 mg/l and 2,000 mg/l respectively. What is the sludge age in days?

(c) The suspended solids content of a mixed liquor is 2500 mg/*l* and sludge index is 80. What should be the percent of return sludge by volume to maintain the condition in (b)?

14.12. Find (a) the allowable BOD loading intensity in kg-hour of a diffused-air activated sludge plant which, is to achieved an efficiency of 80% in BOD removal. The sewage flow is 38,000 m^3 per day and it contains 5-day BOD of 130 mg/l.

(b) Necessary time of aeration and volume of aeration tank if the solid concentration is to be maintained at 0.25% in the aeration unit.

(c) Necessary solid concentration if the detention period available is 6 hours.

(d) The approximate percentages of sludge to be returned under (b) and (c) if the sludge volume index is 80.

(e) The required approximate solids content of the returned sludge under (b) and (c).

Chapter

15

Sludge Treatment and Disposal

15.1 DEFINITION

The solid matter that is separated from sewage in both primary and secondary settling tank is called sludge.

15.1.1 Types of Sludge

According to the treatment process producing it—

(1) Plain sedimentation sludge

(2) Chemical precipitation sludge

(3) Trickling filter humus

(4) Activated sludge

(5) Mixture of any two or more of the above sludge.

15.1.2 Moisture-weight-volume Relationship

If W = Wt. of water in a sludge

W_s = Wt. of solid in a sludge

Then the percentage moisture content of the sludge is

$$p = \frac{100W}{W + W_s}$$

$$\% \text{ of solid} = (100 - p) = \frac{100\, W_s}{W + W_s}$$

The specific gravity of sewage solid, S_s is given by

$$= \frac{100}{S_s} = \frac{p_v}{S_v} + \frac{100 - p_v}{S_f}$$

or
$$S_s = \frac{100\ S_f\ S_v}{100\ S_s + p_v (S_f - S_v)}$$

where S_f – Sp. gravities of volatile and fixed solid respectively,
p_v – Percentage of volatile matter in sludge.

The sp. gr. of volatile solids S_v is closed to 1.0 and that of fixed solid is about 2.5.

The specific gravity of wet sludge, S is Y

$$S = \frac{\text{Wt. of water and dry solid}}{\text{Volume of water and solid}}$$

$$= \frac{p + (100 - p)}{\frac{p}{S_w} + \frac{100 - p}{S_v}} = \frac{100\ S_s\ S_w}{PS_s + (100 - p)S_w}$$

$$= \frac{100 S_s S_w}{100 S_w + p(S_s - S_w)}$$

The sp. gr. of water $S_w = 1$

Volume of wet sludge V = Volume of water + volume of solid

i.e.
$$V = \frac{W_s}{S_s W} + \frac{p W_s}{(100 - p) S_w w}$$

$$= \frac{W_s}{w} \cdot \frac{100 S_w + (S_s - S_w)}{(100 - p) S_s S_w}$$

where w = Wt. per unit volume of water

The volume of partly dewatered sludge V is given by

$$V = V_0 \frac{[100\ S_w + p(S_s - S_w)]}{[100 S_w + p_o (S_s - S_w)]} \frac{(100 - p_0)}{(100 - p)}$$

where V_0 = Original volume of sludge
V = Volume of same sludge after partly dewatering
p, p_o = % moisture content in dewaterel and original sludge.

The above expression of V is valid if all the voids between the solids are filled with water.

Example 15.1. *A plain sedimentation sludge has a moisture content of 95%, 72.2% of the dry solid being volatile matter. Find* (*a*) *Sp. gr. of dry solid if sp. gr. of volatile and fixed solid are 1.00 and 2.5 respectively.*

(*b*) *Sp. gr. of wet sludge, if the sp. gr. of included water is 1.00.*

(*c*) *Relatively decrease in volume, if the moisture content is reduced to 85%.*

Solution

$$S_s = \frac{100\ S_f\ S_v}{100\ S_v + p_v(S_f - S_v)}$$

$$= \frac{(100)(2.5)(1.00)}{(100)(1) + (72.7)(2.5 - 1.0)} = 1.20$$

$$S = \frac{100\ S_s\ S_w}{pS_s + (100 - p)S_w}$$

$$= \frac{(100)(1.20)(1.00)}{(95)(1.2) + (100 - 95)(1)} = 1.008$$

$$\frac{V}{V_0} = \frac{[(100)(1) + 85(1.20 - 1.0)]}{[(100)(1) + 95(1.20 - 1.0)]} \cdot \frac{(100 - 95)}{(100 - 85)}$$

$$= 0.328$$

Let wt. of solid = 5 gm

When moisture content = 95%

$$\text{Volume} = \frac{5}{1.2} + 95 = 99 \text{ ml}$$

When moisture content = 85% .

$\therefore$ 5 gm = 15% of wet solid

$$\text{Wt. of wet sludge} = \frac{5}{15} \times 100 = 33.3 \text{ gm.}$$

$$\text{Wt. of water} = 33.3 - 5 = 28.3 \text{ gm.}$$

$$\text{Volume} = \frac{5}{1.2} + 28.3 = 32.3 \text{ ml}$$

$$\text{Ratio of volume} = \frac{32.3}{99} = 0.328$$

15.2 QUANTITIES OF SLUDGE

It depends upon the composition of sewage and efficiencies of the removal of the treatment process producing it.

Table 15.1 *Average per capita solids and BOD in domestic sewage in gm/ day*

State of solid	*Mineral*	*Organic*	*Total*	*5 day 293° K (20°C) BOD*
1. Suspended	25	65	90	42
(a) Settleable	15	39	54	42
(b) Non-settleable	10	26	36	23
2. Dissolved	80	80	160	12
3. Total	105	145	250	54

Table 15.2 *Relative efficiencies of processes sewage treatment operations and processes in % removal*

Treatment process	*5-day 293° K (20°C) BOD*	*Suspended solids*	*Bacteria*
1. Fine screaming	5–10	2–20	10–20
2. Chlorination of raw or settled sewage	15–30	—	90–95
3. Pain sedimentation	25–40	40–70	25–75
4. Chemical precipitation	50–85	70–90	40–80
5. Trickling filtration preceded and followed by plain sedimentation	80–95	70–92	90–95
6. Activated sludge process preceded to be followed by plain sedimentation	85–95	85–95	90–98
7. Intermittent sand filtration	90–95	85–95	95–98
8. Chlorination of biologically treated sewage	–	–	98–99

The solid of the domestic sewage sludge is commonly observed as follows.

1. **Plain Sedimentation Sludge.** Fresh sludge is composed of 100% settleable solids in the raw sewage. Digestion destroys about 67% of volatile, matter, 25% being converted to fixed solid.

2. **Chemical Precipitation Sludge.** Chemical floc increases the volume of sludge. Each mg/*l* of alum produces 0.66 mg/l of $Fe(OH)_3$ and each mg/*l* of alum produces 0.26 mg/*l* of $Al(OH)_3$. Digestion produce similar change as in plain sedimentation sludge.

Table 15.3 *Proportion of solids in domestic sewage sludge*

Treatment process	*Condition of sludge*	*% solid*
1. Plain sedimentation	(a) Fresh	25–5
	(b) Digested-wet	10–15
2. Chemical precipitation	(a) Fresh	2–5
	(b) Digested-wet	10
3. Trickling filtration	(a) Fresh humus alone	5–10
	(b) Fresh humus mixed will 1 (a)	3–6
	(c) Digested (3b)	10
4. Activation	(a) Fresh—as pumped	0.5–1.0
	(b) Thickened	1–2
	(c) Digested — alone (a or b)	2–3
	(d) Fresh —settled with 1a	4–5
	(e) Digested with 1a	6–8

3. **Trickling Filter Humus.** Humus quantities collected in secondary sedimentation unit are produced by 50 to 60% of non-volatile solid applied to low-rate filter and 80 to 90% for high-rate filter. The humus are generally mixed with primary sludge for digestion and change in composition is same as for plain sedimentation sludge.

4. **Activated Sludge.** Excess activated sludge is composed of non settleable solid 5 to 10% of volatile solid is destroyed during re-circulation. The excess sludge in secondary settling tank may be estimated on the basis of 80 to 90% of the remaining of applied non-settleable solid.

Example 15.2 *Estimate for an activated-sludge plant treating domestic sewage and including primary and secondary sedimentation.*

(*a*) *The weight of dry sludge solid in kg per 1000 persons per day and in kg per million litres of sludge.*

(*b*) *Volume of wet sludge in litres per 100 persons per day and in litres per million litres of sewage.*

Find in each case separate and combined amount of fresh and digested sludge if the primary sludge is digested alone and in combination with secondary sludge. Assume a sludge flow of 450 litres/capita/day and specific gravity of 2.5 for fixed solid and 1.0 for volatile solid. For the data refer to previous tables.

Solution

Primary Sludge-fresh

From Table 15.1, daily per capita production of settleable solid is 54 gram of which 39 gram (or 72.2%) being volatile and 25 gram (or 27.8%) fixed.

(a) Wt. of dry solid per day per 1000 persons

$$= \frac{54 \times 1000}{1000} = 54 \text{ kg.}$$

Since the sewage flow is 450 litres/capita/day the wt. of dry solid is 54 gram in 450 litres

or $$\frac{54}{450} \times \frac{10^6}{1000} = \text{kg. per million litres}$$

i.e. 120 kg per million litres of sewage.

(b) Sp. gr. of dry solid, $$S_s = \frac{100\ S_t\ S_v}{100\ S_t + p_v(S_f - S_v)}$$

$$= \frac{(100)(2.5)(1)}{(100)(1) + (72.2)(2.50 - 1)} = 1.20$$

From Table 15.2 assume the percentage of solid in wet sludge is 5%

Then the sp. gr. of wet sludge

$$S_s = \frac{100\ S_s\ S_w}{pS_s + (100 - p)S_w}$$

$$= \frac{(100)(1.20)(1.0)}{(95)(1.20)(100 - 95)(1.0)}$$

$$= 1.008$$

$$\text{Wt. of wet sludge} = \frac{54}{5} \times 100 \text{ kg} = 1080 \text{ kg/1000 persons}$$

$$\text{Then volume of wet sludge} = \frac{1080}{1.008} \text{ litres}$$

$$= 1070 \text{ lit/1000 persons/day}$$

and volume of wet sludge/million litres/day

$$= \frac{1080}{(1000)(450)} \times 10^6 = 2380 \text{ litres.}$$

15.2.1 Primary Sludge-digested

(a) Assume 67% of the volatile matter is destroyed, 25% being converted to fixed solid as a result of digestion.

Then volatile solid remaining

$$= (1 - 0.67)\ (39) = 13 \text{ gm/person.}$$

$$\text{Fixed solid} = [15 + 0.25\,(39 - 13)] = 22 \text{ gm/person.}$$

$$\text{Then total solid} = (13 + 22) = 35 \text{ gm/person/day,}$$

$$\%\text{ of volatile solid} = \frac{13}{35} \times 100 = 37\ \%$$

Then wt. of dry solid/day/1000 persons

$$= \frac{35}{1000} \times 1000 = 35 \text{ kg}$$

Wt. of dry solid per million litres of sludge

$$= \frac{35}{450} \times \frac{10^6}{1000} = 77.7 \text{ kg}$$

(b) Sp. gr. of dry solid $S_s = \dfrac{(100)(S_f)(S_v)}{100 S_v + P_v (S_f - S_v)}$

$$= \frac{(100)(2.5)(1.00)}{(100)(1) + (37)(2.5 - 1)} = \frac{250}{100 + 55.5}$$

$$= 1.607$$

Assume % of solid = 13% in digested sludge and % of moisture 100 –13 = 87%

Then sp. gr. of wet digested sludge

$$S = \frac{100\, S_s\, S_w}{p S_s + (100 - p) S_w}$$

$$= \frac{(100)(1.607)(1)}{(87)(1.607) + (13)(1.00)} = 1.052$$

Wt. of dry solid/1000 persons/day = 35 kg
which is only 13% of the wet sludge.

$$\text{Then wt. of sludge} = \frac{35}{13} \times 100$$

$$= 270 \text{ kg/1000 persons/day}$$

$$\text{Volume of wet sludge} = \frac{270}{1.052}$$

$$= 256 \text{ liters/1000 persons/day}$$

$$= \frac{256}{(1000)(450)} \times 10^6 = 570 \text{ lit/million lit/day}$$

15.2.2 Excess Activated Sulge—Fresh

From Table 15.1 daily per capita production of non-settleable solid is 26 gm of volatile solids and 10 gm of fixed solid.

Assume that 7.5% of volatile solid is destroyed during activation and 87.5% of the remaining solid is captured in the excess activated sludge.

Then wt. of volatile solid/capita/day

$$= (0.875)\ (0.925)\ (26) = 21 \text{ gm.}$$

and wt. of fixed solid/capita/day in the sludge

$$= (0.875)\ (10) = 9 \text{ gm}$$

Total wt. of solid/capita/day in the sludge

$$= 21 + 9 = 30 \text{ gm}$$

$$\text{Then \% of volatile solid} = \frac{21}{30} \times 100 = 70\%$$

and

$$\text{\% of fixed solid} = \frac{9}{100} \times 100 = 30\%$$

(a) Wt. of dry solid/day/1000 persons $= \dfrac{30 \times 1000}{1000} = 30$ k

Wt. of dry solid/day/million lit $= \dfrac{30 \times 10^6}{450 \times 1000} = 67$ kg

(b) Sp. gr. of dry solid S_s is

$$S_s = \frac{100\ S_f S_v}{100\ S_v + p_v(S_f - S_v)}$$

$$= \frac{(100)(2.5)y(1.0)}{(100)(1.0) + (70)(2.5 - 1.0)}$$

$$= \frac{250}{205} = 1.219$$

Assume solids in thickened fresh sludge = 1.5%

The sp. gr. of wet sludge is

$$S = \frac{100 S_s S_w}{100 S_w + p_v(S_s - S_w)}$$

$$= \frac{(100)(1.219)(1.0)}{(100)(1) + (100 - 1.5)(1.219 - 1)}$$

$$= \frac{121.9}{(100)+(98.5)(0.219)} = \frac{121.9}{100+21.55}$$

$$= 1.002$$

Wt. of dry soild/1000 persons/day = 30 kg
which is only 1.5% of the wet sludge.

Then wt. of wet sludge/1000 persons = $\frac{30}{1.5} \times 100 = 2000$ kg.

Then volume of west sludge/1000 persons/day

$$= \frac{2000}{(S)(\text{wt.of water in kg/liter})} \text{ litres}$$

$$= \frac{2000}{1.002} = 1995 \text{ litres.}$$

Volume of wet sludge/million liters/day

$$= \frac{(1995)(10^6)}{(1000)(450)} = 4440 \text{ litres}$$

15.2.3 Combined Primary and Excess Activated Sludge — Fresh

Combining quantities of solid already calculated for fresh primary sludge and fresh excess activated sludge, we get

Total solid/capita/day = 54 + 30 = 84 grams
Fixed solid/capita/day = 15 + 9 = 24 grams (i.e. 28.6%)
Volatile solid/capita/day = 39 + 21= 60 grams (i.e. 71.4%)

(a) Wt. of dry solids/day/1000 persons = $\frac{84 \times 1000}{1000} = 84$ kg.

Wt. of dry solid/million litres/day = $\frac{84 \times 10^6}{(450)(1000)} = 186.6$ kg.

(b) Sp. gr. of dry solid is

$$S_s = \frac{100\ S_f S_v}{100\ S_v + P_v(S_f - S_v)}$$

$$= \frac{(100)(2.50)(1.0)}{(100)(1.00) + (71.4)(2.5 - 1)}$$

$$= \frac{250}{207} = 1.206$$

Assume solid in combined sludge = 4.5%

Then sp. gr. of wet sludge is

$$S = \frac{100\ S_s S_w}{100\ S_w + P(S_s - S_w)}$$

$$= \frac{(100)(1.206)(1.0)}{(100)(1.0) + (100 - 4.5)(1.206 - 1)}$$

$$= \frac{120.6}{119.7} = 1.007$$

Wt. of dry solid/1000 persons/day = 84 kg

which is only 4.5% of the wet sludge

$$\text{Then wt. of wet sludge} = \frac{84}{4.5} \times 100 = 1867 \text{ kg. /1000 persons/day}$$

$$\text{Then vol. of wet sludge} = \frac{1867}{1.007} = 1850 \text{ litres /1000 persons/day}$$

$$= \frac{(1850)(10^6)}{(450)(1000)}$$

$$= 4120 \text{ litres/million litres/day}$$

15.2.4 Combined Primary and Excess Activated Sludge — Digested

In the combined fresh sludge

Volatile solid = 60 gm/capita/day

and fixed solids = 24 gm/capita/day

Assuming that by digestion 67% of the volatile matter is destroyed, 25% being converted to fixed solid, we get

Volatile solid (1– 0.67) (60)= 20 gm/capita/day

Fixed solid = 24 + 0.25 (60 – 20) = 34 gm/capita/day

Then total solid in digested sludge = (20 + 34) = 54 gm/capita/day

$$\% \text{ of volatile solid} = \frac{20}{54} \times 100 = 37\%$$

(a) $$\text{Wt of dry solid} = \frac{(54)(1000)}{1000}$$

$$= 54 \text{ kg/1000 persons/day}$$

$$= \frac{54 \times 10^6}{(450)(1000)}$$

$$= 120 \text{ kg/million litres/day}$$

(b) Sp. gr. of dry solid is

$$S_s = \frac{100\, S_f\, S_v}{100\, S_v + p_v (S_f - S_v)}$$

$$= \frac{(100)(2.5)(1.0)}{(100)(1.0) + (37)(2.5 - 1.0)}$$

$$= \frac{250}{155.5} = 1.607$$

Assume solid in digested sludge = 7%

i.e. Moisture = 93%

Then sp. gr. of wet sludge is

$$S = \frac{100\, S_s\, S_w}{100\, S_w + P(S_s - S_w)}$$

$$= \frac{(100)(1.607)(1.0)}{(100)(1) + (93)(1.607 - 1.0)} = 1.027$$

Wt. of dry solid = 54 kg/1000 persons which is only 7% of wet sludge.

$$\text{Then wt. of wet sludge} = \frac{54}{7} \times 100 = 771 \text{ kg /1000 persons/day}$$

$$\text{Then vol. of wet sludge} = \frac{771}{1.027} = 750 \text{ litres /1000 persons/day}$$

$$= \frac{(750)(10^6)}{(450)(1000)}$$

$$= 1667 \text{ litres/1000 persons/day}$$

15.3 CHEMICAL CONDITIONING OF SLUDGE

Coagulation of sludge solids is called chemical conditioning of sludge. Chemical conditioning increases the rate at which water can be removed from the sludge by filtration through cloth or by drying on sand beds. The common conditioning chemical for sewage sludges are:

Ferric chloride	—	$FeCl_3$
Chlorinated copper	—	$FeSO_4Cl$
Ferric sulphate	—	$Fe_2(SO_4)_3$
Aluminium sulphate	—	$Al_2(SO_4)_3 18H_2O$
Lime	—	CaO.

Lime is not a coagulating chemical. It is used as a precipitant of bicarbonates present in the liquid fraction of the sludge. Coagulant demand is exerted by both the solid and liquid fraction of the sludge. The coagulant demand by the liquid fraction is due to its alkalinity or content of bicarbonate. Lane can be substituted for that portion of the coagulant which is required to satisfy the coagulant demand of the liquid fraction. Lime does not produce floc, but acts as a precipitant.

The coagulant or conditioner requirements of sludge are generally expressed as percentage ratio of pure chemical to the weight of solids fraction on a dry basis. The total coagulant requirement is divided into two parts (1) Liquid fraction requirement and (2) Solid fraction requirement. The liquid fraction requirement can be determined stoichoimetrically from the chemical reaction equation. But the solid fraction requirement has to be determined from experience. For ferric chloride, A.L. center has developed the following relationship.

$$P_c = \left[\frac{1.08\times10^{-4}AP}{100-P}+\frac{1.6p_v}{p_f}\right]$$

where p_c = Percentage of liquid on dry weight basis
p = Percentage of moisture in sludge
p_v = Percentage of volatile solids
p_f = Percentage of fixed solids
A = Alkalinity of sludge as mg/1.

The first term gives requirement by liquid fraction and the second term gives requirement by solid fraction. The requirement by solid fraction is proportional to the volatile matter content and hence digestion prior to coagulation will reduce the coagulant requirement as digestion destroys the volatile matter to some extent. But since digestion increases the alkalinity of the liquid fraction, the coagulant requirement by liquid fraction is increased. It can be reduced by the use of lime as precipitant or by washing out a part of the alkalinity with water of low alkalinity. The latter process is know as 'elutriation'.

15.4 ELUTRIATION

Washing out of the alkalinity of the sludge liquid with water is called elutriation. Elutriation is done in single or multiple tanks by single or repeated washings the wash water being used serially if desired. During washing the sludge is kept in suspension by air or mechanical agitation.

There are two types of elutriation depending on the method of operation.

(1) *Multiple elutriation* — in which washing is done in a single tank in several stages.

(2) *Counter current elutriation* — in which washing is done in a series of tanks. Wash water being used serially from the last tank to the first tank.

15.4.1 Multiple Elutriation

In this scheme $\frac{1}{11}$th volume of elutriating water is used in each of n-times of washing operation.

Let R = total volume of elutriating water for each volume of moisture in the sludge.

W = alkalinity of elutriating water in mg/l.

A = initial alkalinity of sludge liquid in mg/l.

E = alkalinity of elutriated sludge liquid in mg/l.

Then balance the alkalinity entering and leaving the tank, we get

$$= \left(\frac{1}{n}\right)(RW) + A = \left(\frac{1}{n}\right)(RE_1) + E_1 \text{ for first stage}$$

$$= \left(\frac{1}{n}\right)(RW) + E_1 = \left(\frac{1}{n}\right) RE_1 + E_2 \text{ for second stage and so on}$$

$$= \left(\frac{1}{n}\right) RW + E_{n-2} = \left(\frac{1}{n}\right) RE_{n-1} + E_{n-1} \text{ for } (n^{-1})\text{th stage}$$

$$= \left(\frac{1}{n}\right) RW + E_{n-1} = \left(\frac{1}{n}\right) RE_n + E_n \text{ for } n\text{th stage}$$

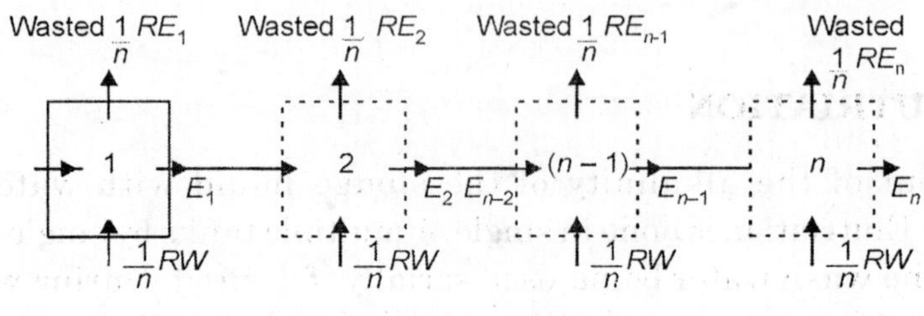

Fig. 15.1

Solving for alkalinity E_n of final sludge, and the ratio R of the wash water to sludge, we obtain

$$E_n = \frac{A + W\left[\left(\frac{R}{n}+1\right)^n - 1\right]}{\left(\frac{R}{n}+1\right)^n}$$

$$R = n\left[\left(\frac{A-W}{A-W}\right)^{1/n} - 1\right]$$

15.4.2 Counter Current Elutriation

In this system a series of tanks is used for washing and the cleaning wash water is first introduced into the last tank which receives partially elutriated sludge from the just previous tank. Then the wash water from the last tank is used to wash the sludge in the next to the last tank and this is followed upto the first tank.

$RW + E_{n-1} = RE_n + E_n$ for the nth stage

$RE + E_{n-2} = RE_{n-1} + E_{n-1}$ for (n–l)th stage

....

$RE_3 + E_1 = RE_2 + E_2$ for 2nd stage

$RE_2 + E_1 = RE_1 + E_1$ for 1st stage

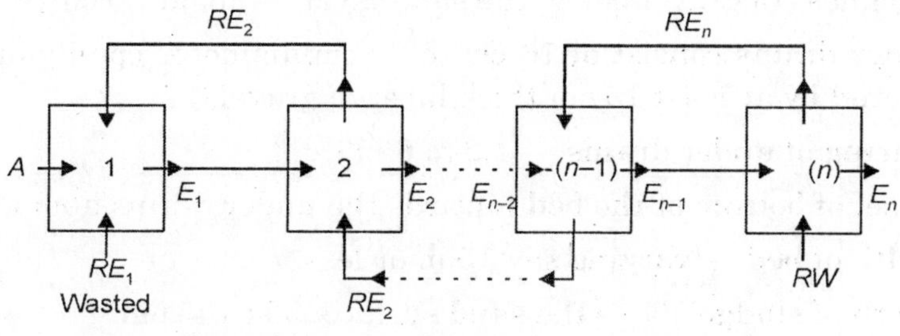

Fig 15.2

Solving for E_n and R, we get

$$E_n = \frac{A(R-1)+WR(R-1)}{R^{n+1}-1}$$

and $$= \frac{R^{n+1}-1}{R-1} = \frac{A-W}{E-W}$$

The ratio of elutriating water to the wet sludge is usually 2 to 1.

15.5 SLUDGE DRYING

Before sludge can be incinerated or disposed of as fill or fertilizer it must be suitably dried. A sludge containing 75 per cent moisture can be moved by shovel and transported in non watertight containers.

15.5.1 Method of Sludge Drying

(i) Air drying on sand beds
(ii) Presses
(iii) Vacuum filters
(iv) Centrifuges
(v) Heat dryers.

For small sewage treatment plants air drying on sand beds is most common.

15.5.2 Air Drying Sand Beds

15.5.2.1 Sand Beds

(i) Thickness of sand — 30 to 60 cm.
(ii) 7.5 cm thick upper layer of gravel should consist of 8 to 10 mm size gravels.

(iii) Thickness of gravel below the sand layer — about 30 cm.

(iv) Under drains consist of 16 cm to 20 cm diameter open joint tile pipe covered by at least 15 cm thick layer of gravel.

(v) Spacing of under drains — 1.2 to 6 m c/c.

(vi) Slope, of bottom of the bed towards the under drains about 2%.

(vii) Width of bed — varying say 15 m or less.

(viii) Depth of sludge above the sand surface 30 to 45 cm.

(ix) Height of side walls above surface of sludge —about 15 cm.

15.5.2.2 Bed areas

It depends on types of treatment, temperature of the locality and whether beds arc open or covered.

Table 15.4 *Air drying bed areas*

Type of treatment	*Bed area in sq. m per 1000 persons*	
	Open bed	*Covered bed*
Primary	94	70
Standard rate filter	118	94
High rate filter	140	118
Activated sludge	164	127
Chemical precipitation	188	140

15.5.2.3 Sludge disposal

There are various methods of sludge disposal. Depending upon the sewage treatment plant and the conditions of the surrounding the method can be adopted. The methods are as follows.

1. Disposal on land
2. Distribution by pipe line
3. Dumping into sea
4. Drying on drying beds
5. Ponding
6. Heat drying
7. Press filters

1. **Disposal on Land.** It can be done in two ways i.e., by ploughing and trenching.

Ploughing. Sludge is mixed with milk of lime or powdered lime and spread over the land. Then the land is ploughed.

Trenching. Trench of 90 cm wide and 60 cm deep are prepared in parallel rows at about 140 cm apart. Then the trenches are filled with sludge, a layer of excavated earth is placed over it.

2. **Distribution by Pipe Lines.** This is most simple and useful method. In this method the sludge is taken away by pipe lines to the nearest farm house to use as a fertilizer.

3. **Damping into Sea.** In this method the sludge is discharged into the sea. If the sea is near by the locality. Care should be taken to see that the sludge is discharged into the sufficiently deep sea. This method is adopted in big cities.

4. **Drying on Drying Beds.** In this method the sludge is spread over the land for drying.

5. **Ponding.** Sludge is discharged into a pond at about 100 cm depth, at the bottom of the pond is filled with ashes at about 15 cm thick, sludges are carried to the pond by tile drain or 100 mm diameter placed at 3 m apart. Embankment can be formed by the excavated earth.

6. **Heat Drying.** Wet sludge is heated to become dry. The dried sludge is usually use as a fertilizer.

7. **Press Filters.** Sludges are taken into jute or cotton bags, and placed between the filter plates. Then the plates are pressed at about 4 to 5 kg/ cm^2 so that all the waters are coming out from the sludge and form a cake. These cakes can also be used as a fertilizer.

15.6 SLUDGE DIGESTION

Process. Sewage sludge contains abandon food for bacteria and other micro-organisms which convert the complex organic substances into simple and more stable substances through their metabolic processes. Oxygen dissolved in fresh sludge is quickly exhausted and anaerobic decomposition, septic action or digestion starts.

In the beginning of anaerobic decomposition, the materials first attacked by the micros are sugars, starches and cellulose and soluble nitrogen compounds. The products of decomposition are organic acids, acid carbonates and gas consisting mainly of carbon dioxide. During this stage intensive acid formation drops the pH to 6 or lower and it is called the acid fermentation.

Acid fermentation which lasts for about 2 weeks at 278°K (5°C) is followed by a relatively lengthy period (about 3 months) of acid regression. During this stage organic acid and nitrogen compounds are attacked and converted to ammonia compounds and acid carbonates. Gas production drops and odor and odor of the sludge becomes extremely offensive. Sludge becomes foamy and scum is formed. pH value rises to about 6.8.

With the recession of acidity alkaline or methane fermentation starts. During this stage more resistant materials like protiens and organic acids are attacked. Gas production increases and it consists mainly of methane besides carbondioxide. pH may rise about 7, the sludge becomes well buffered (i.e., its hydrogens concentration remain unchanged even when relatively large amounts of acid or alkali are added).

In about a month the digestable material is exhaused with gas formation being sluggish and the sludge becomes stable enough for disposal.

The period of acid fermentation and acid regression is known as breaking in or ripening period. A new digestion tank have to pass through this period. Once the methane fermentation is well established fresh sludge solids, if added will be digested rapidly.

Digestion tank can be broken in rapidly by adding a batch of well digested sludge. If acid fermentation cannot be overcome spontaneously, lime can be applied to raise the pH value to help methane fermentation start.

15.6.1 Effect of Sludge Digestion

The results of sludge digestion are listed below.

(1) Conversion of complex organic compounds into stable inoffensive compounds.

(2) Conversion of a portion of sludge solids into liquids and gases resulting in reduction of sludge volume. About 67% of volatile solids are destroyed and converted to fixed solids having one forth weight of the solids which have been converted.

(3) Reduction of percentage moisture of sludge.

(4) Quick release of remaining moisture and rapid drying of the digested sludge.

(5) Reduction of coliforms by about 99.8% by digestion for 30 days at 95 to 100°F

(6) Use of digested sludge as fertilizer and of gas as fuel.

15.6.2 Optimum Conditions for Sludge Digestion

The speed of digestion is indicated by the rate of gas formation and depends on three factors (i) temperature (ii) pH value and (iii) proportion of sludge addition and withdrawal.

Temperature. The relation between the temperature of digestion and the period for 90% digestion of sewage sludge is shown in the curve (Fig.15.3).

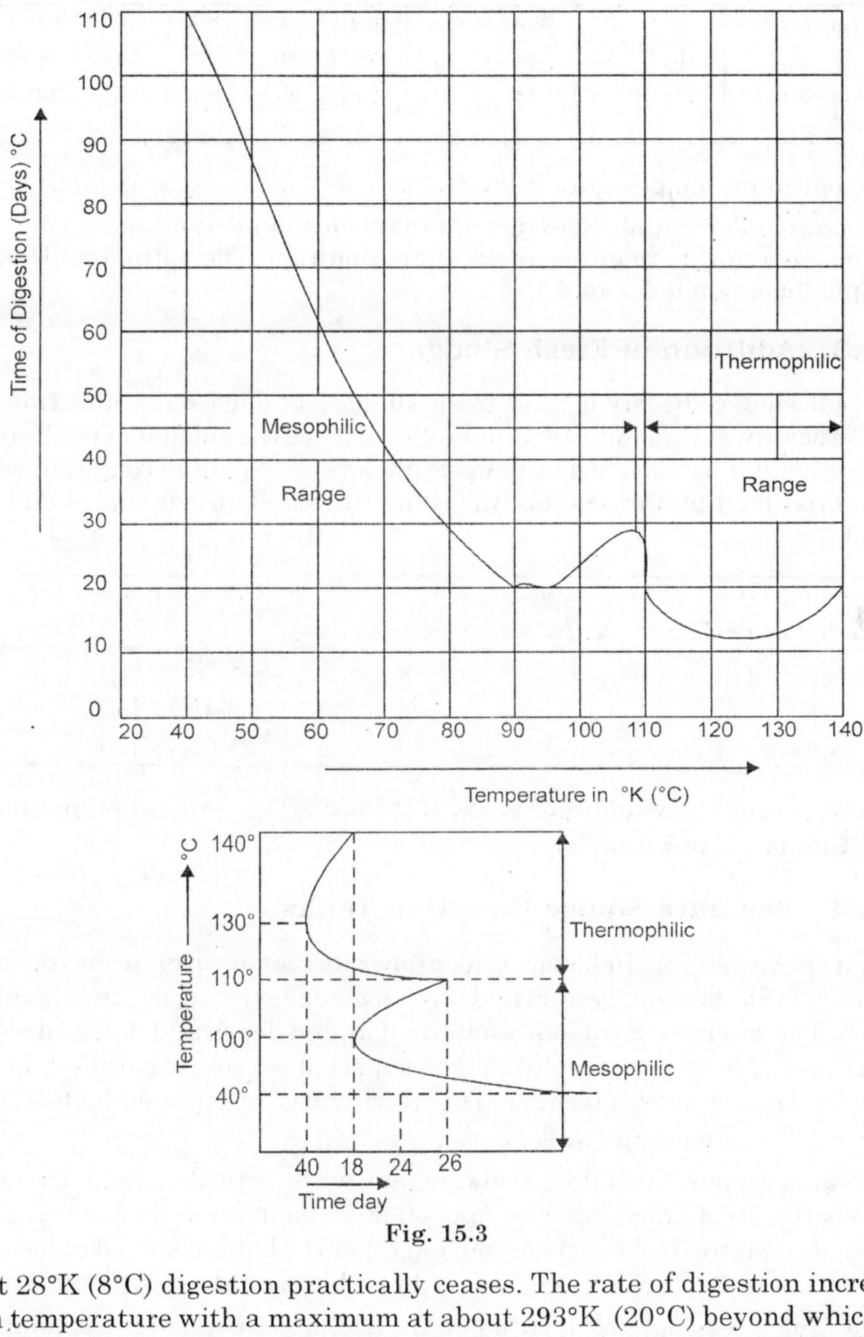

Fig. 15.3

At 28°K (8°C) digestion practically ceases. The rate of digestion increases with temperature with a maximum at about 293°K (20°C) beyond which the rate slows down a little and again reaches a maximum at about 299°K (26°C). Two types of bacteria are active at the two temperature zones, mesophilic and thermophilic at lower and upper zones respectively.

Temperature °K	283	285	287	289	291	293	295	297	299	301
or °C	10	12	14	16	18	20	22	24	26	30
Digestion period days	75	56	42	30	25	24	26	16	14	18

The optimum temperature is 293°K (20°C). $\because C + 273 = K$

pH value. For rapid digestion methane fermentation should be well established and this requires an alkaline condition. The optimum pH value for rapid digestion is 7.2 to 7.4.

15.6.3 Addition of Fresh Sludge

In actual sludge digestion tank fresh sludge is added daily and this will produce acidity during initial decomposition. Hence the amount of fresh sludge addition should be such as not to upset the overall alkaline condition which is essential for rapid digestion. According to Rudolfs fresh sludge addition should be:

Percentage by wt. of fresh solids w.r.t well digested solids added daily on dry basis	*Temperature*
2%	285°K or 12°C
2.5 to 3%	287°K or 14°C
3.5 to 4%	289°K or 16°C

On wet basis daily addition is about 3 to 5%. This proportion maintains the optimum pH of 7.2 to 7.4.

15.6.4 Separate Sludge Digestion Tanks

A separate digester includes the tank proper its roof or cover, a gas collecting system, a tank, heating system and often a sludge stirring or scum breaking system. The fresh sludge usually admitted at mid depth and digested sludge is withdrawn from bottom. Outlets for supernatant are generally 2 to 4 in numbers, spaced about 60 cm apart vertically beginning from 90 to 120 cm below the flow line of the tank.

The tank proper is usually circular in plan with vertical walls and a conical or flat bottom. R.C. tanks may be upto 30 m in diameter. Side water depth is commonly 6 metre. If there is no sludge scraper the bottom should be slopping at 1 vertical to 1 horizontal to 1 vertical to 4 hiorzontal.

The tank cover may be floating or fixed. Since fresh sludge is admitted intermittently, the flow line in the digestor fluctuates considerably. The gas collection system should be protected against entrance of air.

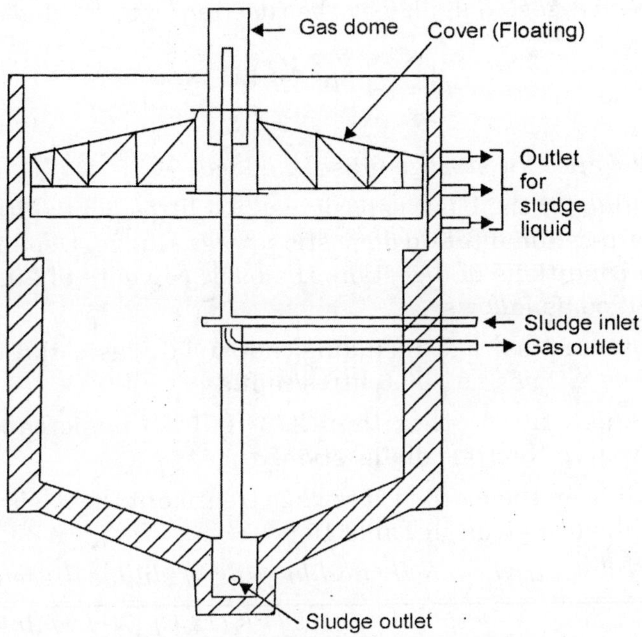

Fig. 15.4

Digestors may be heated by means of (1) heating the sludge outside the tank by a heat exchanger (2) fixed or moveable heating coils inside the tank (3) sludge gas burnt under water inside the tank etc.

15.6.5 Design of a Sludge Digesters

Assuming that the progress of decomposition is parabolic and that the average difference in volume of the fresh and digested solids is about $\frac{2}{3}$rd find difference.

$$C = \left[V_f - \frac{2}{3}\left(V_f - V_d\right)\right]t$$

where C = basic tank capacity per capita

V_f = Volume of fresh sludge/capits/day

V_d = Volume of digested sludge/capita/day

t = Time of digestion in days.

The space for sludge storage will mend on the period of storage.

In another method, the digester volume is calculated by multiplying the average of fresh digested sludge by the time required for digestion.

i.e.
$$C = \left(\frac{V_f - V_d}{2}\right)t$$

To this allowance for storage must be added.

A useful value for the basic capacity is 28.3 litres per capita for digestion at 60° F of plain-sedimentation domestic sewage sludge. Depending upon the variations in conditions of digestion, the basic capacity of 28.3 litres/capita must be modified as follows.

(i) For combined sewage including industrial waste the basic capacity should be 2 times i.e., 56.6 litres/capita.

(ii) For temperatures other than 278°K (5°C) basic capacity vary in proportion to the time of digestion.

(iii) Depending on the method of sewage treatment the basic capacity must be modified as given in Table 15.5

Table 15.5 *Required capacity and loading of sludge digestion tanks*

Treat process producting it	*Basic values at 285°K (12°C)*			*Loading of heated 291°K (18° C) tanks in gm volatile solids per litre per month*	
	litre per capita	*litres per kg of volatile solids daily*	*gram of volatile solids per litre per month*	*Conventional operation*	*High rate operation*
Plain sedimentation	28.3	746	400	800	1765
Plain sedimentation and trickling filtration					
(a) Low rain	36.8	810	369	738	1440
(b) High rate	45.3	870	353	706	1440
Plain sedimentation and activation					
(a) High rate	48.1	870	353	706	1440
(b) Conventional	65.0	1060	288	576	1122

Tank capacity must be provided for second storage at the rate of one-third the basic capacity for each month in excess of required digestion period or equal to the volume of digested sludge for period in excess of the digestion period.

Example 15.3 *Find the capacity of a heated digestion tank in which plain sedimentation sludge is to be digested at 289° K (18°C) for 1000 persons. Assume volume of sludge as 1070 litres when fresh, and 256 litres when digested per 1000 persons per day. Provide storage space for 60 days in excess of digestion period.*

Solution

The digestion period is 30 days at a temperature of 289°K (16° C)

$$\text{Then basic capacity} = \left\{V_f - \frac{2}{3}(V_f - V_d)\right\}t$$

$$= \left[1070 - \frac{2}{3}(1070 - 256)\right] \times 30$$

$$= (1070 - 542.7)(30) = (527.3)(30)$$

$$= 15819 \text{ litres}$$

$$\text{Storage space} = (256)(60) = 15360 \text{ litres}$$

$$\text{Total volume} = 15819 + 15360$$

$$= 31179 \text{ litres/100 persons}$$

$$\text{Sludge withdrawal} = 1070 - 256$$

$$= 814 \text{ litres/1000 persons/day}$$

$$\text{Volume of digestor} = \frac{(1070 + 256)}{2} + 30 \text{ litres/1000 persons}$$

$$\text{Add to it space for storage} = 19{,}890 \text{ litres.}$$

15.7 HEAT REQUIREMENTS OF DIGESTION TANKS

The heat supplied to the digestion must be sufficient (1) to raise the temperature of incoming sludge to that of the tank (2) to compensate the heat lost from the tank through walls, bottom and cover and (3) to compensate the heat lost in piping and other structures between the source of heat and the tank. The amount of heat entering into digestion reactions and used up in the evaporation of water is negligibly small.

The specific heat of most sludges is practically same as that of water.

The heat lost through walls, bottom and cover of the tank is

$$Q = CA\,(\Delta T)$$

where Q = heat loss

C = co-efficient of heat flow

A = area

ΔT = temperature difference between two sides.

For concrete tanks,

e = 2.59 kJ or 0.597 kilo calorie/sq. m/hour for exposure to dry earth.

= 3.68 kJ or 0.895 kilo calorie/sq. m/hour for exposure to air

= 7.37 kJ or 1.790 kilo calorie/sq. m/hour for exposure to wet earth. [∵ 1 kJ = 4.12 KCal]

Heat transfer to the sludge is analogous to heat loss. For stationary heating surfaces the co-efficient of heat flow in Joules or kilo kalorine per sq. m per hour varies from 1.936 to 254.204 or 47.8 to 61.7 for sludges of ordinary thickness and 861.08 to 1108.28 kJ or 209 to 269 Kcal.for thin sludges water.

Example 15.4 *Compute the volume of a sludge digestion tank treating primary sludge, only from the following data:*

(i) *Population = 100,00*

(ii) *Sewage flow = 200 litres/capita/day*

(iii) *Suspended solids = 300 mg/l in raw sewage out of which volatile matter = 250 mg/l and fixed matter = 50 mg/l.*

(iv) *Primary settling tank removes 60% of suspended solids.*

(v) *Present solids in raw sludge = 4%.*

(vi) *Present solids in digested sludge =10%*

(vii) *During digestion only 67% of the volatile solids are destroyed, out of which 25% are converted into fixed solids.*

(viii) *Period of digestion =30 days.*

Solution

$$\text{Daily sewage flow} = 100{,}000 \times 200 \text{ litres}$$

$$= 20 \times 10^6 \text{ litres.}$$

Wt. of volatile solids in the primary sludge

$$= \frac{210 \times 0.6 \times 20 \times 10^6}{10^6} = 2520 \text{ kg}$$

Wt. of fixed solids in the primary sludge

$$= \frac{90 \times 0.6 \times 20 \times 10^6}{10^6} = 1080 \text{ kg}$$

$$\text{Total solids in sludge} = 2520 + 1080 = 3600 \text{ kg}$$

$$\% \text{ of volatile solids, } p_v = \frac{2520}{3600} \times 100 = 70\%$$

Assume sp. gr. of volatile solids $S_v = 1.0$

Assume sp. gr. of fixed solids $S_f = 2.5$

$$\text{Sp. gr. of dry solids, } S_s = \frac{100\, S_f S_v}{100\, S_v + P_v\left(S_f - S_v\right)}$$

$$= \frac{(100)(2.5)(1.0)}{(100)(1.0)+(70)(2.5-1.0)}$$

$$= \frac{250}{205} = 1.22$$

Per cent of solids in raw sludge = 4%

i.e. percent of moisture in raw sludge $p = 96\%$

$$\text{Sp. gr. of the wet sludge } S = \frac{100\, S_s\, S_v}{100\, S_w + P\left(S_s - S_w\right)}$$

where $S_w = 1$

$$= \frac{(100)(1.22)(1.0)}{(100)(1.0)+(96)(1.22-1.0)}$$

$$= \frac{122}{100+(96)(0.22)} = \frac{122}{121.22} = 1.002$$

$$\text{Now, weight of wet sludge} = \frac{3600}{4} \times 100 \text{ kg} = 90{,}000 \text{ kg.}$$

$$\therefore \text{Volume of fresh wet sludge } V_f = \frac{90{,}000}{1.002} \text{ litres.}$$

$$= 89{,}800 \text{ litres.}$$

Volume of digested sludge

Volatile solids remaining after digestion

= (1– 0.67) (2550) kg

= (0.33)(2550) = 840 kg

Fixed solids after digestion

= 1080 + 0.25 (2520 – 840) kg

= 1080 + 420 = 1500 kg

Total solids = 1500 + 840 = 2340 kg

$$\therefore \text{ Percentage of volatile solids } p = \frac{840}{2340} \times 100 = 36\%$$

$$\text{Sp. gr. of dry solids, } S_s = \frac{100\ S_f S_v}{100\ S_v + p_v(S_f - S_v)}$$

$$= \frac{(100)(2.5)(1.0)}{(100)(1.0) + 36(2.5 - 1.0)} = \frac{250}{100 + 54} = 1.625$$

Percent of moisture in digested sludge = 90%

$$\therefore \text{Sp. gr. of wet sludge} = \frac{100\ S_s S_w}{100\ S_w + P(S_s - S_w)}$$

$$= \frac{(100)(1.625)(1.0)}{(100)(1) + 90(1.625 - 1.0)}$$

$$= \frac{162.5}{100 + 56.25} = \frac{162.5}{156.25} = 1.04$$

$$\text{Wt. of wet sludge} = \frac{2340}{10} \times 100 = 23400 \text{ kg.}$$

$$\text{Vol. of wet digested sludge} = \frac{23400}{1.04} \text{ litres}$$

$$= 22{,}500 \text{ litres.}$$

Digestion capacity C is

$$C = \left[V_f - \frac{2}{3}\left(V_f - V_a\right)\right] \times t \qquad [\because t = \text{in days}]$$

$$= \left[89{,}800 - \frac{2}{3}(89{,}800 - 22{,}500)\right] \times 30 \text{ litres.}$$

$$= \left[89{,}800 - \frac{2}{3} \times 67{,}300\right] \times 30 \text{ litres}$$

= (89,800 – 44,867) (30) litres
= (89,800 – 44,867) (30) liters
= 134,7990 liters
= 1348 m^3

Alternatively,

$$C = \left(\frac{V_f - V_d}{2}\right)(30)$$

$$= \frac{(83,800 + 22,500)}{2} \times 30$$

$$= \frac{112300}{2} \times 30$$

$$= 1684,500 \text{ litres} = 1685 \text{ m}^3$$

Capacity for storage of digested sludge

= Volume of digested sludge for 30 days

= 22,500 × 30 = 675,000 litres = 675 m^3

Total volume of digester tank

= 1348 + 675 = **2023 m^3**

15.8 FUEL VALUE OF SLUDGE

Volatile solids content of a sludge is a measure of its ignitiable constituents. From observations the statistical correlation between fuel value of volatile solids of different sludge is given as

$$Q = 1.222\,a\left(\frac{100p_v - b}{100 - p_c}\right)\left(\frac{100 - p_c}{100}\right)$$

where Q = fuel value in kilo-Joules or kilo calorie per kg of dry solids [∵ 1 kJ = 4.12 k calorie)

p_v = % of volatile solids

p_c = % of precipitating or conditioning chemical present

a, b → coefficients of different classes of sludge.

Type of sludge:	*a*	*b*
1. Plain sedimentation sewage sludge (fresh and digested)	131	10
2. Activated sludge —fresh	107	5

Example 15.5. *A chemically precipitated sludge contains 68% volatile matter on a dry basis and the precipitated chemical is 8.3% of sludge weight. Estimate fuel value of sludge.*

Solution

Using values of a and b as for plain sedimentation sludge,

$$Q = (1.222)(131)\left(\frac{100 \times 68}{100 \times 8.3} - 10\right)\left(\frac{100 \times 8.3}{100}\right) \text{ kJ or Kcal/kg}$$

= 9400 k cal per 38728 kJ or kg

15.9 FERTILIZER VALUE OF SLUDGE

The principal fertilizing ingredients in sludge are nitrogen (N), phosphorous and phosphoric acid ((P_2O_5) and potassium as potash (K_2O). Many sludges can be used as soil builders or soil conditioners i.e., as humus. Nitrogen promotes the growth of leaf and stem, phosphoric acid stimulates root growth, hastens ripening and increasing disease resistance, potash encourage vigorous growth, develops the woody part of the stem and the pulp of fruits and is needed for formation of chlorophyll, increases disease resistance, but delays ripening and humus improves water holding capacity of soil.

Type of sludge		*Nitrogen*
Plain sedimentation	....	0.8 to 5% of dry solids
Activated sludge	...	3 to 10% of dry solids
Trickling filter humus	....	1.5 to 5% of dry solids

Digestion reduces nitrogen content of sludges by 40 to 50%.

Phosphoric acid content of most sludges is 1 to 3% and potash about 0.1 to 0.3%.

15.10 FLOW CHARACTERISTICS OF SLUDGE

Hydraulic of sludge flow are dependent on the nature and amount of sludge solids. Solids are of such nature that flow is plastic or pseudo plastic rather than viscous. Again most sludges are thixotropic i.e., their plastic properties are changed by stirring and turbulence.

The friction losses are controlled by temperature, solids content and nature of solids. For thick sludges laminar or transitional flow persists up to a velocity of 0.45 to 1.35 metre/sec in pipes 12.5 to 30 cm in diameter. At turbulent flow all sludges behave more like water.

For laminar flow of plastic liquids the following relationship based on Poiseuille's equation for viscous liquid can be given

$$\frac{h}{l} = \frac{28,800}{g} \cdot \frac{v}{d^2}\left(930\frac{\eta}{\rho} + \frac{1}{5400} \cdot \frac{\tau_\gamma}{\rho} \cdot \frac{l}{v}\right)$$

In a receiving water well defined zones of pollution and recovery are observed after the point of sewage discharge. These zone are as follows.

1. **Zone of Degradation.** This will be established below the point of application of sewage. In this zone water becomes turbid and dark. Dissolved oxygen gradually diminishes, fish will be killed or driven a way; solid matters will settle as sludge at bottom and anaerobic decomposition will start, fungi will be present.

2. **Zone of Active Decomposition.** This zone is marked by the absence of dissolved oxygen and anaerobic decomposition. In this zone gases are formed as a result of decomposition with increase of CO_2 and ammonia, scum is formed on the surface, colour of water will be grayish or black, odours of H_2S and other sulphur compounds will be present Fungi will disappear, anaerobic bacteria will be dominant, fish absent and there will be warms and insect larvae present in the sludge.

3. **Zone of Recovery.** D.O., nitrites, nitrates increase with decrease of CO_2 and ammonia as a result of reaeration occurring at a high rate than that of BOD. Turbidity is decreased, anaerobic bacteria decreases in numbers with aerobes taking the place. Fungi, blue algae and green algae appear successively. Algae helps in reaeration by consuming CO_2 and releasing oxygen. Protozoa rolifera and crustacea will appear. Fish may also appear water becomes clear.

4. **Clear water zone.** In this zone the stream will assume its normal natural appearance. The amount of D.O. will be near saturation value. Normal fish life will be restored.

15.11 FACTORS AFFECTING SELF-PURIFICATION

1. **Dilution.** If the volume of the diluting water is large and if it contains considerable amount of D.O. the degradation of stream condition may be prevented. The minimum desirable amount of D.O. at all times is 4 mg/*l*.

2. **Currents.** Rapidly flowing turbulent river speeds up reaeration at the water surface and purify themselves in a shorter time than a sluggish river. But longer stretch is affected in case of a swiftly glowing river than a sluggish river.

3. **Sedimentation.** Sewage solids settle and accumulate on the river bed. Anaerobic decomposition takes place, currents mix up the decomposition products will overlying water.

4. **Sunlight.** If acts as a disinfectant against pathogenic bacteria. It is effective in self-purification through photosynthesis by which chlorophyll bearing organisms absorb carbon-dioxide and give off oxygen.

5. **Temperature.** Temperature effect self-purification by its influence on biological activities and on the solubility of oxygen. Higher temperature enhances biological activities and reduces solubility of oxygen.

6. **Biological Activity.** Biological activities are mainly responsible for self- purification. Biological forces include bacteria which attack organic matter and convert them into simple chemical substances through their metabolic process, algae and large aquatic plants absorbs carbon-dioxide and produce oxygen through photo-synthesis; protozoa some preyon bacteria and some act like algae; rotifers and crustaceans which consume algae and protozoa, macroscopic animals, which like an aquatic plants insect larvae.

Dissolved Oxygen. The most important parameter indicating the condition of a river is its dissolved oxygen content. If the quantity of D.O. in the receiving water is sufficient to satisfy the, BOD of the sewage, there will be no nuisance. The D.O. concentration in any water depends on the temperature and the amount and nature of dissolved substances. W.D. Hatfield gives solubility of oxygen in quiet water as

$$\text{D.O} = \frac{0.678(P - V_p)(1 - S \times 10^{-5})}{T + 35}$$

where

D.O. = Dissolved oxygen in mg/l
P = Barometric pressure in mm of mercury
V_p = Vapour pressure in mm of mercury
S = Chloride content in mg/l
T = Temperature in °K or °C

Formula given by G. A. Truesdale (1955) is

$$\text{D.O} = 14.161 - 0.3943\ T \text{ or K} + 0.007714\ T^2 \text{ or K}^2 - 0.0000646\ T^3 \text{ or K}^6 - 5 \times 10^{-5}\ (804101500.256\ T \text{ (or K)} + 0.00374\ T^2) \text{ (or K}^2)$$

where nomenclature is same as before.

15.12 OXYGEN-DEFICIT

It is the difference between the actual D.O. content and saturation content expressed in ppm or milligrams per litre.

15.13 OXYGEN BALANCE

According to H.W. Streeter, it is the difference between dissolved oxygen content and the total first stage BOD at a point in a stream.

15.14 DEOXYGENATION

Deoxygenation result from the withdrawal of oxygen from the D.O. reserve to meet the BOD of sewage solids. We have already obtained an expression for the amount of oxygen needed to satisfy BOD for a time t as

$$X = La\ (1 - 10^{-kt}) = La\ (1 - e^{-k't})$$

$$[\because K' = 2.30258 \text{ K}]$$

where x = Oxygen required in time t
La = Initial first stage BOD
K = Deoxygenation constant

L_a and K are dependent on temperature and may be given as

$$(La)_T = (La)_{20}\ [0.02T + 0.6]$$

and $$(K)_T = (K)_{20}\,[1.047^{(T-20)}]$$

where $(La)_T$, $(K)_T$— respective values at T° C (or K°) [∵ C + 273 = K]

$(La)_{20}$ $(K)_{20}$—values at 20°C (or 293° K)

But $K_{20} = 0.1$

Then the rate at which oxygen is used up i.e., the rate of deoxygenation is

$$\frac{dx}{dt} = -(La)(-K')e^{-k't} = (La)\,K'e^{-k't} \quad \text{... (1)}$$

This is, in other words the rate at which the oxygen deficit increases. If D is the oxygen deficit at any time, then

$$\frac{dD}{dt} = La\,K'e^{-k't} \quad \text{... (2)}$$

15.15 REOXYGENATION

Oxygen is absorbed from atmosphere by a water containing D.O. less than saturation quantity. And this rate of oxygen absorption is proportional to the oxygen deficit. In other words, rate of decrease of oxygen deficit is proportional to the oxygen deficit at any time. If D is the oxygen deficient, then

$$\frac{dD}{dt} = -K''\,D$$

i.e., $$\frac{dD}{D} = -K''\,dt$$

Integrating

$$\log_e D = -K''\,t + C$$

where C = Constant of integration.

Let $D = Da$ at $t = 0$ i.e., Da = Initial oxygen deficit.

Then

$$\log_e Da = C$$

$$\log_e D = -K''\,t + \log_e Da$$

$$\log_e \frac{D}{Da} = -K''t$$

i.e. $$\frac{D}{Da} = e^{-K't} = 10^{\frac{-K''}{2.30258}t} = 10^{-k_2 t}$$

i.e. $$D = Da\,e^{-K't} = Da = 10^{-k_2 t} \quad \text{... (3)}$$

where $K_2 = \frac{K''}{2.30258}$ ... (3)

where K'' and K_2 are called reoxygenation constant

K_2 is dependent on temperature and is given as $(K_2)_T = (BC_2)_{20}\ (1.0153^{T-20})$

Where$(K_2)_T$, (K_2) 20 — Values of K_2 at T°C and 20°C

Table 15.6 to shows the values of K_2

Table 15.6

S. No.	Type of stream	Values of K_2 at 293°K or 20° C
1.	Small ponds and backwater	0.05–0.10
2.	Sluggish streams and large lakes	0.10–0.15
3.	Large streams of low velocity	0.15–0.20
4.	large streams of normal velocity	0.20–0.30
5.	Swift streams of normal velocity	0.30–0.50
6.	Rapid and water falls	0.50 and higher

In most practical causes the co-efficient K_2 of reoxygenation is determined from observations of disolved oxygen and BOD at different points of the stream.

The rate of reoxygenation r is given as

$$r = -\frac{dD}{dt} = K''\ D = 2.3\ K_2 D \quad \text{... (4)}$$

r is maximum when D is maximum i.e., D is equal to oxygen saturation at a given temperature i.e.,

$$r_{max} = 2.3\ K_2\ D_{max}$$

According to H.M. Streeter,

$$(r_{max})_{T_2} = (r_{max})_{T_1}\left[1.016^{(T_2 - T_1)}\right] \quad \text{... (5)}$$

The average rate of reaeration, r_m between two stations A and B is given as

$$\left.\begin{aligned} r_m &= x_1 + (O_b - O_a) = (L_a - L_b)(O_b - O_a) \\ &= x_1 + (O_a - O_b) = (O_b - L_b)(O_a - L_a) \end{aligned}\right\} \quad \text{... (6)}$$

where

O_a, O_b = Dissolved oxygen *at* A and B respectively

L_a L_b = BOD at A and B respectively

D_a, D_b = Oxygen dficit at A and B respectively

x_t — Total oxygen demand exerted between A and B i.e., $x_t = (L_a - L_b)$

15.16 DEOXYGENATION AND REOXYGENATION

In a polluted stream deoxygenation and reoxygenation proceed simultaneously. The amount of dissolved oxygen increases with the absorption of atmospheric oxygen and at the same time some oxygen is used up to satisfy BOD or in other words the oxygen deficit decreases by reaeration and increases by BOD satisfaction. The rate of change of oxygen deficit at any time is given by summation of (2) and (3).

$$\frac{dD}{dt} = La\,K' e^{-kt} - K''D$$

i.e.

$$\frac{dD}{dt} = K'D = La\,K'\,e^{-k't}$$

Multiply both sides by integrating factor

i.e. $e^{\int k''dt}$ i.e. $e^{k''t}$

$$e^{k''t}\frac{dD}{dt} + K''\,De^{k''t} = La\,K'e^{-k't}\cdot e^{k''t}$$

i.e.

$$\frac{d}{dt}(De^{k''t}) = La\,K'e^{(k''-k')t}$$

$$De^{k''t} = \int K'La\,e^{(k''-k')t}dt$$

$$= K'La\,\frac{e^{(k''-k')t}}{(K''-K')} + C$$

where C = Constant of integration.

Let $D = Da$ at $t = 0$

Then

$$Da = \frac{K'La}{K''-K'} + C$$

whence

$$C = Da - \frac{K'La}{K''-K'}$$

Then

$$De^{k''t} = \frac{K'La}{K''-K'}\cdot e^{(k''-k')t} + Da - \frac{K'La}{K''-K'}$$

i.e.

$$De^{k''t} = \frac{K'La}{K''-K'}e^{k''t}\cdot e^{-k't} + Da - \frac{K'La}{K''-K'}$$

whence $D = \frac{K'La}{K''-K'}e^{-k't} + Dae^{-k't} - \frac{K'La}{K''-K'}e^{-k't}$

$$= \frac{K'La}{K''-K'}\left(e^{-k't} - e^{-k''t}\right) + Da''e^{-k't}$$

$$= \frac{\frac{K'}{2.3}La}{\frac{K''}{2.3} - \frac{K'}{2.3}}\left[10^{-kt} - 10^{-k_2t}\right] + D10^{-k_2t}$$

$$D = \frac{KLa}{K_2 - K}\left[10^{kt} - 10^{-k_2t}\right] + Da\ 10^{-k_2t} \qquad \text{... (7)}$$

where K = Co-efficient of deoxygenation

K_2 = Co-efficient of reoxygenation.

For curve A, $D = \frac{KLa}{K_2 - K}\left[10^{-kt} - 10^{-kt}\right] + Da\ 10^{-k_2t}$

For curve $B_{2,}\log \frac{L}{La} = -K_1t$

or $X = La(1-10^{-kt})$

The curve is shown in Fig. 15.5

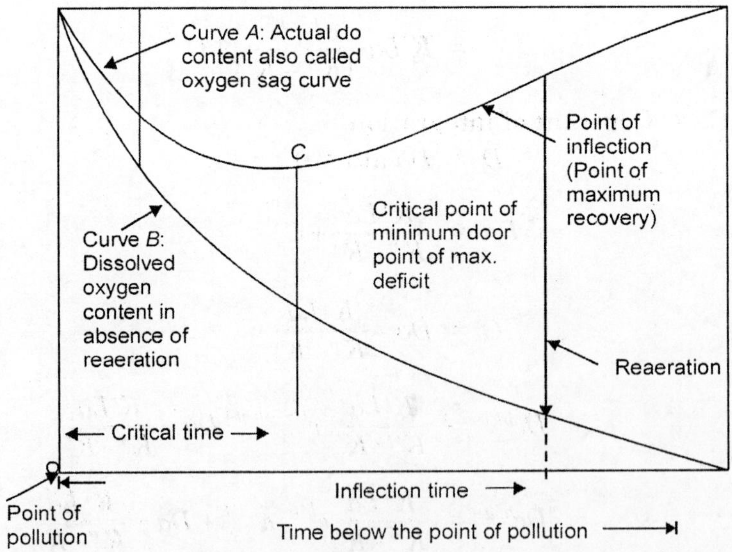

Fig. 15.5

15.17 CRITICAL POINT

Critical point is point of minimum D.O. or point of maximum deficit and obained from the condition $\frac{dD}{dt} = 0$ and $\frac{d^2D}{dt^2} < 0$.

where D is oxygen deficit.

Putting $f = \frac{K''}{K'} = \frac{K_2}{K}$ Equ. (7) can be written as

$$D = \frac{La}{f-1}\left[e^{-k't} - e^{-kft}\right] + Da\, e^{-k'ft}$$

$$= \frac{La}{(f-1)}\left[e^{-k't} - e^{-k'ft} + (f-1)\frac{Da}{La}e^{-k'ft}\right]$$

$$= \frac{La}{(f-1)}\left[e^{-k't} - e^{-k'ft}\left\{1-(f-1)\frac{Da}{La}\right\}\right] \quad \text{... (8)}$$

$$\frac{dD}{dt} = \frac{La}{(f-1)}[(-K')e^{k't}(-K'f)e^{-k'ft}\left\{1-(f-1)\frac{Da}{La}\right\}] = 0$$

i.e., $e^{k't} - fe^{k'ft}\left\{1-(f-1)\frac{Da}{La}\right\} = 0$

Multiplying by $e^{k'ft}$

$$e^{k't}e^{-k'ft-}\left\{1-(f-1)\frac{Da}{La}\right\} = 0$$

i.e., $$e^{k't(f-1)} = f\left\{1-(f-1)\frac{Da}{La}\right\}$$

Then $$K'(f-1)t = \log_e\left[f\left\{1-(f-1)\frac{Da}{La}\right\}\right]$$

$$= \log_{10}\left[f\left\{1-(f-1)\frac{Da}{La}\right\}\right](2.30)$$

i.e., $$t_c = \frac{2.30}{K'(f-1)}\log_{10}\left[f\left\{1-(f-1)\frac{Da}{La}\right\}\right]$$

$$= \frac{1}{K(f-1)} \log_{10}\left[f\left\{1-(f-1)\frac{Da}{La}\right\}\right] \quad \text{... (9)}$$

where $\frac{K'}{2.30} = K$

This is the expression for critical time.

Putting the value of t_c in Eq. (8)

$$D_c = \frac{La}{f}e^{-k't_c} = \frac{La}{f}10^{-k'tc} \quad \text{... (10)}$$

where D_c = Critical depth.

15.17 POINT OF INFLECTION IN OXYGEN SAG CURVE

The point of inflection is the point of maximum recovery and obtained from the condition $\frac{d^2D}{dt^2} = 0$

The time t_i and oxygen deficit D_2 for the point of inflection are given as

$$\left.\begin{aligned} t_i &= \frac{1}{K(f-1)} \log_{10}\left\{f^2\left[1-(f-1)\frac{Da}{La}\right]\right\} \\ &= \frac{1}{K(f-1)} \log_{10} f + t_c \end{aligned}\right\} \quad \text{... (11)}$$

$$D_i = \frac{La}{f^2}(f+1)10^{-kt_i} \quad \text{... (12)}$$

If $f = 1$, Eqs (9) and (11) for times become indetenninate but can be obtained from

$$t_e = (0.434)\left(1-\frac{Da}{La}\right)\frac{1}{K} \quad \text{... (13)}$$

$$t_i = (0.434)\left(2-\frac{Da}{La}\right)\frac{1}{K} \quad \text{... (14)}$$

PROBLEMS

15.1. A plain-sedimentation sludge has a moisture content of 95%, 72.2% of the dry solids being volatile matter. Find

(a) Specific gravities of dry solid, if the sp. gr of volatile and fixed solids are 1.00 and 2.50 respectively;

(b) Specific gravity of the wet sludge, if the specific gravity of the included water is 1.00 and

(c) Relative decrease in volume, if the moisture content is reduced to 85%.

15.2. (a) What is meant by "Sludge digestion"? Describe the process of sludge digestion in a digestion state the effects of sludge digestion. What are the optimum condition for sludge digestion?

(b) Compute the wet volume of digested sludge per 1000 persons after digestion of primary sludge. The primary sludge contains 39 gin of volatile solids and 15 gm of fixed solids per person before digestion. During digestion 67% of the volatile solids are destroyed. Out of which 25% are converted into fixed solids. The specific gravity of volatile and fixed solids are 1.00 and 2.50 respectively. Assume digested sludge contain as 13% solids.

15.3. Compute the volume of a sludge digestion tank treating primary sludge only from the following data.

(i) Population = 150,000

(ii) Sewage flow = 250 litres/capita/day

(iii) Suspended slides = 300 mg/ lit in raw out of which volatile matter = 250 mg/lit and final matter = 50 mg/lit

(iv) Primary settling tank removes 60% of suspended solids

(v) Percent solids in raw sludge = 4%

(vi) Percent solids in digested sludge = 10%

(vii) During digestion only 67% of the volatile solids are destroyed, out of which 25 % are converted into fixed solids

(viii) Period of digestion = 30 days.

15.4. (a) What is meant by "Elutriation"? Describe the various processes of Elutriation.

(b) A digested sludge has 40% volatile solids on dry basis, 90% moisture and an alkalinity of 2500 mg/lit. Find the percentage of ferric chloride that must be added to this sludge (a) If this sludge is not elutriated (b) If the sludge is elutriated in three tanks by counter current operation in a.ratio of 2 to 1 with water containing 25 mg/lit of alkalinity. Find also (c) the elutriation ratio that will reduce alkalinity of the sludge to 200 mg/lit by counter current elutriation and two stage multiple elutriation.

15.5. Define sludge. What are the design consideration of sludge digestor?

15.6. Write notes on:
(i) BOD,
(ii) Elutriation
(iii) Sludge Index
(iv) Time of concentration
(v) Sludge digestion
(vi) Population equivalent
(vii) Sludge Bulking
(viii) Surface area loading.

15.7. How the sludge can be conditioned. Describe the method with a sketch.

15.8. Why sludge is necessary to dry? How the sludge can be dried.

15.9. Explain clearly the optimum condition of sludge digestion. What does it mean the "Optimum condition"?

15.10. Explain the term sludge disposal. At what condition sludge is suitable for disposal? What should be the proper way to dispose off the sludge.©

Chapter 16

Septic Tank

16.1 INTRODUCTION

A septic tank is a water tight single storeyed tank in which sewage is retained for a sufficiently long period to permit sedimentation of suspended solids and partial digestion of settled sludge by anaerobic bacteria.

In unsewerid areas where adequate water supply is available from pipe, well or any other source for both conveyance of domestic wastes to treatment unit and also functioning of the unit septic tank system is a recommended method of disposing of wastes.

The septic tank combines two processes in a single unit sedimentation in the upper portion and digestion in the lower portion. At the bottom of the tank digested sludge accumulates whence from it is withdrawn at intervals. Lighter particles including oil, grease etc. rise to the surface and form flotating scum. The close contact of sewage with scum and sludge together with the agitation caused by the rising gasses produced as a result of anaerobic decomposition makes the effluent highly odourous, high in BOD high in finally divided solids and dark in colour. There is also every possibility that the effluent contains pathogenic organisms.

Design consideration for small tanks (up to 50 users) [IS : 2470 (Part-I) 1963].

16.2 SEDIMENTATION

1. Peak discharge.

No. of users	*Maximum, discharge*	*Fixture units*	
		Actual number	*Probable number discharging simultaneously*
5	9	1	1
10	18	2	2
15	18	3	2
20	27	4	3
50	63	10	7

Note. Fixture units is a quantity in terms of which the load producing effect of different plumbing fixtures on the plumbing system ere expressed on some arbitrarily chosen scale. Here it is 9 litres per minute.

2. The surface area of the tank assumed at 0.82 m for tank serving 10 persons or less at a flow rate of 9 litres per minute.
3. Depth of sedimentation assumed at 23 cm to 30 cm.

16.3 SLUDGE DIGESTION

1. For Indian conditions per capita contribution of drthy solids assumed at 70 gm per day.

2. The capacity required for sludge digestion at 0.033 m^3 per capita at 25°C.

16.4 SLUDGE STORAGE SPACE

1. Volume of digested sludge assumed as 0.00021 m^3 per capita per day. Sludge storage space to be provided depending in interval of cleaning which may be 6 months, 1 year or 2 years.

16.5 DIMENSIONS OF SEPTIC TANKS

1. Min. width = 75 cm
2. Min.depth = 1 m
3. Min. liquid capacity = 1 m^3
4. L : W = 2 : 1 to 4 : 1.

16.6 CONSTRUCTION

1. Walls of septic tanks may be built out of brick work, stone masonry, concrete or other suitable material.

Thickness of wall made of brick = 20 cm min. and both sides should be plastered with 12 mm thick cement mortar not weaker than 1:3. If made of stone masonry then it should be 37 cm thick.

2. *The flooring.* The floor should be cement concrete of proportion not lesser than 1 : 2 : 4.

3. *Bottom slope.* The bottom slope should be 5 to 10% towards inlet or sludge outlet pipe.

4. *Cover.* Septic tank should be covered by concrete slabs of adequate strength in which opening should be provided for purposes of desludging or inspection. The opening should be covered by suitable cover.

5. *Ventilating pipe.* Each septic tank should be provided with a ventilation pipe of at least 50 mm diameter. The top of the pope should be provided with a cage of wire mesh. The ventilating pipe should extend to a height of about 2 m when the septic tank is at least 15 m away from the nearest building and to a height of about 2 m above the top of the building when it is closer than 15 m.

16.7 DISPOSAL OF SEPTIC TANK EFFLUENT

The disposal of septic tank effluent by soil absorption depends upon the position of ground water level and subsoil conditions. The effluent may be treated in a thick ling filter or chlorinated where necessary.

16.8 SOIL ABSORPTION SYSTEM

There are mainly three types.

1. Seepage pit or soak pit.
2. Leaching cess pool.
3. Dispersion trench.

Leaching Cess Pool. It functions as a septic tank for portion at the bottom and as a soil absorption system for a portion at the top.

It is an underground structure. It may be circular or rectangular. This may be constructed with brick or cement concrete. Inside surface should be plastered and an air tight cover should be provided.

This is adopted to get the advantage of water carnage system. The sewage is carried to the cess pool through properly laid sewers.

The bottom should be water tight where highest water table is below the bottom of the cess pool to retain sewage and sludge while the upper portion shall be provided with open jointed lining to allow the supernatant effluent to disperse into the surrounding soil. Depth of bottom may be calculated as septic tank while the depth of top portion may be determined as a soil absorption system where the water table is above the bottom, it need of be water tight at bottom.

Dispersion trench:

Depth of trench = 50 to 100 cm

Width of trench = 30 to 100 cm

Diameter of pipes = 75 to 100 mm

Length of trench = 30 m maximum

Spacing of trenches = 1.8 m minimum

Design of Soil Absorption System. The allowable rate of application of effluent per unit area of dispersion trench or soak pit is limited by the percolation rate of soil.

Percolation rate (Minute)	*Maximum rate of effluent application litres/m²/ day*
1 or less	204
2	143
3	118
4	102
5	90
10	65
15	53
30	37
45	33
60	24

Note: (1) Absorption area of dispersion trench is the trench bottom area.

(2) Absorption area for soak pit or Leaching cess pool is the effective side wall area (effective depth is measured from 15 cm below the invert level of inlet pipe to the bottom of the pit).

(3) If the percolation rates exceed 30 mins the soil is unsuitable for soak pit or leaching cess pool. If the percolation rate exceeds 60 mins the soil is unsuitable for any soil absorption system.

Percolation Rate. It is the time required in minutes for water to fall 25 mm in the test hole.

To determine percolation rate, a vertical hole of horizontal diameter about 10 cm to 30 cm is dug or bored up to the depth of the proposed dispersion trench. Coarse sand or fine gravel is added for a depth of 5 cm.

Water is poured up to a min. depth of 30 cm over the gravel and left for 30 cm over the gravel and left for 24 hours. 24 hours after the water is added, water depth should be adjusted to 15 cm over the gravel. Then from fixed reference point, the drop in water level should be measured over a period of 30 mins. This drop should be used to calculate the percolation rate.

If no water remains after 24 hours water is added for a depth of 15 cm over the gravel. From fixed reference point drop is measured at 30 min. intervals for 4 hours, refilling 15 cm over the gravel as necessary. The drop that occurs during the final 30 min. period should be used to calculate percolation rate.

Example 16.1 *To design a septic tank with a soak pit, the following informations are available.*

Water supply = 150 litres/capita/day.

Population = 1000

It is assumed that the total amount of water supply is available in the form of sewage.

Solution

Sewage per day = 150 × 1000 = 150,000 litres.

Assuming the detention period is 12 hours.

$$\text{Capacity of the septic tank} = \frac{150{,}000}{24} \times 12 = 75{,}000 \text{ litres.}$$

Considering that the tank will be cleaned every one year and also considering that the sludge capacity per person will be 15 litres.

∴ The capacity for sludge storage

= 1000 × 15 = 15,000 litres

Then the total capacity of the bank

= 75,000 +15,000

= 90,000 litres.

Allowing 5,000 litres additional capacity for future extension.

Then the total capacity = 90,000 + 5,000

= 95,000 litres = 95 m^3

Assuming the depth of tank = 2 m

$$\text{Then the plan area} = \frac{95}{2} = 47.5 \text{ m}$$

Considering the size of the tank = 6 m × 8 m × 2 m

Also considering 30 cm as free board, then the overall size of the septic tank

= **8 m × 6 m × 2.3 m.**

Check: Design size = 8 m × 6 m × 2.3 m = 110.4 m^3 which is more than 95 m^3

16.9 DESIGN OF SOAK PIT

Let the absorption value of the soil is taken as, 1500 litres/m^3

Then the volume of the soak pit = $\frac{95 \times 1000}{1500} = 63.4 \text{ m}^3$

The depth of pit is taken as 4 m, then the area $= \frac{63.4}{4} = 15.85 \text{ m}^2$

Now $$\frac{\pi d^2}{4} = 15.85$$

$$d^2 = \frac{15.85 \times 4}{\pi} = 20.19$$

$$d = 4.48$$

Say $$d = 4.5 \text{ m.}$$

The diameter of the soak pit = 4.5 m.

16.10 RATIONAL METHOD OF SEPTIC TANK DESIGN

This method is more or less the same as the standard method. But only the difference lies in calculating the volume. The method is as described below:

(i) Volume for settling zone should be kept minimum 30 cm deep.

(ii) Volume for digestion zone should have the capacity to digest at the rate of 0.032 m^3 per capita.

(iii) Volume for storage zone should have the capacity to store at the rate of 0.002 m^3/head/day.

So the volume of this tank mainly depends on the interval of cleaning generally the cleaning interval is used between 1 year to 2 years.

For 10 or less than 10 persons the surface area of this type of tank is used at the rate of 9.0 m^2.

Construction. It can be constructed with bricks, stone or concrete. Thickness should be provided to see the tank has enough strength and perfectly water tight. On top of the tank manhole covers are used to inspect and for desludging when necessary. In case of circular manhole cover minimum clear opening is needed to have 50.0 cm dia and for rectangular manhole cover need the clear opening of 60 cm × 45 cm. Every tank must have a ventilation pipe of 5 cm. dia (minimum). The top of the ventilation pipe should have a wire mesh cover to check the mosquito breeding and fly breeding ventilation pipe should be so long that it should not create any smell nuisance to the sarroundings. Generally the height of the pipe is used as 2 m but it may be adjusted as required.

It is very urgent to see the water tightness of the tank before use after construction. It can be done by filling the tank with water and wait to see the water level after 24 hours. After 24 hours the water level in the tank should not fall more than if $1\frac{1}{2}$ cm.

Example 16.2 *Design a septic tank for 50 persons by using rational method.*

Solution

(i) *Settling zone.* Minimum liquid depth is 30 cm so the depth of liquid is taken as 30 cm.

Minimum surface area = 0.9 m^2 for 10 persons

Take 1.0 m^2 for 10 persons

$\therefore$ For 50 persons surface area

$$A = 1\times\frac{50}{10} = 5 \text{ m}^2$$

$$\therefore \quad \text{Vol. of settling zone} = 0.3 \times 5 = 1.5 \text{ m}^3$$

(ii) *Volume of digestion zone*

$$\text{Per capita volume} = 0.032 \text{ m}^3$$

$$\therefore \quad \text{For 50 persons} = 0.032 \times 50 = 1.6 \text{ m}^3$$

(iii) *For storage zone*

It is at the rate of 0.002 m^3/capacity/day

Cleaning Interval is assumed as one year

$$= 1 \text{ year} = 365 \text{ days}$$

$$\therefore \quad \text{Storage vol.} = 0.002 \times 50 \times 365$$

$$= 36.5 \text{ m}^3$$

$$\text{Total vol. required} = 15 + 1.6 + 36.5 = 39.5 \text{ m}^3$$

$$\text{Use depth} \quad \frac{39.5}{10} = 3.95 \text{ m}$$

Breadth = 2m

Length = 5 m

Depth = 3.95 m + 30 cm free board.

Check:

Flow is assumed to be BOD lit/day

Total flow = 700 × 50 = 35,000 lit/day

= 35 m^3/day

For detention period of 24 hours

Vol. required = $35 \times 1 = 35$ m^3 day

Vol. provided in the tank = 39.5 m^3

The volume will be **39.5 m^3**

16.11 IMHOFF TANK

This type of tank was developed by Dr. Karl Imhoff of Germany. Like septic tank, it combines sedimentation and digestion. But unlike septic tank it is a two storeyed tank separating sedimentation and digestion, sedimentation takes place in the upper compartment while digestion takes place in the lower compartment since there is no intimate mixture of sewage and the digesting sludge, the effluent is of better quality than that of septic tank.

16.11.1 Constructional and Operation Feature

Sewage enter the flow chamber overflowing the inlet weir. To prevent direct current from inlet to outlet, hanging baffle walls are provided before inlet and outlet weirs extending 37.5 cm below the water surface. The flow chamber and digestion chamber are separated by means of slopping walls. The solids settling in the flow chamber slides down the slopping wall through the slit at bottom of the flow chamber into the digestion chamber. After being digested in the digestion chamber, the sludge can be withdrawn through the sludge pipe under hydrostatic pressure. The gasses produced as a result of digestion escape through the gas vent. The lighter solid like oil, grease, etc. will rise to form scum which accumulate in the scum chambers.

In order to prevent the rising gas and solids from entering the flow chamber the slot is trapped by over lopping of the slopping walls which are sometimes called septum walls. The slope of the septum walls should be 50° to 60° to horizontal.

The bottom of the tank is made slopping to from a hopper or series of hoppers divided by partition walls, partition walls are provided with openings at bottom to help sludge flow to maintain a common level in the hoppers. These partition walls serve to support the septum walls and side walls and tend to break up bottom currents of sewage. In large tank these walls should be spaced 4.5 m apart.

Inlets and outlets are similarly designed and arrangements are made to reverse the flow. The object of this is to maintain a uniform accumulation of sludge in hoppers.

The space extending 45 cm below the plane of slot is called, 'Neutral Zone' and not included in the space for digestion.

16.11.2 Design Considerations

(1) Average sewage detention period in flow chamber = 1 to $2\frac{1}{2}$ hours.

(2) Depth of flow chamber from weir line to plane of slot should not exceed 2.7 m to avoid great depths.
(3) Surface area loading should not exceed 24,500 litres/day/m^2.
(4) Length: width of flow chamber = 3 : 1 to 5: 1.
(5) Length should not exceed 30 m.
(6) Sludge digestion capacity 28 to 56 lit/capita/day.
(7) Minimum total depth = 4.5 m and maximum total depth may upto 10 m.
(8) Depth of neutral zone below the plane of slot is 45 cm.
(9) Gas vent area = 15% to 25% of the total surface area.
(10) Slope of slopping septum walls = 50° to 60° to horizontal i.e., 1.25 V : 1 H to 1.7 V: 1 H.
(11) Width of slot = 20 cm minimum.
(12) Overhang of septum at slot = 20 cm minimum.
(13) Slope of sludge hopper = 30° to 40° with horizontal i.e., IV : 1.4 H to 1V : IH
(14) Diameter of sludge pipe = 20 cm minimum.
(15) Vertical distance between the sludge outlet and the sewage level in tank = 1.5 m minimum.

Advantages

The following are the advantages to an Imhoff tank.
(1) Sludge obtained from the imhoft tank can be easily dried.
(2) Comparing to the septic tank the process of sludge digestion is better in imhoff tank.
(3) Imhoff tank delivers better effluent than that received from the plain sedimentation tank.
(4) No need to have skilled supervision constantly.
(5) It is quite capable of removal of BOD and settleable solid. The BOD removed upto 30 to 40 percent.

Disadvantages

The following are the disadvantage of the Imhoff tank.
(1) In hard soil the construction of Imhoff tank may be uneconomical due to its big depth.
(2) If the tank is improperly operated, offensive odours coming out from the tank.

Table 16.1 *Advantage and Disadvantages of Septic tank and Imhoff tank*

S.No.	Septic Tank	Imhoff Tank
	Advantages	**Advantages**
1.	It does not need skilled supervision and easy to construct.	No need to have skilled supervision constantly but the construction is little costly.
2.	Considering to the advantage, the cost is reasonable.	Considering to the advantage, the cost is little high.
3.	Function of a septic tank is good. About 90% BOD is removed and 80% suspended soild is also removed.	Function is good. About 30 to 40% BOD can be removed, 60% suspended solid is also removed.
4.	There is no moving parts.	Some moving parts are present.
5.	Volume of sludge is reduced. Comparing the sludge in plain Sedimentation tank it is about 60% less in volume and 30% less in weight.	No such noticeable reduction of sludge volume and weight.
6.	Effluent and scum obtained in septic tank can easily be disposed without any trouble.	Sometimes scums may create trouble they can change the quality of the effluent.
	Disadvantages	**Disadvantages**
1.	Improper functioning of septic tank, foul smell will come out and the effluent will be dark.	When not working satisfactorily offensive odours will come out.
2.	For serving large number of people it requires very large size.	It can serve large number of people with reasonable size.
3.	Leak through the top of septic tank will cause the air pollution.	It has a tendency to foam or to boil, which will effect the working of sedimentation chamber.
4.	Removal of sludge occasionally the maintenance cost gets high as well as tedious job.	Occasional removal of sludge is not required if the tank is in order.
5.	Working condition of septic tank is nonuniform and unpredictable.	The working condition of this tank is more unifom than that of the septic tank
6.	The depth of septic tank may not be too deep.	The depth of this tank is very deep and costly.
7.	Construction of septic tank is not difficult.	It is difficult to construct.

(3) It has the tendency to boil or foam. This is very harmful for the tank, becauses it leads the scum to go up to the top of the tank. It also try to force the sludge to enter the sedimentation chamber through the slot. This will effect the quality of effluent.

(4) At high acidic condition the tank will not operate properly, it will be unsuitable at this condition.

Note : For large sewage treatment works Imhoff tank is not suitable. But it is most suitable for small cities and institutions where the separate digestion tanks are not possible to install.

Example 16.3 *To design an imhoff tank the following informations are available.*

(1) Population to be served = 20,000 persons.

(2) Water consumption per person = 135 lit/capita/day. Assume the other suitable data required.

Solution

Sedimentation chamber

Assume detention period = 2 hours

$$\text{Chamber capacity} = 20{,}000 \times 135 \times \frac{2}{24} = 225000 \text{ lit.}$$

Assume the effective depth of the chamber = 2 (m)

(Including part of the bottom slopping)

Assume the width of the chamber = 4 m

$$\text{Chamber length} = \frac{225}{2 \times 4} = 28.125 = 28.5 \text{ m say.}$$

This seems to be too long for a single tank. It will be better to have two tanks of 14.25 m each.

$$14.25 \text{ Ratio of length to width} = \frac{14.25}{4} = 3.81$$

This is satisfactory because the value lies between 3 and 5.

$$\text{Surface loading} = \frac{20{,}000 \times 135}{2 \times 14.25 \times 4} = 23{,}700 \text{ lit/m}^2\text{/day}$$

This is also with in the permissible limit.

$$\text{Height of slopping} = \frac{1}{2} \times 4 \times \frac{5}{4} = 2.5 \text{ m}$$

(Considering 4 m width and bottom side slope as 5 : 4).

Height of the vertical portion below the liquid surface

$$2 - \frac{1}{2} \times 2.5 = 0.75 \text{ m}$$

(Considering average effective depth 2 m)

For free board 0.45 m is added. Then the total depth of the sedimentation chamber upto the bottom at the entrance slot will be 0.45 + 0.75 + 2.5 = 3.7 m.

Depth of neutral zone considered = 50 cm.

The length 14.25 m is divided in further 4 units. Each unit will have a cross-section of 3.56 m × 4.00 m.

Sludge Digestion Chamber

4 m width of sedimentation chamber and having 15 cm thick wall of scum chamber, 6 m = 6– (4 +2 × 0.15) = 1.7 m.

This is 25% of the width of sludge digestion chamber. Vent width 0.85 m will be safe.

Assume 0.05 m^2 sludge per capita

Sludge digestion chamber capacity = 20000 × 0.05 = 1000 m^3

$$\text{Because of 8 units, capacity for each unit} = \frac{1000}{8} = 125 \text{ m}^3$$

Assume hopper depth = 2 m with 1 : 1 slope

and bottom cross-section as 2 m × 1 m

$$\text{Then the volume or hopper} = \frac{1}{2} (6 \times 3.56 \times 3 - 2 \times 1 \times 1) = 20.7 \text{ m}^3$$

Because the cross-section of hopper at top = 6.00 × 3.56 m

Sludge digestion chamber (balance volume) = 125.0 – 20.7 = 104.3 m^3

$$\text{Straight portion chamber height} = \frac{104.3}{6 \times 3.56} = 5 \text{ m (say)}$$

∴ Total depth of imhoft tank

= 3.7 + 0.5 + 2.0 + 5.0

= **11.2 m which is O.K.**

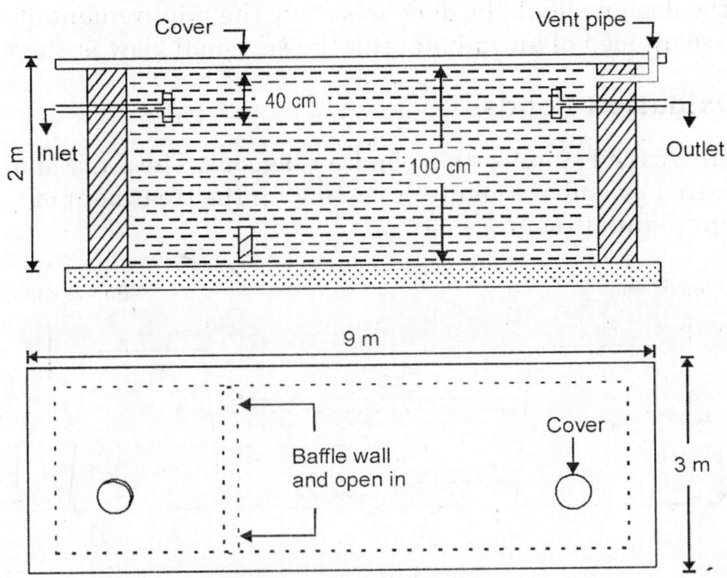

Fig. 16.1 *Sectional view of septic tank*

Plan and sectional elevation with standard dimensions are shown for the septic tank.

There is no such hard and fast rule to have the same dimension. It may be changed according to the requirement needed.

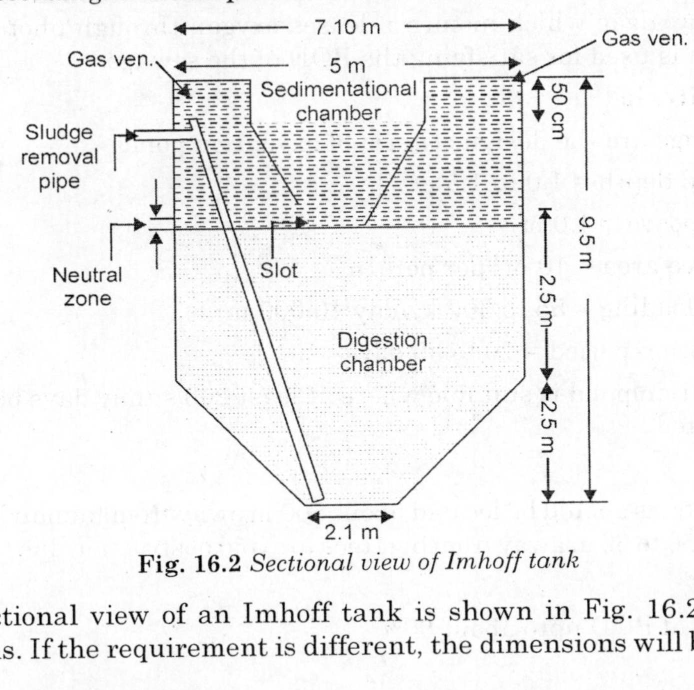

Fig. 16.2 *Sectional view of Imhoff tank*

The sectional view of an Imhoff tank is shown in Fig. 16.2 with some dimensions. If the requirement is different, the dimensions will be varied. In that case the design should be done to satisfy the requirement of the Imhoff tank. To give an idea of an Imhoff tank the sectional view is shown.

16.12 Oxidation Ponds

These are also called stabilization ponds or lagoons. These are shallow ponds or pools created by embankments or cutting for the treatment of raw sewage or treatment plant effluent.

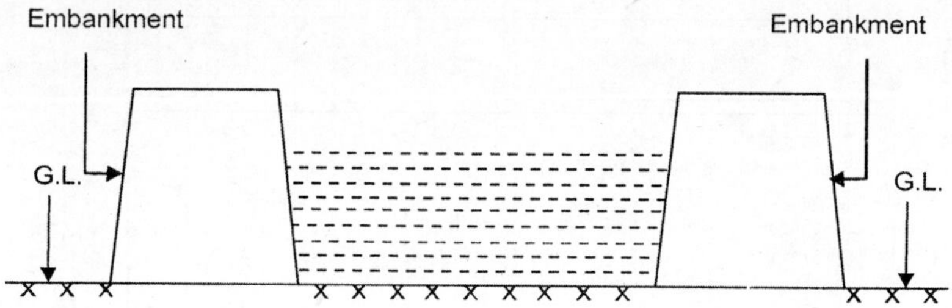

Fig. 16.3

The treatment of this pond is effected by sunlight, algae, evaporation, percolation, sedimentation and oxidation. It results in biological stabilization of influent, oxidation, nitrification and reduction of liquid volume. Oxygen is obtained from air or as a result of photosynthesis.

Bacterial decomposition of organic matters releases carbon di-oxide, which is utilised by algae which in turn releases oxygen through photosynthesis. This oxygen is used for satisfying the BOD of the sewage.

Design Criteria

The followings are the design criteria of oxydation pond.

(1) Liquid depth = 1.0 to 1.5 m

(2) Free board = 1.0 m

(3) Surface area = 10 m^2 per person

(4) BOD loading = 55 to 300 kg/day/10,000 m^2

(5) Detention period = 20 to 30 days.

(6) Oxidation pond is suitable where at least 200 sunny days per year are expected.

Location

Oxydation ponds should be located about 500 m away from human habitation. It should be 50 to 60 m away from big trees to avoid obstruction due to shadows.

Efficiency

(1) Removal of BOD upto about 90%

(2) Removal of chloroform upto about 99%.

Advantages

(1) Successful disposal of waste liquids.

(2) Reduction of volume due to evaporation and percolation.

(3) Effectiveness as an equalising basin to absorb rapid fluctuations in flow.

(4) Low cost.

Disadvantages

(1) Large area required

(2) Hazards to public health under unformable condition.

(3) Possible pollution of ground-water.

(4) Breeding of mosquitoes and other flying insects.

Example 16.4 *Sewage flow = 4 × 10^5 lit/day.*

BOD = 300 mg/l. Allowable BOD loading is 100 kg/10,000 m^2/day.

Find the hydraulic loading.

Solution

Assume liquid depth = 1.0 m

$$\text{BOD loading} = \frac{3000 \times 4 \times 10^5}{10^4} \text{ kg} = 120 \text{ kg.}$$

$$\therefore \text{ Surface area required} = \frac{120}{100} \times 10{,}000 \text{ m}^2$$

$$= 12{,}000 \text{ m}^2$$

$$= 1.2 \text{ hectares.}$$

$$\text{Volume of the pond} = 12000 \times 12000 \text{ m}^3$$

Daily flow = 400 m^3

$$\therefore \quad \text{Detention period} = \frac{12000}{400} = 30 \text{ days}$$

$$\text{Hydraulic loading} = \frac{400}{12000} \text{ m}^3\text{/day/m}^2$$

$$= \frac{1}{30} \text{ m}^3\text{/day/m}^2$$

16.12.1 Use of Oxidation Ponds

This pond is generally suitable for small towns. This is useful in warm climate. It is better to have the natural depression of ground to form the pond. These ponds are used in Europe on small communities and becoming popular in some parts of America. Indian climate and condition is the most oxidation ponds. Shallow oxidation ponds are proved to be very effective for sewage. Treatment in many countries with warm sunny climates.

PROBLEMS

16.1. With a net sketch describe the relevant parts of an Imhoff tank.

16.2. Discuss briefly the design considerations of an Imhoff tank.

16.3. What are the advantages and disadvantages of an Imhoff tank?

16.4. Draw net sketch of a septic tank and explain it properly.

16.5. What are the advantages and disadvantages of a septic tank?

16.6. Explain why and where the septic tank is suitable.

16.7. Discuss briefly the design consideration of a septic tank.

16.8. Compare the advantages and disadvantages of Septic tank and Imhoff tank.

16.9. Discuss briefly the oxidation pond.

16.10. Write the advantages and disadvantages of oxidation pond.

16.11. What are the conditions needed to have an oxidation pond? State its use.

16.12. Why oxidation ponds are constructed with shallow depth.

16.13. Why the cover is provided at the top of a septic tank.

16.14. What are the conditions when oxidation ponds will not produce any effluent?

16.15. Design an Imhoff tank for teating sewage of a town of population 25,000, if the flow of sewage is 150 litres/capita/day.

Assume reasonable data if required

16.16. State the design considerations for small septic tanks and compute dimensions of a septic tank for 50 users, using reasonable data.

16.17. What is population rate? How is it obtained?

Chapter

17

Refuse Sanitation

17.1 DEFINITION

Refuse means solid and liquid wastes which should be properly collected and transported to the suitable spot for disposal. The cost of collection and transportation to the disposal spot covers the major cost of the sanitary project. The refuse formed should be quickly, conveniently and safely taken to the disposal site to maintain the environment clear.

17.2 REFUSE CARRYING METHODS

Generally there are two ways for carrying the refuse.

1. Conservancy System.
2. Water Carriage System.

1. **Coservancy System.** In this system refuse of different types are collected separately, then each type is taken to the suitable disposal spot.

The dry garbages are collected from streets and roads, using pans and baskets. It is then transported by cars or trucks to the suitable spot. This garbage is classified into two classes. Mainly flammable and inflammable. The flammable is burnt down to ashes, and the inflammable is buried into low lying areas for reclamation of soil.

The night soil is collected from pan by manual labour to carts or trucks for transportation. It is then buried into the ground and thus the buried material is converted into manure.

The storm water and sullage are collected and transported. Generally they will be disposed to rivers or streams.

In modern cities nowadays this system i.e., conservancy system is out dated. But it is still adopted in small towns, villages, underdeveloped areas, where there is water is not much available easily. So there difficulty arises to adopt the water carriage system.

Disadvantages of the System

1. Cost : At the beginning it is cheap, but maintenance and establishment costs are high.
2. Building design : Lavatories should be separated from the main building. It will not to have a compact structure.
3. Conditions of Insanitary: After 4 or 5 hours later the sewage starts to decompose. Around the residential area insanitary condition develop, so it is necessary to remove the sewage within 24 hours. It also leads to fly nuisance.
4. Problems of Labour : It completely depends on the hands of labour. Same reason if the labour do not do the jobs for few days then the entire public health system will be in great trouble.
5. Spot for disposal : The night soil needs considerable land for disposal and which should be quite away from the habitation.
6. Carts for night soil : The night soil cart should not pass through the main road of a town or city.
7. Drains : If the storm water and sullage are carried though the open drain it will create insanitary atmosphere.
8. Water pollution : Every possibility lies there to soon the liquids waste from lavatories to the ground, which may lead to the pollution of water.
9. Epidemic : There is every possibility to out break the epidemic, if the proper disposal of the sewage is not done.

2. Water Carriage System. Water is used to carry the sewage to the point of treatment or disposal. The amount of water mixed with the amount of soil is quite sufficient, the dilution ratio of solid matter and water is so high that the mixture behaves more or like water. Sewage the easily carried through the designed sewers.

The garbage may be collected and carried away as in the conservancy system. Sometimes storm water may be allowed to flow with the sewage or carried out separately.

The initial installation cost is very high for this system. In poorly financial area, it will be very difficult to adopt this system. The water carnage system is the modern method of carrying the sewage. This can be done in stages as the township develops.

Advantages

1. *Compact design.* There will be no difficulty to have the lavatories inside the building itself.

2. *Hygienic.* Night soil, foul matter etc. are carried away by the water through the closed conduits, so it is hygienic out break of the epidemic is highly reduced.

3. *Land for treatment works.* Land requires for this system for treatment works is less than that of for the conservancy system.

4. *Methods of treatment.* Modern methods of sewage treatment can be adopted to this system. Sewage from all the places can be collected to the treatment plant, after treatment it can be suitably disposed off.

5. *Water requirement.* This system requires more water for flushing of solid matter. But actually in practice, no extra amount of water is required for this system other than the water supply.

It is fact that this system i.e., the water carriage system become the universal technique of collection and transport of domestic or sanitary waste of the locality. From the generating point of sewage to the final point of disposal, through the close conduit sewer, the system generates the liquid waste approximately 100 times the actual domestic sewage. So the water carriage system generate, huge amount of liquid waste, which have high pollution potential. Ultimately this will lead to the pollution of land and water.

Table 17.1 *Comparison between conservancy and water carriage system*

Conservancy system	*Water carriage system*
1. It should be separated from compact design structure.	It can be placed inside the compact structure.
2. It is dumped on the ground it can be seen but non hygienic.	It will pass through the under ground, so it is not visible and hygienic.
3. It needs very small amount of water about 30 to 40 litres per capita per day.	It needs large amount of water about 110 to 1201itrea per capita per day.
4. Putrefaction exists.	Putrefaction is not existed.
5. This system is suitable for rural condition.	This system basically meant for an urban system.
6. Much labour requires.	Very few labour requires.
7. Segregation is present.	No segregation is there.
8. Initially cheap, but maintenance cost is high.	Initially costly, but the maintenance cost is low.
9. Chances of epidemic out break is very high.	Chances of epidemic out break is very low.

Contd.

Contd. Table 17.1

Conservancy system	Water carriage system
10. No need to have skilled or technical personnels.	It needs the help of skilled and technical personnel for laying, maintenance and running the treatment unit.
11. Town looks dirty and foul smelling.	Town looks neat and clean.
12. Underground source of water may get polluted due to the soaking of liquid wastes.	Underground sources of water will not be polluted, because there will be no soaking wastes are carried by closed sewers.

17.3 INSTITUTIONAL SANITATIONS

Institutional sewage is nothing but the domestic sewage. It is obtained from lavatory basins, urinal and water closets of homes, offices and institutions except from hospital. This sewage is highly offensive on account of human night soil and urine are present in it. Night soil start putrifying and gives out offensive smell. This sewage will have good amount of disease producing germs and bacteria due the presence of external waste. Care must be taken to handle and disposal of this sewage.

This sewage comes from the lateral sewer to the branch sewer. There the branch sewer carry this so the main. The main sewer carry this sewage up to a certain point and then handed over to the out fall sewer.

The out fall sewer is a sewer which receives the entire discharge from the system and carries it to the point of disposal. It is that length of the main or trunk sewer between the connection of the lowest branch and the final point of disposal. It thus carries the sewage either to the point of final disposal.

17.4 PUBLIC BATHING PLACE SANITATION

Public bathing place may be bath tub, swimming pool, public showers etc. In case of bath tubs, before goes to bath it should be properly cleaned and the room where the bath tub is, should be well ventilated. After having completed the bath the used water should pass through the hole at the bottom of the tub and pass on to the lateral sewer. Lateral sewer will take this is to the branch sewer and branch will take it to the main sewer. In case of a public shower the same way the waste water will go the main for final disposal. Bath tubs and showers should be cleaned every day with some germ killer.

In the case of a swimming pool, the water should be changed once a week, if this is expensive, in that case the water should be made germ free. But the water of pool must change at least once a month care should be taken that, before using the pool, they should take a fresh bath in the shower, and also after the use of the pool they should take a fresh bath in the shower too. In case any one needs toilet, they should not use the pool but they should use the outside toilet. Otherwise the water in the swimming pool will be spoiled.

If the pool is in the open space, care should be taken to have enough sun light in the pool. If it is in the covered place, sometimes ultraviolet ray is used to disinfect the water in the pool.

At the time of changing the water at the bottom there will be an outlet value. The value should be opened slowly. So that all the water coming out from the pool should go to the lateral sewer which is placed immediately below the value. Then from lateral it will go to main via branch to the point of final disposal.

17.5 WATER SUPPLY AND SANITARY ARRANGEMENT IN A RESIDENTIAL AND PUBLIC BUILDING

The water supply in residential purposes will be as follows:

(i) *Drinking.* Average consumption of water of a man per day for drinking purpose is about 2 liters a day. This amount is very little as compared to various other uses of water.

(ii) *Cooking.* It depends upon the size of the family or the community. In any case, the average requirement of water for this purpose is about 5 litres per capita per day is recommended.

(iii) *Bathing.* It depends upon the climatic condition of the country. But in Indian climate, it is taken as 30 to 44 litres per capita per day it may also be recommended that the requirement will be 50 to 80 litres per capita per day.

(iv) *Washing hands face etc.* It depends upon the habit of the people, however roughly it is taken as 5 to 10 liters per capita per day.

(v) *Household sanitary purposes.* Under this heading, the water is required for washing cloths, floors, utensils etc. and it may be assumed to be about 50 to 60 litres per head per day.

(vi) *Private gardening and irrigation.* In case of developed cities, there will be practically no demand of water for this purposes.

(vii) *Road washing.* The roads with heavy amount of dust are to be sprinkled with water to avoid inconvenience to the users. Even in case of dust-proof roads, periodical washing is necessary. On an average, the quantity of water required for this purposes may be taken as about 5 liters per head per day.

(viii) *Sanitation purpose.* In this heading, water is required for cleaning public sanitary blocks, large markets etc. and for carrying liquid wastes from houses. The quantity of water required for this purpose will depend on the growth of civilization and may be assumed to be about 2 to 3 liters per head per day.

To fulfil the above mentioned demands for residential and public purpose, the water should be supplied is such a manner, that all the requirements can be satisfied. Water supplied from water works in such a manner, that all the facilities can be easily available. At the same time the sanitation arrangement should be well designed to meet the demand of proper sanitation for residential and public buildings. All the used water, and wastes should be drained immediately so that there should not be any cloged water any where.

17.6 PLANNING FOR WATER SUPPLY AND SANITATION

Planning for Water Supply: Before going for the proper planning of the scheme, different aspects should be carefully noticed with different angle.

The planning is done mainly by the combination of field observation and office work. It is generally built up stage by stage. First part for the developed part of the city, then the next stage for the other part of the city.

The following are the points of importance for any water supply planning.

1. Financial side.
2. Population to be served.
3. Quality of water required.
4. Consumption rate.
5. Sanitary survey of area.
6. Water supply sources.
7. Area topography.
8. Possibilities of future development.

17.6.1 Financial Side

At the beginning of the planning, the available fund should be checked carefully. Then the work should be done according to the fund permits. Step should be taken to see that the planning must be done as much economical as possible.

17.6.2 Population

Present population is decided by the census of previous years, it is the general practice to make the planning in such a way that it should accommodate the population after three or four decades. This three or four decades are calculated by any suitable method. Failure to provide the future expansion will cause great trouble in future.

17.6.3 Quality of Water Required

The quality of available water will be the guideline to decide the treatment of water. If the water is more pure, the treatment cost will be less, but if it is not so, cost will be high. So water sample available should be analysed and various results should be studied carefully to decide the economical treatment method for the locality.

17.6.4 Consumption Rate

Water demand depends on various types of uses. Such as domestic, industrial, public, trade etc. The consumption rate per capita should be decided by considering all these uses. Rate multiplying with the population, will give the total quantity of water required for the water supply.

17.6.5 Sanitary Survey of Area

This survey should be done in the surrounding area of water available. This job should be done carefully. This helps to estimate the possibility of pollution or contamination of water. This survey means to find out the information regarding the possible source of water pollution.

17.6.6 Water Supply Sources

This planning depends on the good source of water supply. It should be remembered that the source should have adequate quantity of water and also it should be cheap too.

17.6.7 Area Topography

The topographical map of the area should be available. For this map low lying area, ridges, density of population should be marked. Studying all these factors, the cheap planning of water-supply should be made.

17.6.8 Possibilities of Future Development

Possibilities of future development should be predicted and properly adjusted with this planning. Such possibilities may be of different forms such as possibility of new industry, public recreation centres, public institutions, residential blocks etc.

17.7 PLANNING FOR SANITATION

Before planning the sanitation project, few important points should be considered, because, it is purely dependent on water supply planning. Generally some advantages are given to the sanitary project in stages. For this reason the total locality is divided in to zones and then the installation of sanitory facilities are given to one by one.

Following are the important points:

1. Cost of the planning.
2. Pulsation to be served.
3. Existing disposal method.
4. Type of sewage to be handled.
5. Rain fall.
6. Amount of sewage.
7. Sources of sewage.
8. Area topography.
9. Methods of treatment.
10. Possibilities of town development.

17.8 COST OF PLANNING

To complete any sanitary project needs considerable fund. Considering the condition and taking the full advantage of the natural conditions, the planning should be made economical as much as possible. In case of alternative design, the costs of sewage treatment, both capital and operating should be worked out and the one which gives better connection between cost and performance should be chosen.

17.8.1 Population to be Served

Generally the planning of sanitary project is done in such a way, that it could accommodate the population after three or four decades. The present population is calculated from the current census report and the future population is calculated by any suitable method.

17.8.2 Existing Method of Disposal

The present method of waste disposal should be studied carefully, and by to make maximum use of this method as much as possible for the new planning.

17.8.3 Type of Sewage to be Handled

After proper analysis and knowing the properties of the sewage to be handled, line of treatment method should be decided.

17.8.4 Rainfall

There is an aim in sanitary planning to remove the storm water by suitable disposal method. So the local rainfall data will help to know the quantity of storm water may have to be handled.

17.8.5 Amount of Sewage

A reasonable amount of sewage should be assumed for the design of various sanitary units. Usually the assumption is made in this connection is that the rate of sewage is equal to the rate of supply of water.

17.8.6 Source of Sewage

The possible sources of s wage should be studied carefully, that will give some idea of the possible amount of sewage available and also help to get an idea of the type of sewage too.

17.8.7 Area Topography

Study of the topographical map of the area must be done carefully to decide low lying areas, ridges, density of population nature of soil layers, site of treatment plant etc. Study of topographical map is essential to make cheap planning.

17.9 METHODS OF TREATMENT

An outline of the treatment methods should be prepared and at the same time the methods of disposal of the final product of the treatment should be carefully decided.

17.9.1 Possibilities of Town Development

Possibilities of town development in future should be predicted first and then adjusted in the planning. Such possibilities may take various forms such as possibilities of new industries, public recreation centre, public institutions, new residential colonies etc.

PROBLEMS

17.1. Define refuse. State briefly how the refuse sanitation is done.

17.2. What is meant by institutional waste?

17.3. Describe briefly the method of sanitation of a public bathing place.

17.4. How the water supply and sanitation is done in a residential building?

17.5. Describe briefly the arrangement made for water supply and sanitary in a public building.

17.6. What are the important points are to be considered for a proposed water supply planning?

17.7. Discuss briefly the important points which should be in water supply planning.

17.8. What are the important points are to be included in proposed sanitation planning.

17.9. Write notes on:

(1) Refuse.

(2) Institutional waste.

(3) Area topography.

(4) Possibility of town development.

(5) Sanitation planning.

(6) Water supply planning.

Chapter

18

Industrial Waste Water Treatment

18.1 TREATMENT AND PROPER DISPOSAL OF INDUSTRIAL WASTE WATER UNDER INDIAN CONDITION

18.1.1 Introduction

The objective of this topic is to dealt with the industrial waste only. In the industrial waste, there will be organic impurities, inorganic impurities, and both organic and inorganic together are present. First of all the treatment method is given to be discussed. Then the general method of treatment and separately dealt with organic impurities, inorganic impurities, both organic and inorganic present.

18.1.2 Industrial Waste Treatment

1. **Factors Involving in the Treatment.** If the industrial waste and sewage are mixed together, generally they are treated as a domestic sewage. So in that case the treatment method will be the same as the domestic treatment method. If the separate treatment is needed to treat the Industrial wastes. Some units of the plant differ considerably in design from the units used in sewage treatment As for example generally the municipal sewage flow is calculated in millions of litres per day. But for the treatment of industrial waste the flow is taken thousands of litres per day. It is not unusual for Industrial waste to have BOD of round about 4000 mg/*l*. For strong industrial waste may have to be treated with recirculating ratio of 10 :1. With high rate filter, some time the ratio may be as high as 12 :1.

2. **Treatment Preparation.** Nearly all industrial waste are in an organic nature; they can be treated by biological methods. Some time a waste may have to be adjusted and corrected by primary treatment so that the effluent may be treated by biological methods.

There are number of recognised methods, any one method can be used in the Industrial Plant. The methods are as follows.

(1) Particularly for strong waste which are very difficult to treat, some operations can be modified to treat that waste.

(2) To reduce the organic content from a very strong waste an aerobic digestion method is adopted.

(3) To produce a combined waste various wastes are blended so that it is easier to treat

(4) Valuable products which are allowed to go to the waste, measures are taken to recover them.

On the basis of the local conditions the methods are generally selected. It is to reduce very high concentrations of waste, so that combined waste can be treated as domestic sewage. If it is not possible at all the waste should be treated by having some modifications of process used for domestic sewage. Figure 18.1 is a flow diagram usually used for the treatment of strong industrial waste. Only for the strong waste all the steps should be followed. But for low BOD concentration some steps may be rejected.

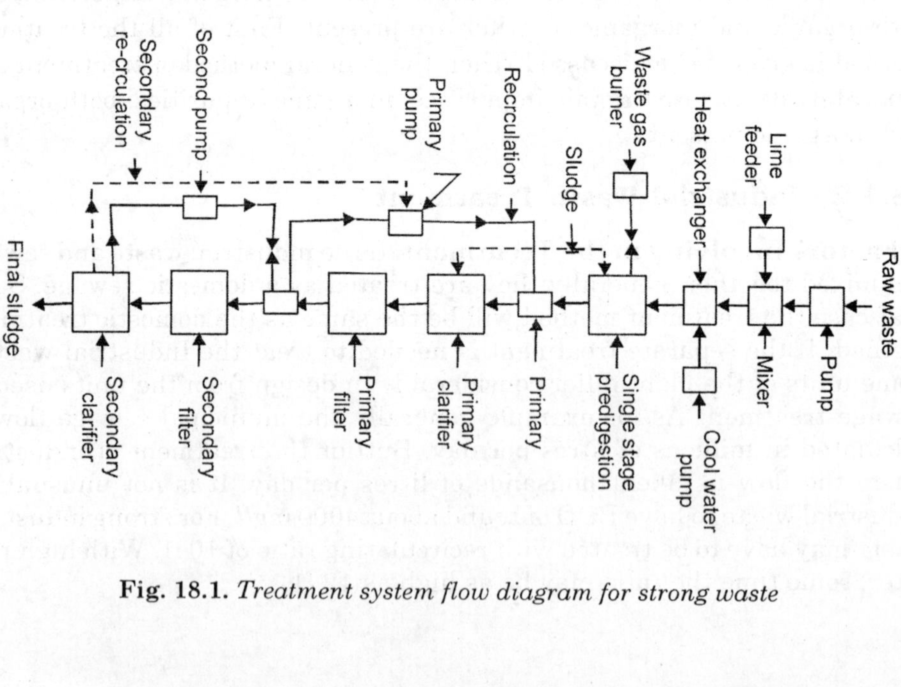

Fig. 18.1. *Treatment system flow diagram for strong waste*

18.2 TREATMENT METHODS IN GENERAL

In general the methods are like as used. Screening grit removal (if needed), sedimentation (sometimes with chemical precipitation), biological treatment (such as filters or activated sludge) and end with chlorination. The methods are briefly described below.

Screening. Screening is necessary to get rid at the unwanted materials present in the waste. The material presents in the waste if they allow to go in to the treatment process they may cause trouble even they may damage the treatment system. Sometimes materials like hair, bricks, pulps etc., are present in the waste.

Skimming. To remove grease, oil, tar from various industrial wastes, skimming tanks are used. Sometime chlorination and aeration may be helpful to remove these materials. Sulphuric acid is used to separate the soap and grease mixture.

Sedimentation and Coagulation. Setting tanks, Lagoons, setting basins and setting ponds are used to settle the materials present in the industrial wastes. In certain industries conventional setting tanks may be very useful.

It will be better to take the full advantage of the natural flocculation that will help not to mix up waste from different plant. Chemical coagulating agents are normally used to have quick sedimentation.

Biological Treatment. Waste like saw-dust etc., are having very good absorbing capacity, due to this property they absorb same fine material from the waste. Many industrial waste have good amount of organic matter and very thick waste. In that case an aerobic decomposition with high temperature is required. Trickling filter can handle the toxic acid very easily without any trouble. If the BOD exceeds 800 mg/lit recirculataon is recommended without checking the BOD from Pilot plant, activated sludge treatment should not be used.

Irrigation. Industrial wastes are not very popular for irrigation cropping. From canning and beat sugar industries seasonal waste are useful by spray irrigation on cut over land and this is also an economical process.

Sludge Disposal. Lots of industrial sludges are in stable conditions, so they can be discharged directly. But some strong sludges needs digestion before disposal. Sometimes the neutralization of industrial waste water could produce large volume of sludges.

18.3 CLASSIFICATION OF INDUSTRIAL WASTE

Industrial waste can be classified as follows:

1. Industrial waste with organic matters.
2. Industrial waste with mineral particles.

3. Industrial waste with both organic and mineral matters.
4. Industrial waste with radioactive matters.

18.3.1 Industrial Waste with Organic Matters

Mainly the organic wastes are produced from the following industries.

1. Milk-Processing Plants.
2. Breweries.
3. Canneries.
4. Meat-Packing Plant

1. Milk-Processing Plants. This plant generally produce waste like skim milk, whole milk, buffer milk and whey.

To eleminate the waste it is better to utilize them as much as possible Preliminary treatment is done in the floor by using sand and grease trap. The treatment methods are as follows.

1. By using lime and iron dosages up to 450 mg/lit.— chemical precipitation.
2. By adding small amount of iron coagulant—activated sludge treatment.
3. Filtration—by using trickling filter.

The trickling filtration is very common. It can handle large volume of waste at a time approximately about 15 to 60 million liters per hectare per day even the liquid coming from the filter will have the BOD less than 350 mg/lit.

The concentrated dirty waste is going to be best treated by using 40% excess lime. This will neutralize the acid which may have very high concentration.

2. Breweries. Spent grain is the main waste from the breweries. Waste from breweries have high food value, they can be recovered and used as a food for cattle, also this can be used as a liquid soap. In this waste suspended particles are in small amount; but the organic matters are in dissolved state. The suspended matters are precipitated by sedimentation and the dissolved organic matters are recovered by either biological treatment or by evaporation. Brewery waste contains high BOD.

3. Canneries. Properties of this waste vary according to the material used. Wastes obtain from this industry, are first treated by screening and then by sedimentation. Sometime coagulation may be needed. If the secondary treatment is needed, then the trickling filters is used. Many canneries used oxidation pond for the disposal of their wastes. Prevention of bad odour is done by adding sodium nitrate.

4. Meat Packing Plant. Waste from meat packing plant contains, blood, excreta, grease etc. Treatment like screening, sedimentation and grease removal actually removes 25% to 35% of BOD. By using low rate sometimes by high rate filter, 80% of the BDO can be removed. To reduce the strength of meat packing plants wastes anaerobic digestion is the suitable one.

18.4 INDUSTRIAL WASTE WITH MINERAL PARTICLES

1. Copper Bearing Waste. Copper bearing wastes are produced by the copper and copper alloys after bot working. This shorts of wastes disturb the life stream as well as the biological functions in the sewage treatment plant. Less than 1 mg/lit, copper concentration waste will good enough to effect the biological activities. Copper itself is a costly metal. The recovery from the wastes is always advisable. Generally the methods are adopted for this perposes are (1) sedimentation by time neutralization, (2) copper sulphate crystallization, (3) electrolytic treatment used to the pickling liquor.

2. Waste from Gas and Coke Plant. Wastes from this industry considerable amount of phenols, and cresols, ammonia, sulphides and cyanides are mixed with wastes. This wastes have the peculiar odour which is very difficult to get rid off it and having the high BOD. Mainly the phenol produce the odour. To recover from the wastes, the following methods are generally used. Solvent extraction, activated carbon absorption and distillation.

Phenol is removed from the solvent by adding caustic soda or by distillation. Benzene, Pyridine are received from the solvent. Treatment used for gas and coke plant wastes are trickling filters and some time they are diluted with water and treated as a domestic effluents.

3. Refinery Wastes (Petroleum). Brine is a common associate of petroleum. Different oil bearing are have the different brine contact. Generally the average of 3.0 barrels of brine per barrel of oil is taken. The brine is rich in sulphate and chlorides alkali and also have the alkaline earth metals. It is much salty than sea water. If this liquid is mixed with, stream of sea must be controlled carefully otherwise this will destroy water life and fish. Heavy concentration of solid contents can be removed by distillation with a heavy expense. The economical disposal of brine may be disposed into lagoon or in a pond. It should not dispose into sewers or streams, only the natural evaporation and seepage is the method of disposal.

Refinery waste is created by the chemical treatment of crude oil and washing. In the refinery wastes there are phenol, hydrogen sulphide and other compounds. Sedimentation and skimming treatment will be sufficient to remove the oil and solid particles. Before disposal to the stream the refinery wastes should carefully purified and distilled.

18.5 INDUSTRIAL WASTES WITH BOTH ORGANIC AND MINERAL MATTERS

Textile industries, laundries, tanneries and paper mills are treated here.

Waste from Textile Industries. The waste water from this industry is very polluted in nature. It will be better to treat this in two ways.

(1) Manufacturing Process.

(2) Finishing Process.

In manufacturing process it goes through the bleaching retting and boiling. This can be safely treated by biological treatment. Whereas in the finishing process it has to go through the dying, printing, stiffening and loading, wasting etc. This waste can be treated chemical precipitation method.

Apart from that various treatment methods are applied as for example:

For manufacturing process. Pecipitation with copper and lime, screening, coagulation with copperas and lime aeration. Discharge of wastes into lagoons and filters etc. It depends on the materials used for manufacturing.

For Finishing Process. Chemical precipitation is the safe and best from the point of economical point of view.

Laundry Wastes. The wastes from laundry contains spent soap, bleaches, synthetic detergents, dirt and grease. This waste is treated mainly by acidification following the treatment of chemical precipitation. If needed trickling filtration may be added to this. The chemical precipitation alone reduces the BOD upto 85%.

Waste from Tannery. Tannery wastes generally contain toxic chemical such as sulphides and chromates. Elemination of highly putrescible elements are done in the primary process. Approximately half of BOD concentration is found in liquors in the primary process. In this industry the treatment methods are screening flocculating, sedimentation and biological filtration.

Wastes from Paper Making Industry. Paper making is divided into two pans, first the paper pulp is prepared and then paper is manufactured. Wastes from sulphite mills is the most damaging one and is very difficult to treat ponding digestion and evaporation have been tried in patented processes. Waste treatment is followed by settling tanks, fine screen, rotary filters, floatation tanks, it is necessary to have the chlorination treatment to the waste to keep down the lime growths.

18.6 DISPOSAL OF INDUSTRIAL WASTE WATER

Some of the industrial waste water is harmful, before disposal of industrial waste water. Few points should be kept in mind such as (1) flammable or toxic waste matters, (2) corrosive acid and other chemicals, (3) hot fluids, that weaken pipe joints and accelerate unwanted chemical reaction, (4) heavy sludges and greases reduce the flow, (5) organic and fluid load is high, (6) interference is created by the toxic acids to the biological treatment, (7) due to this wastes, residues of these and other substances are not removed in the treatment works. There are existing rules about the industrial wastes. What type of waste water is allowed in the sewer system. High flows or excessive loads of suspended solids organic contaminants measured as BOD or chlorine consuming chemicals are accepted by some municipalities at a surcharge over the normal systems-use fee.

Industries wanted to be connected with the municipality, some of them may have to do pretreatment, simple or elaborate to more their waste water acceptable by the municipal authority. It is simple matter to neutratize acids, skim off oils, strain out oarse solids or settle finer ones, and cool hot liquids. Even when more complicated techniques are necessary the cost of purification is generally less than for direct discharge to natural water courses.

Industrial plants situated in rural areas or in small town with treatment plant that cannot handle the hydraulic or pollutional loads of the manufactured materials. They must be prepared to collect, treat and dispose of their waste-water themselves. In large cities too, may require separate treatment plant for technical and political reasons. Industries near tide water may be able to discharge their waste waters to the sea or big river after having some treatment before disposal.

Waste water can be disposed in the underground but it is costly. This method may dispense the Geological formation not only that this may contaminate the fresh water sources. The waste water must be clean enough not to clog the underground formations, be relatively free from unneutralized acids and other chemicals that might attack well equipment or the minerals composing the geological formations and be unable to supply nutrients to biological growths.

Strong industrial waste waters can be disposed off into the lagoon or large earth basins, left it for natural evaporation and seepage. It should not be disposed into the stream or rivers without having any treatment.

Considering the Indian condition, the industrial waste water should be properly treated by the industries themselves. Most of the industry in India are in the big cities or in towns are situated in the thickly populated areas, so the industrialist should keep in mind about the health of the public. Good amount of pollution starts from the various types of industrial waste water, if it is not properly treated. Initially the treatment plant may cast a little high but it will pay in the long run.

18.7 TREATMENT FOR ORGANIC MATTER AND WASTE IN INDIA

18.7.1 Biological Methods

In biological waste treatment, oxygen is always supplied in excess of the requirement. As for example 10% of the oxygen content of compressed air is absorbed by the mixed liquor of activated sludge units. But in the tricking filters, the gelainous film absorbs only 5% of air, sweeping through the filters. In the treatment of waste on natural soils and sands the oxygen helps the biological activities as high as 50% of the supplied oxygen.

Bacteria and the other associated organisms with the biological treatment of waste need food supply continuously. For the maintenance of maximum activity for maximum population of bacteria to achieve best result if the food supply to the bacteria should be maintained. Sufficient organic matters are available in the flowing wastes which froms food in the organisms. As more food in the form of organic matters becomes available in the concentrated wastes, therefore the biological activity is increased in such waste. Absorption and other inter-facial phenomena are therefore increased by high concentrated organic matters in the flowing wastes. That is why the work done is greatest in the upper strata of tricking filters and during the first hour of activated sludge treatment. So increased organic load in sewage do not reduce the efficiency of treatment process until very high concentrations of organic matters are reached.

So we can easily understand that the supply of oxygen and food is the most important to get the successful biological treatment The wastes coming out from the biological treatment is not very clear but the suspended matters in suspension are well stabilized and flocculated. For cleaning the effluent the wastes are allowed to settle in the secondary settling tanks. Thus the effluent gets clarified of colloidal and dissolved organic matters. The sludge obtain from the secondary settling tanks is highly putresible and needs to be treated again before its disposal. The effluent is free of all the polluting elements and be disposed early without any difficulty.

There are mainly two biological treatment methods.

(1) Biological beds or Trickling filter.

(2) Activated sludge process.

18.7.2 Economical Methods

Recently the concept of using natural ecosystem, such as wet lands for waste water treatment has received considerable attention. Applications of waste water to artificial and natural wetlands and the use of waste water in agriculture are explained here.

Wet lands which constitue 3 percent of the land area are the intermediate area in a hydrological sense they have too many plants and too little water to be called lakes, yet they have enough water to prevent most agriculture or silvicultural uses. The term wet lands is used here to encompass areas also known as marshes, bogs, wet meadows, peat lands and swamps. The ability to wet lands to influence water quality is the reason for much current research on the use for wastewater treatment.

Three categories of wet lands are now used for waste water treatment, artificial wet lands, existing wet lands and peat lands. Peat land is explained separately due to highly organic soil can be drained and managed in a manner similar to that used in irrigation.

Two types of artificial wet lands treatment systems have developed long Island in New York. Both of them are wet lands-pond system. In one system the wet land consists of wet meadows merging into marsh followed by a pond (meadow marsh). In the other system the wet meadows are eliminated.

These wet lands were formed in a sandy soil by making an impervious plastic liner under the soil. Operating methods have evolved from the original recycling to the once through approach with increasing loading rates. In April 1976 the rate of about 63 cm/week (25 in/week) was established. Waste water is aerated for short periods prior to application. Typical averaged results for July through September 1975 for operation with one-to-one recycling of pond effluent is noticed.

The wet lands area occupies 0.08 ha (0.2 acre) and is flooded to a depth of about 0.1 m (0.5 ft). Small recommends a 0.3 m (1 ft) depth or more to prevent volunteer weed growth and to prevent washout during storms. Cattails were planted and duckweed (Lemna minor) is prevented.

Peat lands or organic soils for land application has also been studied. A system has been designed using sprinkler application. This system was designed for 33.8 cm/week (13.3 in/week) and is under drained at a depth of about 1 m (3 ft). Secondary effluent was applied. Because of high loading rate the nitrogen uptake of the grass planted on the peat surface was exceeded. Although the peat pH was 4 the effluent pH was consistently between 6.5 to 7.5.

The above mentioned methods will be the most suitable and economical method for Indian industries.

18.7.3 Chemical Units Operations

Considering the application of the chemical unit operation, it is important to remember that one of the inherent disadvantages remains with the use of chemical unit operation as compared with the physical unit operations is that they are additive process. In most cases, something is added to the waste water to achieve the removal of something else. As a result, there is usually a net increase in the dissolved constituents in the waste water. As for example, where chemicals are added to enhance the removal efficiency of plain sedimentation the total dissolve solids concentration of the waste water is always increased. If the treated water can be reused, this can be a significant factor. This additive aspect is in contrast to the physical units operations and the biological unit process, which may be described as being subtractive, in that the material is removed from the wastewater. Another disadvantage of the chemical unit is that they are all intensive in local operating costs. The costs of some of these chemicals are tied to the costs of energy and can be expected to increase similarly.

Application of chemical unit operation in waste water treatment is as follows:

Chemical precipitation. Removal of phosphorus and enhancement of suspended solids removal in primary sedimentation facilities used for physical chemical treatment.

Gas transfer. Addition and removal of gas.

Absorption. Removal of organics not removed by conventional chemical and biological treatment methods. Also used for dechlorination of waste water before final discharge of treated effluent.

Disinfection with chlorine. Selective destruction of disease-causing organisms. Chlorine is the most commonly used chemical.

Disinfection. Removals of total combined chlorine residual that exists after chlorination (can be achieved in various ways).

Disinfection with ozone. Selective destruction of disease-causing organisms

Others. Various other chemicals can be used to achieve specific objectives in waste water treatment.

Chemical used in waste water treatment. Alum, ferrous sulfate, ferric chloride, ferric sulfate.

Disinfection with Ozone. Ozone is considered as a strong oxidising agent that has similar bactericidal properties to chlorine, has been found to be equal or superior to chlorine in its vircidal effects (viral inactivation).

When ozone is added to water, it rapidly reverts oxygen as shown.

$$2O_3 \rightarrow 3O_2$$

For this phenomenon no chemical residue exists in the treated effluent: that may require removal as is the case with chlorine residuals. Ozonation dose not produce dissolved solids and is not affected by the ammonium ion or pH influent to the process. Aeration of the waste water deriving from the use of

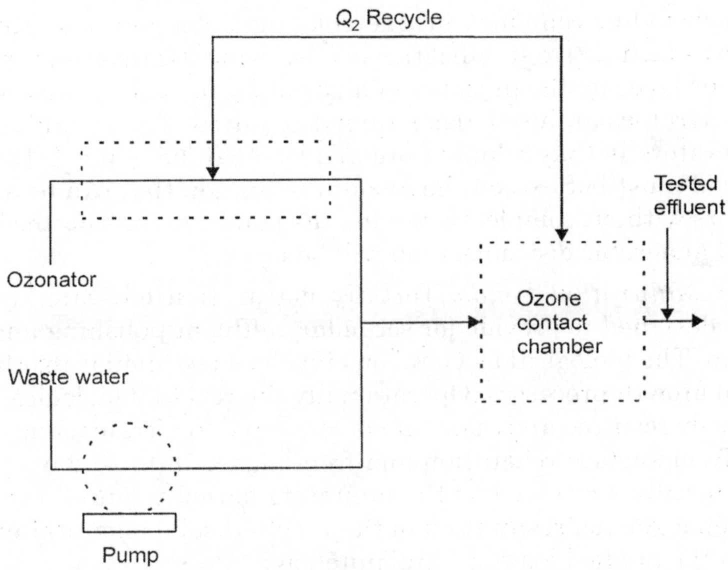

Fig. 18.2 *Ozonation system*

ozone for disinfection is and added benefit. For this reason, ozonation is considered a good alternative to chorinatior or hypochiorination especially where dechlorination may be required.

18.7.4 Natural Treatment Systems

Waste water can be treated naturally by using Oxidation ponds, Anaerobic ponds. Facultative lagoons (ponds) and Tertiary-Maturation ponds.

Oxidation Pond. This has been discussed earlier Anaerobic ponds—these are used for high strength organic waste water treatment that also contain a high concentration of solids. Usually an anaerobic pond is a deep earthen pond with appropriate inlet and outlet piping. To conserve heat energy and to maintain anaerobic conditions, anaerobic ponds have been constructed with depths up to 6 m (10 ft). The added wastes are settled at the bottom of the pond. The partially clarified effluent is usually disposed to another treatment process for further treatment. These ponds added wastes are all through their depth except for extremely shallow depth at the surface zone. Stabilization takes place by a combination of precipitation and anaerobic conversion of organic waste to CO_2, CH_4 other gaseous end products organic acids and cell tissues. The conversion of BOD is up to 70 percent and obtain routinely. At the maximum operating condition the removal efficiencies upto 82 percent may be possible.

Facultative Lagoons (ponds). The ponds in which the stabilization of waters is brought about by a combination of aerobic, anaerobic and facultative bacteria are known as facultative stabilization ponds. In practice oxygen is maintained in the upper layer by the presence of algae or by the use of surface aerators. If surface aerators are used algae are not required. The advantage of using surface aerators is that a higher organic load can be applied. However the organic load must bot exceed the amount of oxygen that can be supplied by the aerators without completely mixing the pond contents on the benefits to be derived anaerobic decomposition will be lost.

Tertiary-Maturation Ponds. Tertiary maturation low-rate stabilization ponds are designed to provide for secondary effluent polishing and seasonal nitrfication. The biological mechanisms involved are similar to other aerobic suspended growth processes. Operationally the residual biological solids are endogenously respired and ammonia is converted to nitrate using the oxygen supplied from surface reaeration and from algae. 18 to 20 day of detention time is generally suggested as the minimum period required to provide for complete endogenous respiration of the residual solids maintaining aerobic conditions the applied loadings are quite low.

18.7.5 Advanced Waste Water Treatment

Lots of substances found at waste water are not affected or are little affected by conventional treatment operation. These substances are ranges from relatively simple organic ions, such as calcium, potassium, sulphate nitrate and phosphate organic compounds. As the effects of these substances on the environment becomes more clearly understood. It is anticipated that treatment requirements will become more stringent in terms of the allowable concentration of many of these substances in the effluent from waste water treatment plants. In other words, this will need advance waste water facilities which are not used widely.

Most domestic waste waters also have a wide variety of trace compounds and elements though they are not regularly measured. If industrial wastes are discharged to the domestic sewers, the distribution of the materials will vary considerably some of the substances found in the waste water and the concentrations that may cause problems when discharge to the environment. Typical chemicals that may be found in waste water and their effects are as follows.

18.7.5.1 Inorganic

Ammonia. Chlorine demand is increased, toxic to fish, converted to nitrate in the process oxygen resources can be destroyed, with phosphorus. Undesirable aquatic growth may be developed.

Calcium and magesium. Total dissolved solids and hardness is increased.

Chloride. Salty test is created, industrial and agricultural processes will be disturbed.

Mercury. Human and aquatic life gets toxicated.

Nitrate. Aquatic growth like algal get stimulated infants may be affected by methemoglobinemia (blue babies).

Phosphate. Aquatic growth and algae get stimulated, coagulation will be interfered; lime-soda softening will be interfered.

Sulphate. Cathartic action.

18.7.5.2 Organic

Hexachoride Petrochemicals Phenolic Compounds. Development of cancer may have the relation to this, test and odour problem of water may also be the cause.

Surfactants. Interference with coagulation and may cause foaming.

Development of treatment operation depends on the following points: (i) The use to be made of the treated effluent, (ii) Type of waste water, (iii) Compactness of the various operations, (iv) Availability to dispose of the ultimate contaminants, (v) The economic feasibility of the various combinations. In extreme conditions economic feasibility may not be a controlling factor in the design of an advanced waste water treatment system.

The advanced waste water treatment operations and processes are as follows.

18.7.5.3 Physical unit operations

Air stripping or ammonia. Effluent of the secondary treatment will be treated. Removal of ammonia nitrogen.

18.7.5.4 Filtration

Multimedian. Effluent after secondary treatment will be treated—suspended solids will be removed.

Microstrainers. Effluent from biological treatment will be treated—Removal of suspended solids.

Distillation. Effluent after secondary treatment nitrified and go through filtration— Removal of dissolved solids.

Electrodialysis. Effluent after secondary treatment will go through the filtration and carbon absoption— dissolved solids will be removed.

Flotation. Effluent from primary treatment go through the secondary treatment—suspended solids will be removed.

Foam fractionation. Effluent after secondary treatment—refractory organics, surfactants and metals will be removed.

Freezing. Effluent after secondary treatment goes through filtration—dissolved solids will be removed.

Gas. Phase separation—Effluent after secondary treatment—ammonia nitrogen will be removed.

Land application. Effluent from primary treatment goes through the secondary treatment. Nitrification, denitrification is done with the removal of ammonia. Nitrogen and phosphorus.

Reverse osmosis. Effluent after secondary treatment goes through the filtration—dissolved solids are removed.

18.7.5.5 Chemical unit operation

Break point chlorination. Effluent after secondary treatment goes through filtration — ammonia, nitrogen will be removed.

Carbon absorption. Effluent from primary treatment goes though the secondary treatment may have filtration—dissolved organic, heavy metals and chlorine will be removed.

Chemical precipitation. Effluent from biological treatment—Precipitation of phosphorus, heavy metals, colloidal solids are removed.

Chemical precipitation in activated sludge. Effluent from primary treatment —phosphorus will be removed.

Ion exchange. Effluent after secondary treatment goes through the filtration-nitrate nitrogen and ammonia will be removed.

Electrochemical treatment. Untreated effluent—dissolved solids will be removed.

Oxidation. Effluent after secondary treatment—refractory organic will be removed.

18.7.5.6 Biological unit operation

Bacterial assimilation. Effluent from primary treatment — ammonia nitrogen will be removed.

Denitrification. Agricultural return water reduction of nitrogen.

Harvesting of algae. Effluent from biological treatment—ammonia nitrogen will be removed.

Nitification. Desitification — Effluent from primary treatment goes thourgh biological treatment — oxidation of ammonia.

Nitrifica:ion. Denitrification — Effluent goes through biological treatment —removal of total nitrogen.

18.7.5.7 Plant design of waste water treatment

The first step is to determine of the physical facilities needed. The size of the plant depends upon the process design criteria are adopted. As for example, if the hydraulic detention time in the aerated grit chamber is to be 4 mm at the peak flow rate, for a peak flow rate of m^3/sec, the corresponding volume would be 2/5 m^3. Sometimes the designer may increase the size on account of nonideal inlet and outlet conditions. Same method follows to determine the size of each unit operation. When the calculation have been completed, all the key data should be listed in a tabular manner. A typical example of such a table of a given setup shown in the figure is given here. Typically, these basic data include the population served the per capita contributions of various waste contaminants, average and peak flow rates and daily average of BOD loading and suspended solids. As most of the treatment plant is designed for the period of 15 to 25 years. Design criteria are given for the time when the facilities will first put into operation and for the end of the design period. The latter will, be influenced by projection of the population to be serve and the economic studies of cost effectiveness for various design periods.

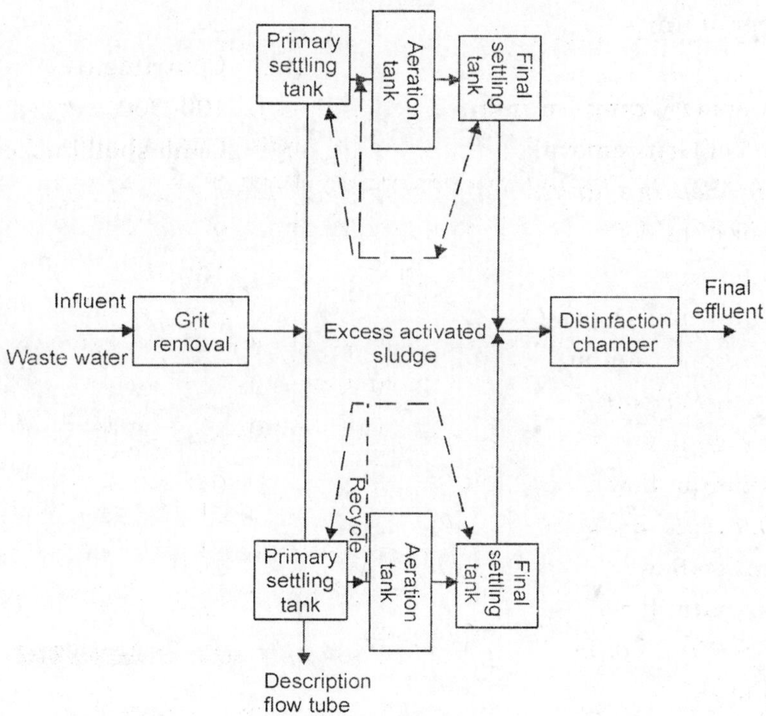

Fig. 18.3 *Sludge for processing*

The following design description is given for approximately population 20,0000 persons.

Item	*Description*
Flow measuring	Flow tube
equipment type size (mm)	600
Flow Range m^3/d	3,000–50,000
Grit Chamber	
Number of units	1
Length (m)	8–10
Width (m)	3.5–4
Average water depth (m)	3–3.5
Detention time at maximum hour flow (min)	3.0
Air-supply range m^3/min	10.30
Air blowers	
Number of units	2
Type	Centrifugal
Unit Capacity range $m^3/min.$	100–300
Method of Grit removal	Clam shell bucket
Primary Settling tanks	
Number of tanks	2
Diameter (m)	16–18
Sidewall water Depth (m)	2.0–3.0
Bottom slope (mm/m)	150
Detention Period (h)	2.5
At average flow	0.6
At maximum flow	0.6 to 0.8
Overflow rate $m^2/m^3.d$	
At average flow	29.0
At maximum flow	88.0
Final Settling Tanks	
No. of tanks	2
Diameter (m)	24
Side wall water depth (m)	3.0
Bottom slope (mm/m)	20

Contd.

Contd.

Item	*Description*
Detention Period (h)	
At average flow	6
At average flow	2
Overflow rate $m^2/m^3.d$	
At average flow	15
At maximum flow	46
Chlorination System Chlorinators	
No. of units	2
Unit Capacity range (kg/d)	440
Dosage (mg/l)	
At average flow	30
At maximum flow	10
Aeration tanks	
BOD applied (kg/d)	2.170
Average	3.250
Peak (1.5 × average)	
No. of tanks	
Length (m)	2
Width (m)	25
Average water depth (m)	3.0
No. of tanks in operation	2
Detention Period (h)	
At average flow	7.0
At maximum daily flow	3.0
Return sludge range (m^3/d)	0–15,000
Mixed liquor suspended solid concentration (mg/*l*)	2.500
Mixed liquor volatile suspended	
Solid concentration (mg/*l*)	2000
Food to micro-organism ratio (kg/BOD/ kg Mixed liquor volatile suspended solids	0.25

Contd.

Contd.

Item	*Description*
BOD to volume ratio kg/m^3.d	0.05
Aeration equipment	
Type	Machanical
No. of aerators	8
Unit power (kW)	15
Total power (kW)	120
Oxygen transfer capacity (kg/d)	5,000
BOD applied (kg/kg)	
Average	2.0
Peak	1.0
Chlorine contact tank	
No. of units	2
Length (m)	12
Width (m)	6
Depth (m)	7
Detention period at max. flow (min.)	
Contact tank	11.0
Out fall	5.0
Total	16.0
Sludge de-watering units	
Type	Vacuum filters
No. of units	2
Filtration area (m^2)	36
Filtration capacity (kg/h)	650
Total filtration capacity (kg/h)	1.250
Estimated operation period (h/week)	24

In the above the design detail is given for a waste water treatment plant.

18.7.8 Physical Properties of Air and Water

18.7.8.1 Physical properties of air

It is very important to know the composition of air. Air in atmosphere mainly consists of oxygen and nitrogen. But there are same other gases are present in the atmosphere. Though their presence is not remarkable still they exist in the atmospheric air. The composition of clean air dry atmospheric air is given below:

Component	*Concentration (ppm)*
Nitrogen (N_2)	780,900
Oxygen (O_2)	209,500
Argon (Ar)	9,300
Carbon dioxide (CO_2)	320
Neon (Ne)	18
Helium (He)	5.2
Methane (CH_4)	1.5
Krypton (Kr)	1.0
Hydrogen (H_2)	0.5
Nitrous oxide (NO_2)	0.2
Carbon monoxide (Co)	0.1
Xenon (Xe)	0.08
Ozone (O_3)	0.02
Ammonia (NH_3)	0.006
Nitrogen oxide (NO_2)	0.001
Nitric oxide (NO)	0.006
Sulphur dioxide (SO_2)	0.002
Hydrogen sulphide (HS_2)	0.0002

In the pure regions like above the sea and mountain the air may contain 21,999 percent of oxygen whereas in the over crowded halls it may be as low as 20.65 percent. Carbon dioxide is produced by combustion, putrefaction, fermentation, and respiratory processes of animal life. A balance is maintained by vegetation, rain and high winds by decreasing the quantity of carbon dioxide from the atmosphere. If the percentage of CO_2 increases upto 1.5 it will produce nausea, depression and headache. If the percentage of carbon dioxide becomes 2.5 it will extinguish the candle and 5 percent by volume may cause the fatal accidents.

It has been noticed that the amount of moisture which the air carry depends on the temperature. At higher temperature more quantity of humidity is required to saturate the air. If the temperature is then lowered. Some moisture condenses to the saturation point called the dew point for the lower temperature. The percentage ratio of the moisture content present in the air at any temperature to the moisture content required to make the saturated at the temperature is known as the relative humidity. This is index of the capacity of air to evaporate water. The relative humidity is measured through a comparison of the wet bulb and dry-bulb temperature system. This system consists of a bulb covered with a thin piece to wet cloth is allowed to move through the air rapidly. Due to evaporation the wet bulb temperature goes down initially. But later this comes to a state of equilibrium. In the same method the dry bulb temperature is a measure of relative humidity. From a psychrometric chart the relative humidity can be directly obtained.

The effective temperature may be defined as the temperature that will produce the same degree of comfort as obtained by the temperature of saturated still air. The temperature of air has an effect on humidity. Air temperature does vary according to seasons and also according to the altitudes of the place. Specially the tropical countries have the temperature of the air in atmosphere is higher than that of the non-tropical countries like Europe.

Some other physical properties are also noticed in air, they are as follows.

1. *Variation in composition.* It has been shown that the composition of air varies from place to place. In the cities the percentage of oxygen is slightly less than the rural areas. Similarly the percentage of oxygen drops as we go to higher altitudes.

2. *No specific properties.* Air has no specific properties. As a matter of fact the properties of air are the average properties of its constituents. The vapour density of air is 14.4.

3. *Liquid air has no definite boiling point.* All compounds in air have a definite boiling point. However air has no definite boiling pont and hence it is a mixture.

4. Separation of constituents of air is possible by physical means.

5. All components of air do not dissolve equally in water. If air is dissolved in water and the composition of air from the water is studied, it is found that it consists of 34% by volume of oxygen and 64% by volume of nitrogen. This suggests that air is a mixture, because if a compound dissolves in water then the concentration of compound in water is the same throughout.

6. *No chemical change is involved in mixing nitrogen and oxygen.* When four parts of nitrogen are mixed with one part of oxygen. No energy is absorbed or released, Moreover the constituents remain their individual properties there by that suggestion that air is a mixture and not a compound.

7. *No chemical formula is possible.* Knowing that air contains 78% by volume of nitrogen and 21% by volume of oxygen, if we make its formula by weight it will turn out $N_{15}O_4$. Using the formula we know that the actual vapour density is 137. However we know that the actual vapour density of air is 14.4. Thus the formula for air cannot be $N_{15}O_4$. As a matter of fact no formula for air is possible. Hence air is mixture and not a compound.

18.8 PHYSICAL PROPERTIES OF WATER

The important physical properties of water in hydraulic and quality management are its molecular structure, density, viscosity, vapour pressure, surface tension, resistance to diffusion, power of solution and suspension, light absorption, heat capacity and heat absorption. Nearly all the physical properties of water as well as its chemical and biological properties are dependent on temperature. Their variation with temperature is so great that we often speak of temperature itself, rather than temperature-dependent properties as controlling the behaviour of water.

Molecules of water occupy a volume of $2.97 \times 10^{-11}\ \mu^3$ (i.e. 29.7 cubic angstrom units Å^3). They occur singly or in groups of molecules of H_2O and as hydrogen and hydroxyl ions. Although they are in intimate contact with each other their arrangement is no directional and they are free to move about. The mean pore space between molecules is about 38%. In ice the molecules occupy, uniquely, a slight larger volume, namely, $3.23 \times 10^{-11}\ \mu^3$ (3.23 A^3). In the vapour state the molecules arc widely separated. Their equivalent size is about $3.3 \times 10^{-4}\ \mu$ (i.e. 3.3 Å) and they move at high velocity, exerting a pressure $P = \frac{1}{3}Nmv^2$ in conformance with the kinetic theory of gases. Here N is the number of molecules in a unit volume of gas, m the mass of each molecule and v^2 the mean square of the velocity of the molecules.

Density is expressed in one of the three ways: (1) as mass density ρ or mass per unit volume. (2) specific weight r or weight (force) per unit volume. (3) the specific gravity $s = \rho/\rho_0 = \frac{r}{r_0}$ (dimensionless). The suffix zero denotes the density at a standard or reference temperature such as at the temperature of maximum water density, 4°C (39.2° F) when waxer weight 1 gr per ml. The pressure has little effect on the density of water. The bulk modulus of elasticity of water is about 3×10^{5} Psi. The density of sea water is a function of its salinity which varies considerably for different oceans, seas and salt lakes. The normal specific gravity of sea water is 1.025.

Viscosity is the resistance to deformation and very kin to internal friction. The viscosity of water is expressed as follows: (1) Absolute viscosity μ, or mass per unit length and time ($ml^{-1}\ t^{-1}$) and (2) Kinematic viscosityv $v = \mu/\rho$ or length squared per unit time ($ml^2\ t^{-1}$) Fluidity is the reciprocal of absolute viscosity (m^{-1}/t). In cgs system the usual measure of absolute viscosity is the centipoise. 1 centipoise = 10^{-2} poise or 10^{-2} (dyne)(sec)/cm^2 = 10^{-2} (gram mass)/(cm)(sec). Effects of viscosity are so common and significant that it is worth while to remember Hazen's approximation of the variation of viscosity with temperature with in the normal temperature range form 32° to 80° F.

Vapour pressure is a controlling factor in evaporation whether from a free surface or from sewage sludge being air dried. Air or gas in immediate contact with water is soon saturated with water vapour. Vapour molecules then condense at the water surfaces as fast as they evaporate and the vapour reaches equilibrium with the water. The pressure exerted by the water vapour is known as partial pressure and equals its relative volume in the atmosphere. Among other things the vapour pressure of water reduce the suction lift of pumps in proportion to this partial pressure.

Surface tension governs the capillary rise of water it is an important link in the exchange of substances from and to water and determines its wetting ability. Surface tension depressants soap and synthetic detergents for example improve the wetting and deterging effect of water. Dust, Pollen other foreign particles are held at the water surface by surface tension.

Even without mechanical mixing the concentration of substances in true solution in water, both molecules and ions, will eventually become uniform. However this equalization process or diffusion is extremely low.

How a solid dissolves in water is not comparable to the solution of ideal gases. Solubility is a function of temperature, the nature and structure of the solid and the nature of concentration of water impurities. As a general rule the solubility of solids increases with temperature, but it may vary considerably within different temperature ranges. Moreover, some substances are broken down by heat into component constituents. The solubility of these components then govern their maintainable concentration. The precipitation by heat of calcium carbonates from calcium bicarbonates is an example, carbon dioxide and water or carbonic acid being released. On the other hand pressure has little effect on the solution of solids.

Absorption of solar energy by water is important in three ways:

(1) Chlorophyllaceous organisms utilize radiant energy and increase the amount of cell substance (Photosynthesis) (2) bacteria and other living organisms are killed, and colour, especially natural colour, is bleached by actinic i.e., chemically active wavelengths, more particularly ultraviolet light and (3) absorbed energy is converted into heat. Not all the solar energy directed on to a body of water penetrated the water surface. Some is reflected, its amount increasing as the incident angle becomes more acute. Accordingly the period of day light is much shorter in water that in the atmosphere. Reflection increases as the water surface is disturbed by wind.

Heat capacity of a substance is the amount of heat required to rise the temperature of a unit mass of a substance by 1 degree. The specific heat is the ratio of the heat capacity at a given temperature to that of water at a standard temperature. For most of the engineering calculation specific heat and heat capacity may be considered numerically equal. The units of measurement of heat capacity are calories/ gram/degree centigrade. The heat capacity of water is great. Much heat is required to warm water; and much cold to cool water. Variation in heat capacity with temperature is small.

Most of the energy absorbed from the sun by a natural body of water is converted into heat. If the sun's ray were monochromatic and warming took place only by radiation, while the absorptive capacity itself remained uniform, the temperature of water would decrease logarithmically from the surface to the bottom. Selective absorption however steepens the gradient within the upper layers. Even more radical shifts are induced by conduction and convection. Radiant energy absorbed by the bottom is released in the form of longer wavelengths. These are trapped by the overlying water. Bottom sediments themselves are cooler than the overlying water in summer and warmer in winter; the deeper the deposit, the greater the difference.

18.8.1 Solubility of Dissolved Gas in Water and Air

The solubility of gas in water depends upon the following factors.

(1) Upon the partial pressure in the atmosphere in contact with the water (2) upon the water temperature and (3) the concentration of impurities in the water. The rate of solution and precipitation is controlled by (a) the degree of under saturation or super saturation of water; (b) the water temperature and (3) the interfaecial area of gas contact and water exposure, including the prevention, by movement of the atmosphere and the water of the built-up of stationary gas and water films at the gas water interface. Rate of gas dispersion in water depends upon the rate of (i) molecular diffusion; (ii) eddy diffusion by convection; and. (iii) eddy diffusion by agitation. Actually the gas does not react chemically with water. From the atmospheric gases, oxygen, nitrogen and carbon-di-oxide and the rare gases of this kind although carbondioxide does react to the extent of about 1 % to form carbonic acid. Among other gases of significances in natural and treated water and waste waters and products of waste waters such as digesting sludge methane and hydrogen are inert; hydrogen sulfide less; and chlorine strongly reactive.

It is found that the rate of gas absorption is proportional to its degree of under saturation (or saturation deficit) in the absorbing liquid it can be written then,

$$dc/dt = K_g(C_s - C_t) \quad ...(1)$$

where dc/dt is the change in concentration or rate of absorption transport or transfer at time t, C_s, the saturation concentration at a given temperature, C_t the concentration at time t and K_g a proportionality factor for existing conditions of exposure. Integration between the limits C_0 at $t = 0$ and C_t at $t = t$ then the Eq. (1) becomes

$$C_1 - C_0 = (C_s - C_0)[1 - \exp(-K_g t)] \quad ...(2)$$

where K_g increases with temperature and the degree of mixing of the gas and liquid i.e., the rate of renewal of the gas and liquid interface and the degree of eddy diffusion. A common assumption is that the rate of gas transfer across the gas-liquid interface, rather than the rate of diffusion of the dissolved gas within the liquid, is the controlling factor.

In contrast of absorption, the rate of desorption, precipitation, release or dissolution of gas from a liquid becomes proportional to its degree of over saturation in the liquid or the saturation surplus. It follows that the equation of rates of absorption (i.e. Eqs. 1 and 2) apply also to rates of dissolution. As the fact that the saturation concentration C_s will be less than the observed concentration C_0 and C_t makes for negative difference.

Dissolved oxygen indicates the amount of oxygen found in sewage in dissolved form. Solubility of oxygen in water depends upon the temperature atmospheric pressure, percentage of oxygen in atmosphere oxygen deficiency in water turbulence at the surface area of exposed surface to atmosphere and other conditions. Solubility of oxygen is less in saline water than in the fresh water and the solubility in the sewage is about 95% of that in fresh water.

Hatfield formula for the solubility of oxygen in fresh or saline water as follows:

Dissolved oxygen (D.O)

$$= \frac{0.678(P - V_p)(1 - 5 \cdot 10^{-5})}{T + 5}$$

P = barometric pressure in mm of mercury

V_p = vapour pressure in mm of mercury

S = chlorine content in ppm or mg/litre

T = temperature in 0°C

Dissolved oxygen may be expressed in mg/litre or ppm or as percentage of saturation. Commonly it is expressed in mg/litre. It is always desirable to state the temperature whichever way it is expressed, because the temperature has a great effect on the solubility. For solubility at pressure other than those as shown below multiply the figure by $\frac{P}{760}$, where P is the atmospheric pressure in mm or mercury. Solubility of oxygen under normal atmospheric pressure of 760 mm of mercury is shown below. It is assumed that dry atmosphere contains 21% of oxygen. Solubility is expressed in mg/litre.

Table 18.1 *Solubility of oxygen with respect to temperature*

Tem. in °K/0°C	*From Truesdate exp. (mg/lit)*
0	14.16
279° K or 1	13.77
275° K or 2	13.40
276° K or 3	13.05
277° K or 4	12.70
278° K or 5	12.37
279° K or 6	12.06
280° K or 7	11.76
281° K or 8	11.46
282° K or 9	11.19
287° K or 10	10.92
289° K or 11	10.62
285° K or 12	10.43
286° K or 13	10.20
287° K or 14	9.98
288° K or 15	9.76

18.8.2 Solubility of Air

Air is a mixture not a compound. Air is slightly soluble in water. The solubility of air is at 273° K or 0°C is 4.3 $H \times 10^{-4}$ atm/mol fraction.

where $H = 8.04 \times 10^4$ as the temperature rises the solubility increases as shown.

Temp. at	*Solubility of air*
°K or 0°C	4.32 $H \times 10^{-4}$ atm/mol fraction
283° K or 10°C	5.49 $H \times 10^{-4}$ atm/mol fraction
293° K or 20°C	6.64 $H \times 10^{-4}$ atm/mol fraction
303° K or 30°C	7.71 $H \times 10^{-4}$ atm/mol fraction
313° K or 40°C	8.76 $H \times 10^{-4}$ atm/mol fraction
323° K or 50°C	9.46 $H \times 10^{-4}$ atm/mol fraction
333° K or 60°C	10.1 $H \times 10^{-4}$ atm/mol fraction

Air has some special properties which differentiate from the other states of matter i.e., solids and liquids on account of following reasons.

1. The distance between any two molecules of air is far larger than that of solids and liquids with the results that the force of attraction between them is almost negligible.

2. The molecules of air more constantly in a straight line unless they collide with the other molecules. These collisions are random and hence air molecules move in all possible directions in straight line.

3. *No fixed shape and volume.* The force of attraction between the molecules of air is negligible. Thus the molecules of air can fill the entire space available to them. Hence they can attain the shape and size of containing vessel. In other words, air has no definite shape and volume, but take the shape and volume of containing vessel.

4. *Air exert same pressure in all direction.* The molecules of air constantly bombed the sides of containing vessel and hence exert same force per unit area on the sides of the containing vessel which is called pressure. It has been found that at a given temperature the number of molecules striking the walls of containing vessel is same. Thus we can say that air exert same pressure in all directions.

5. *Compressibility.* The high compressibility of air is due to the fact that there are very large in spaces. On applying pressure these molecules simply come close to each other thereby decreasing volume of air. Such a compressibility is not possible in case of solid and liquids which have very small intermolecular space.

6. *Expansibility.* The volume of a given mass of air can be increased by (1) decreasing pressure (2) increasing temperature.

When the pressure is reduced from air its molecules simply move apart thereby increasing molecular space and hence volume increases.

When air is heated the kinetic energy of its molecules increases. Thus the molecules move faster and further from each other. This in turn results in the increase in volume of air.

Other properties of air is given below:

Molecular wt.	–	28.96
Normal boiling point	–	33.56 – 335.40° K or 63.56° C – 62.48°C
Critical point	–	317.26°K or 44.26° C
Vapor Density	–	0.287
Liquid Density	–	54.46
Specific heat	–	0.240
Specific heat ratio cp/cv	–	1.40
Heat of vaporization	–	88.2

As air is mixture of different compounds like nitrogen, oxygen, water vapour etc. absorbtion air mean, the absorbtion oxygen, nitrogen etc. Similarly dissolved air means dissolved of other components of air only.

PROBLEMS

18.1. Define industrial waste? What is the difference between the industrial waste and domestic waste?

18.2. Explain briefly the general methods of treatment of industrial waste. How the industrial wastes can be disposed off?

18.3. What is the condition that the industrial wastes are not mixed up with the domestic wastes?

18.4. Explain the nature of wastes from the following industries and what should be the common methods of their treatment?

(a) Wastes from Oil Refineries.

(b) Wastes from Brewery.

(c) Wastes from Canneries.

(d) Wastes from Dairies.

18.5. Explain the general features of the wastes from the following industries:

(a) textiles, (b) milk-processing plant, (c) wastes from paper mill, (d) wastes from meat packing industry.

18.6. How the industrial waste waters should be properly disposed.

18.7. What are the factors should be considered before disposing the industrial waste waters?

18.8. Why the biological treatment is required? At what stage this treatment is done? Give a brief description of Biological treatment method.

18.9. What is aerobic treatment and what is anaerobic treatment? What shorts of bacteria are living in aerobic treatment?

18.10. What type of treatment method is economical? Why is it economical?

18.11. What are the things required in natural treatment method? Why is it called the natural treatment method?

18.12. What is the difference between the ordinary treatment method and advanced treatment method?

18.13. What benefit generally be available by using the advanced waste water treatment method?

18.14. What are the essential requirement of waste water plant design?

18.15. What is recirculation means? Why is it required? How can it be detected from the effluent that the effluent needs re-circulation?

18.16. Describe briefly the physical properties of air.

18.17. Give some reason to prove that air is a mixture not a compound.

18.18. What are the physical properties of water?

18.19. What are the differences of the physical properties of air and water?

18.20. Define solubility. What gases are soluble in water?

18.21. Define dissolved oxygen. What is the main role played by this dissolved oxygen in sewage?

18.22. What is the relation between the temperature and solubility of air?

BIBLIOGRAPHY

1. Public Health Engineering *by* P.C.G. Issac (CEFFN. Spon. London).
2. Public Health Engineering (Part I and Part II) *by* E.B. Phelps and C.J. Velj, (John Wilye & Sons New York).
3. Water Supply Engineering *by* H.E. Babbitt & J.J. Donald Mc Graw Hillbook Co. New York.
4. Public Water Supplies *by* F.E. Turneaure and H.L. Rusell, John Wiley & Sons, New York.
5. Water Supply Engineering *by* S.R. Kshirsagar, Roorkee Publishing House, Roorkee, India.
6. Water for Cities *by* N.M. Blake, Syracase University New York.
7. Water and Man's Health *by* A.P. Miller, C.E. and F.N. Spon. London.
8. S.N. Garg and R. Garg *by* Khanna Publishers, New Delhi, India
9. Manual on Water Supply *by* Expert Committee of G.O.I, Ministry of Urban and Treatment Development, New Delhi (3rd Ed)..
10. Water Wealth of India *by* Orient Longman Ltd., New Delhi.
11. Water Supply Waste Disposal and Environmenal Pollution Engineering *by* A.K. Chatterjee, Khanna Publishers, New Delhi.
12. Hydrology – Crager, Justin and Hid, John Wiley & Sons, New York.
13. Ground Water and Tube Well *by* Dr. S.P. Garg, Oxford & IBH Publishering Co. Pvt. Ltd.
14. IS : 4097–1967 Gravel for Use *by* Bureau of Indian Standards. as Dock in Tube wells, Manak Bhawan New Dehli-2.
15. Sewerage and Sewage Disposal *by* L.H. Escritt.
16. Waste Water Engineering *by* Metealf & Eddy.
17. Municipal and Rural Sanitation *by* Victor M. Elhlers and Ernest. W. Steel.
18. Journals of the Sanitary Engineering Division. The American Society of Civil Engineers U.S.A. (August 1964 and February 1963).
19. News letters of Indian Association for Water Pollution Control, Nagpur, India.
20. The World's Water by M.I. Lvovich.
21. Journals of the Public Health Engineering division of the Institution of Engineers (India) Kolkata, India.
22. Indian Journals of Environment, Health of National Environmental Engineering Research Institute, Nagpur, India.
23. Sewage Treatment by Karl Imhoff.
24. Sewers *by* Edward Vaughan Bevan and Bernard Trevelyan Rees.
25. Fundamentals of Water Supply and Sanitary Engineering *by* S.C. Rangwala.
26. Waste Water Engineering *by* Metealf and Eddy INC.
27. Water Supply and Sanitary Engineering *by* Gurcharan Singh.
28. The Work of the Sanitary Engineering *by* L.B. Escrift and Sidney F. Rich.